HUMAN RIGHTS AND CLIMATE CHANGE

As the effects of climate change continue to be felt, appreciation of its future transformational impact on numerous areas of public law and policy is set to grow. Among these, human rights concerns are particularly acute. They include forced mass migration, increased disease incidence and strain on healthcare systems, threatened food and water security, the disappearance and degradation of shelter, land, livelihoods and cultures and the threat of conflict.

This inquiry into the human rights dimensions of climate change looks beyond potential impacts to examine the questions raised by climate change policies: accountability for extraterritorial harms; constructing reliable enforcement mechanisms; assessing redistributional outcomes; and allocating burdens, benefits, rights and duties among perpetrators and victims, both public and private. The book examines a range of so-far unexplored theoretical and practical concerns that international law and other scholars and policy-framers will find increasingly difficult to ignore.

STEPHEN HUMPHREYS is Research Director at the International Council on Human Rights Policy (ICHRP), an independent research institution based in Geneva, Switzerland. He holds a Ph.D. from the University of Cambridge and has published widely on human rights.

HUMAN RIGHTS AND CLIMATE CHANGE

Edited by

STEPHEN HUMPHREYS

with a Foreword by

MARY ROBINSON

CAMBRIDGE
UNIVERSITY PRESS

CAMBRIDGE UNIVERSITY PRESS
Cambridge, New York, Melbourne, Madrid, Cape Town, Singapore,
São Paulo, Delhi, Dubai, Tokyo

Cambridge University Press
The Edinburgh Building, Cambridge CB2 8RU, UK

Published in the United States of America by Cambridge University Press, New York

www.cambridge.org
Information on this title: www.cambridge.org/9780521762762

First published 2010

Printed in the United Kingdom at the University Press, Cambridge

A catalogue record for this publication is available from the British Library

ISBN 978-0-521-76276-2 Hardback

CONTENTS

AUTHORS' BIOGRAPHIES

SAM ADELMAN teaches in the School of Law at the University of Warwick. Among the courses he teaches are Legal Theory and Comparative Human Rights at undergraduate level and Theories and Histories of Human Rights on the LLM in International Development Law and Human Rights. He has degrees from the University of the Witwatersrand, Harvard University and Warwick University. A South African national, he was banned and exiled by the regime during the struggle against apartheid. A legal theorist, his writings cover globalisation, governance, development, climate change and human rights. He is currently completing a book on sovereignty. He is engaged in development projects in Africa.

JON BARNETT is a Reader and Australian Research Council Fellow at the University of Melbourne. Jon is a human geographer whose research concerns the impacts of environmental change on social systems. This includes research on environmental security, climate change and water. His research draws on and contributes to a range of disciplines. In recent years he has conducted fieldwork in the South Pacific, China and East Timor. His ARC Discovery project aims to identify the risks climate change poses to social systems in the Pacific Islands, including examining their capacity to manage these risks. Jon is a member of the Scientific Steering Committee of the Global Environmental Change and Human Security (GECHS) Project. He is on the editorial boards of *Geography Compass*, *Global Environmental Change* and the *Journal of Human Security*.

SIMON CANEY is Professor in Political Theory, Oxford University and a Fellow and Tutor in Politics, Magdalen College. His research interests are primarily in contemporary political philosophy, especially the application of political philosophy to problems in global politics. He has published extensively on global justice, climate change, poverty, human rights and international institutions. He is currently an ESRC

Climate Change Leadership Research Fellow and engaged in a three-year ESRC-funded project on 'Equity and Climate Change'. He is the author of *Justice Beyond Borders: A Global Political Theory* (2005) and articles in politics, law and philosophy journals. Professor Caney is currently writing two books: *Global Justice and Climate Change* (forthcoming 2009, co-authored with Dr Derek Bell) and *On Cosmopolitanism* (forthcoming 2009). He is also the co-editor (with Stephen Gardiner, Dale Jamieson and Henry Shue) of *Ethics and Climate Change: A Reader* (forthcoming 2009).

PHILIPPE CULLET is a Reader in Law at the School of Oriental and African Studies, University of London, where he teaches environmental law and intellectual property. He is also a Founding Director of the International Environmental Law Research Centre, Geneva and the Editor-in-Chief of the *Journal of Law, Environment and Development*, a peer-reviewed online journal available at: www.lead-journal.org. His current research interests include biodiversity law, biosafety, climate change and water, as well as the socio-economic aspects of intellectual property rights. He is the author of *Differential Treatment in International Environmental Law* (2003), *Intellectual Property and Sustainable Development* (2005) and *Water, Law and Development in the Context of Water Sector Reforms in India* (forthcoming 2009). More information about edited books and other contributions can be found at www.ielrc.org/about_cullet.php.

STEPHEN HUMPHREYS is Research Director at the International Council on Human Rights Policy (ICHRP), an independent research institution based in Geneva, Switzerland. He holds a Ph.D. in law from the University of Cambridge and has published widely on human rights, international law and climate change, including in the *European Journal of International Law*, the *Cambridge Review of International Affairs*, the *Modern Law Review*, the *International and Comparative Law Quarterly* and New York University's Hauser Global Law Working Paper series. Prior to joining the ICHRP, he was head of publications at the Open Society Justice Initiative at the Open Society Institute (OSI) in New York, where he was lead author of *Legal Remedies for the Resource Curse* (OSI, 2005) and edited the journal *Justice Initiatives*. Before that, he oversaw a project monitoring minority rights in ten Eastern European countries for OSI's EU Monitoring and Advocacy Programme (EUMAP), based in Budapest, Hungary, and edited *Monitoring the EU Accession Process: Minority Rights* (OSI, 2002).

Between 1997 and 1999 he worked with ENDA Tiers Monde's Programme Energie, in Dakar, Senegal, researching and publishing on climate change and development. In 2005–2006 he was a fellow at New York University's Hauser Global Law program.

PAUL HUNT is Professor in the Department of Law, University of Essex. He was the UN Special Rapporteur on the Right to Health from 2002 to 2008. He was Director of the Human Rights Centre at Essex University from 2001–2003. He served as an independent expert on the UN Committee on Economic, Social and Cultural Rights (1999–2002). In 2005, at the request of the UN Office of the High Commissioner for Human Rights, he co-authored draft human rights guidelines for poverty reduction strategies. He has practised as a solicitor in London and held the posts of Legal Officer and Acting General Secretary with the National Council for Civil Liberties (Liberty) in the United Kingdom. He has also lived, and undertaken human rights work, in the Middle East, Africa, Europe and the South Pacific. Professor Hunt has published widely on human rights issues, focusing particularly on economic, social and cultural rights. His publications include *Reclaiming Social Rights: International and Comparative Perspectives* (1996), (co-editor) *Culture, Rights and Cultural Rights: Perspectives from the South Pacific* (2000) and (co-editor) *World Bank, IMF and Human Rights: Including the Tilburg Guiding Principles on World Bank, IMF and Human Rights* (2003). He is also an Adjunct Professor at the University of Waikato, Aotearoa/New Zealand and holds an honorary doctorate from the Nordic School of Public Health.

RAJAT KHOSLA is a Senior Research Officer in the Right to Health Unit at the Human Rights Centre, University of Essex. He is a human rights lawyer researching and writing on issues related to economic, social and cultural rights. He has previously worked as a researcher for the Centre for the Study of Developing Societies (CSDS) in India. Mr Khosla has also worked as an advocate in the Supreme Court of India on several public interest cases involving socio-economic rights.

KYE MESA BARNARD is a second year student in the Master of International Affairs program at Columbia University's School of International and Public Affairs concentrating in environmental policy studies. His work focuses on the links between climate-related events, development and social vulnerability. Kye also works at the International Research Institute for Climate and Society on projects that integrate seasonal climate information with water resources management in developing countries. Previously, Kye served with the International Alliance of Indigenous and Tribal Peoples of the Tropical Forests as the Thailand-based Coordinating Editor of a collection of case studies documenting the impacts of climate change on indigenous peoples across the world.

JOHN MUTTER is Professor at Columbia University in the Departments of Earth and Environmental Sciences and of International and Public Affairs. He is Director of Graduate Studies for the Ph.D. program in Sustainable Development at the School of International and Public Affairs. More recently he has turned his attention to the issue of the role of Earth systems in sustainable development. He is interested in the way in which natural systems interact with human well-being, especially the role of natural extremes, such as earthquakes and hurricanes in reducing development opportunities. At the Earth Institute, where he was Deputy Director until July 2007, he oversees the Fellows program and the establishment of a disasters and development initiative and chairs the committee for the State of the Planet Conference. The Institute is taking a comprehensive approach to understanding the nexus between the Earth's environment and the nature of the human condition.

PETER NEWELL is Professor of Development Studies at the University of East Anglia and James Martin Fellow at the Oxford University Centre for the Environment. He has held positions as Principal Research Fellow at Warwick University's Centre for the Study of Globalisation and Regionalisation, Research Fellow at the Institute of Development Studies, University of Sussex and Associate Professor at FLACSO Argentina and the Instituto de Estudios para la Sustentabilidad Corporativa. He has also worked as a researcher and lobbyist for Climate Network Europe in Brussels and Friends of the Earth London and is a trustee of the NGO, One World Trust. He is author of *Climate for Change: Non-State Actors and the Global Politics of the Greenhouse*, co-author of *The Effectiveness of EU Environmental Policy*, and co-editor of *Development and the Challenge of Globalisation*, *The Business of Global Environmental Governance* and *Rights, Resources and the Politics of Accountability*. He has published widely in journals on issues of environment and development and is currently Associate Editor of the journal *Global Environmental Politics* and sits on the editorial board of the *Journal of Environment and Development*.

FRANCES SEYMOUR is Director General of the Center for International Forestry Research (CIFOR), Bogor, Indonesia. Previously she founded and directed the Institutions and Governance Program at the World Resources Institute (WRI) in Washington DC where she guided the launch of The Access Initiative, a global civil society coalition promoting citizen involvement in environment-related decisions. She also authored and contributed to WRI publications critically examining the role of public and private international financial institutions in promoting sustainable development. She previously served as Director of Development Assistance Policy at WWF, and spent five years in Indonesia with the Ford Foundation, where her grant-making focused on community forestry and human rights. She has served on numerous boards and advisory committees, including those of World Neighbors, the International NGO Forum on Indonesian Development and the African Centre for Technology Studies. She is a member of the Council on Foreign Relations.

DINAH SHELTON holds the Manatt/Ahn Professorship in International Law at the George Washington University Law School. From 1996–2004 she taught international law and was director of the doctoral program in international human rights law at the University of Notre Dame Law School. From 1987 to 1989, she was the director of the Office of Staff Attorneys at the US Court of Appeals for the Ninth Circuit. She is the author of three prize-winning books, *Protecting Human Rights in the Americas* (co-authored with Thomas Buergenthal), *Remedies in International Human Rights Law* and the three-volume *Encyclopedia of Genocide and Crimes against Humanity*. She is a member of the board of editors of the *American Journal of International Law* and is a counsellor to the American Society of International Law. Professor Shelton also serves on the boards of many human rights and environmental organisations. She has also served as a legal consultant to UNEP, UNITAR, WHO, the EU, Council of Europe and the OAS.

ACRONYMS

ATCA	Alien Tort Claims Act
CAFOD	Catholic Agency for Overseas Development
CBDR	common but differentiated responsibilities
CDM	clean development mechanism
CERD	Committee on the Elimination of Racial Discrimination
CER	certified emission reduction
CESCR	Committee on Economic, Social and Cultural Rights
CIFOR	Centre for International Forestry Research
COP	Conference of the Parties (to the UNFCCC)
CRC	Convention on the Rights of the Child
ECtHR	European Court of Human Rights
EPA	Environmental Protection Agency
ETS	emissions trading system
FAO	Food and Agricultural Organisation
FIELD	Foundation for International Environmental Law and Development
GATT	General Agreement on Tariffs and Trade
GCI	Global Commons Institute
GDR	greenhouse development rights
GEF	global environmental facility
GHG	greenhouse gas
IACHR	Inter-American Commission on Human Rights
IBRD	International Bank for Reconstruction and Development
ICCPR	International Covenant on Civil and Political Rights
ICESCR	International Covenant on Economic, Social and Cultural Rights
ICHRP	International Council on Human Rights Policy
ICJ	International Court of Justice
IELRC	International Environmental Law Research Centre
ILC	International Law Commission
IMF	International Monetary Fund
IOM	International Organization for Migration
IPCC	Intergovernmental Panel on Climate Change
IUCN	International Union for the Conservation of Nature and Natural Resources
LDCF	Least Developed Country Fund

LDC	least developed country
MDG	millennium development goal
NAPA	National Adaptation Programme of Action
NGO	Non-governmental organisation
NHTSA	National Highway Traffic Safety Administration
OAS	Organization of American States
ODA	official development assistance
OECD	Organisation for Economic Co-operation and Development
OHCHR	Office of the High Commissioner for Human Rights
PRC	People's Republic of China
REDD	reduced emissions from deforestation and degradation
SCCF	Special Climate Change Fund
SPA	special priority on adaption
TVPA	Torture Victims Protection Act
UDHR	Universal Declaration on Human Rights
UNCED	United Nations Conference on Environment and Development
UNCLOS	United Nations Convention on the Law of the Sea
UNDP	United Nations Development Programme
UNEP	United Nations Environment Programme
UNESCO	United Nations Educational, Scientific and Cultural Organisation
UNFCCC	United Nations Framework Convention on Climate Change
WHO	World Health Organization
WTO	World Trade Organization

ACKNOWLEDGEMENTS

The present collection of articles on the human rights dimensions of climate change had its genesis at a meeting convened by the International Council on Human Rights Policy, held on 12–13 October 2007, which brought together experts in climate change, human rights and/or their inter-relation. Participants included: Sumudu Atapattu, Simon Caney, Philippe Cullet, Stefanie Grant, Rajat Khosla, M. J. Mace, Benito Müller, John Mutter, Peter Newell, Gonzalo Oviedo, Dimitrina Petrova, Wolfgang Sachs, Frances Seymour, Dinah Shelton, Marco Simons, Youba Sokona, Jorge-Daniel Taillant, Martin Wagner and Michael Windfuhr.

A variant of the Introduction to this book, and of the first chapter and Conclusion, were originally published as parts of a report by the International Council on Human Rights Policy entitled *Climate Change and Human Rights: A Rough Guide* (2008). That report benefited from extensive comments from the participants at the October 2007 meeting, whose significant contribution needs special mention. In addition, the following individuals gave generously of their time in commenting on aspects of the original report, and/or in discussing the project with the editor: Bill Adams, Jon Barnett, Nathalie Bernasconi-Osterwalder, Mohamad El Ashry, Scott Jerbi, Sébastien Jodoin, Yves Lador, Siobhán McInerney-Lankford, Nathalie Mivelaz, Marcus Orellano, John Quigley, Kate Raworth, Jesse Ribot, Margot Salomon, Margaret Young and Mona Younis.

Robert Archer, Executive Director of the International Council on Human Rights Policy, played a pivotal role in guiding the meeting and also for editorial suggestions on much of the text reproduced in the Introduction, Chapter 1 and Conclusion of the present volume. Anna Piekarzewski deserves special thanks for close perusal and copy-editing of the book in its entirety and some excellent suggestions on structure. Thanks must also go to those at Cambridge University Press who shepherded the book through its final stages, particularly Finola O'Sullivan, whose encouragement and advice from the outset were invaluable, and Lyn Flight, whose editing was consistently courteous, clear and competent.

This publication, and the meeting and report from which it partly derives, were made possible in part through the direct support of the German Agency for Technical Cooperation (GTZ) and Federal Ministry for Economic Cooperation and Development (BMZ), as well as funding from the UK Department of International Development, the Ford Foundation, the Dutch Ministry of Foreign Affairs, the Swedish International Development Cooperation Agency, the Swiss Agency for Development and Cooperation and the Catholic Agency for Overseas Development. The views expressed here are not necessarily those of the organisations mentioned here.

FOREWORD

MARY ROBINSON

In summer 2008, natural disasters in China and Myanmar brought tragedy to hundreds of thousands of individuals and families. After centuries of technological and industrial advance, it is sobering to reflect on how poorly equipped we were to manage the devastation wrought by these catastrophes. We still lack early warning systems, efficient response mechanisms, and – as the suffering of the Burmese people in particular cruelly reminds us – the global solidarity and coordination needed to deliver help where and when it is most needed.

We know there will be more catastrophes of this kind in future, we know that their number and intensity is likely to grow and we know that they will increasingly be due to the acts of human beings. But they will not always involve horrific headlines and torrid photographs of hurricanes and tsunamis. More commonly, they will be cumulative and unspectacular. People who are already vulnerable will be disproportionately affected. Slowly and incrementally, land will become too dry to till, crops will wither, rising sea levels will undermine coastal dwellings and spoil freshwater, species will disappear, livelihoods will vanish. Occasional cataclysms will exacerbate these trends. Mass migration and conflicts will result. Only very gradually will these awful consequences reach those whose lifestyles and activities are most to blame. Climate change will, in short, have immense human consequences.

We have known this for a long time. We also know that these events are no longer 'natural' in the ordinary sense of the word: they are in part man-made, a product of the very same technological progress that has improved many lifestyles, but has not so far succeeded in protecting the vulnerable from its own worst consequences. The human impacts of climate change also have a human source – and this causal relationship makes the climate phenomenon peculiarly appropriate for human rights analysis.

The present collection of papers is not by any means the first to draw attention to the urgency of the many human impacts that climate change will entail, nor to broach the difficult justice questions that it raises, nor to inquire into its long-term implications for development. Each of these concerns has been discussed repeatedly since the United Nations Framework Convention on Climate Change (UNFCCC) was opened for signature in May 1992, and, indeed, before then. What this collection does for the first time, however, is think through the human rights implications of climate change and ask how the substantial body of international human rights law and experience relates to that phenomenon. Where does international human rights law coincide with or confront obligations under the nascent climate regime? Where must climate change policies contend with human rights imperatives?

Human rights law is relevant because climate change causes human rights violations. But a human rights lens can also be helpful in approaching and managing climate change. The human rights framework reminds us that climate change is about suffering – about the human misery that results directly from the damage we are doing to nature. Many communities already feel the adverse effects of warming temperatures – yet few remedies are so far available to them. While we cannot say precisely who will be affected in future, or how severely, the signs are nevertheless clear. Where information is still lacking, as it often is, we know where and how to gather it. As the present collection progressively clarifies, if we build human rights criteria into our future planning, we will better understand who is at risk and how we should act to protect them.

Attending to human rights also means recognising that as we take steps to address climate change, we must not do so at the cost of the most vulnerable. It is surely possible to repair our environment while still assuring our fellow human beings a path out of poverty and insecurity. Making certain that good information exists – and that it is in the hands of those most affected – can ensure greater participation in efforts to prevent and manage climate change. Beyond that, we must design with care global and regional programmes that substitute fuels, preserve forests, apply new technologies, or redesign markets. At each step we must ask where the heaviest burden will fall and how it should be divided.

Finally, government obligations do not stop at their own borders. As human rights lawyers have long observed, states have a special responsibility to monitor and, where necessary, regulate the behaviour of private entities within their purview, including those operating abroad. This is especially important in the case of climate change, where the immediate causes are generally found in private acts. Yet, as things currently stand, large emitters may fall through the net of a global system that imposes different obligations on rich and poor countries. It is surely right that countries with little or no history of emitting greenhouse gases should not have to take on targets that would impact their development. But it would surely be wrong if private actors were permitted to take advantage of this dispensation to produce carbon-intensive goods cheaply and export the benefits back to wealthy countries. Better to harness private ingenuity toward equitable solutions that can target and head off unacceptable human harms.

Climate change already threatens the livelihoods of peoples in distant corners of the world, from North Alaska to the Pacific islands. It is contributing to rising prices for grains and staples that are undermining food security for millions, particularly in countries with unstable weather patterns. It poses a profound threat to development in states that currently lack the resources to fulfil basic human rights.

The scope of these problems – and of the action required to treat them – reaches beyond previous human challenges. Yet in the seventeen years since the UNFCCC was signed, global negotiations have proceeded at a glacial pace. We have collectively failed to grasp the scale and urgency of the problem. Climate change shows up countless weaknesses in our current institutional architecture, including its human rights mechanisms. To address it effectively will require a transformation of global policy capacity – from information-gathering and collective decision-making to law enforcement and resource distribution.

Climate change is a story about desperation and hope. It can kill us or it can save us. Climate change will test us, threaten us and force us to change. And change, the unknown, is daunting. However, it does not have to be. On the contrary, there are reasons to be optimistic. Because, in fact, we have the know-how, the tools, the technology and the economy to mitigate climate change and ensure climate justice.

Why is the human context so important? In 1820, the United Kingdom was the richest country in the world. The average income per person was three times greater than that of people in the poorest region, sub-Saharan Africa. Today the United States is the richest country in the world, with a per capita income that is roughly twenty times larger than that of the poorest region – still sub-Saharan Africa. Most of the expected rise in global population of 2.6 billion persons by 2050 will come from the poorest regions in the world. These are regions which have no convergent economic growth, are the most unstable politically and will be those hardest hit by climate change.

It is clear that we will not be able to mitigate climate change unless we address poverty and ensure climate justice. Ultimately, achieving sustainability and a low-carbon economy will not only depend on technological innovation, but will require far-ranging social and political innovation. Let us not forget that technology does not have the ability to eliminate poverty, respect human rights, stop climate change and build a sustainable society – people do.

According to the Universal Declaration of Human Rights, 'everyone is entitled to a social and international order in which [their] rights and freedoms … can be fully realized'. Climate change disrupts that order. But perhaps it is also an opportunity, if we are willing to grasp it, to create the kind of international and social order of which the framers of the Universal Declaration dreamt.

Mary Robinson
President, Realizing Rights: The Ethical Globalization Initiative

Introduction: human rights and climate change

STEPHEN HUMPHREYS*

Two starting points inform this collection of articles on human rights and climate change. The first is that, as a matter of simple observation, climate change will undermine – indeed, is already undermining – the realisation of a broad range of internationally protected human rights: rights to health and even life; rights to food, water, shelter and property; rights associated with livelihood and culture; with migration and resettlement; and with personal security in the event of conflict.[1] Few dispute that this is the case.

Moreover, the interlinkages are deep and complex. The worst effects of climate change are likely to be felt by those individuals and groups whose rights protections are already precarious.[2] This is partly coincidence. As it happens, the most dramatic impacts of climate change are expected to occur (and are already being experienced) in the world's poorest countries, where rights protections are too often weak for a variety of reasons. But the effect is also causal and mutually reinforcing. Populations whose rights are poorly protected are likely to be less well-equipped to understand or prepare for the effects of climate change, less able to lobby effectively for government or international action and more likely to lack the resources needed to adapt to expected alterations

* Research Director, International Council on Human Rights Policy. See the Acknowledgements for background to this chapter and for a list of those who contributed comments to sections of it. Special thanks to Robert Archer for valuable editorial and substantive suggestions.

1 On the rights of indigenous peoples under conditions of climate change, see IUCN (2008). On migration, see IOM (2008). On gender, see IUCN (2007). On conflict, see German Advisory Council on Global Change (2008); European Council Doc. 7249/08 Annex, *Climate Change and International Security*, Paper from the High Representative and the European Commission to the European Council (March 2008).

2 The literature on climate change vulnerability is vast and raises significant human rights concerns. See, for example, Brooks *et al.* (2005); Guèye *et al.* (2007); Ribot (1995).

in their environmental and economic situation. A vicious circle links precarious access to natural resources, poor physical infrastructure, weak rights protections and vulnerability to climate change-related harms.

At another level, the close relation between climate change and human rights vulnerability has a common economic root. Rights protections are inevitably weakest in resource-poor contexts. But resource shortages also limit the capacity (of governments as well as individuals) to respond and adapt to climate change. Worse, where governments are poorly resourced, climate change harms will tend to impact populations unevenly and unequally, in ways that are *de facto* discriminatory because the private capacity of individuals to resist it and adapt to it differs greatly.

The construction of an international climate change regime, too, has rights implications. Mitigation policies have clear human rights dimensions. On the one hand, any strategy (or mix of strategies) that is successful at a global level will tend to determine the long-term access that many millions of people will have to basic public goods. On the other hand, choices made in the shorter term – such as whether and where to cultivate biofuels or preserve forests – will affect food, water and health security and, by extension, the cultures and livelihoods of particular persons in particular places.

Adaptation policies raise comparable human rights concerns. International funding for adaptation may be thought of as a compensatory or corrective response to potential or actual climate change-related human rights violations. Adaptive interventions before or during climate change impacts reduce the likelihood that rights infringements might result from those impacts; adaptation actions after the fact may provide redress where rights protection has already suffered. Indeed, discussions of adaptation at international and government level (as opposed to autonomous local measures) already assume a rights basis for policy construction, even if it is rarely articulated in those terms. At the same time, adaptation actions can themselves affect human rights; for example, if communities or individuals are forcibly removed from disaster or flood-prone areas, or, less forcibly, expected to conform to new economic policy imperatives (by adopting different cash crops or energy sources, for instance).

A second starting point is the observation that, despite the obvious overlaps outlined above, the mainstream climate change literature and debate has, until very recently, given little or no attention to human rights concerns.[3] This has been so even though the reports of the Intergovernmental Panel on Climate Change (IPCC) have examined the human impacts of climate change – in particular, on food, water and health – and have progressively expanded their sphere of reference to include the social as well as the physical sciences. Moreover, perhaps unavoidably, climate change analyses generally remain aggregated at continental or sub-regional level: the available information is still not sufficiently nuanced to cover the situation of individuals and communities who experience climate impacts directly as rights infringements. This, too, reflects the resource asymmetries that everywhere inform climate change discussion and research. Information is far more detailed for those areas likely to experience lesser impacts than for those where the consequences will be most devastating.

The paucity of rights-specific information is not, of course, merely a *cause* of the negligible analysis of the human rights dimensions of climate change, it is also a *consequence*. Given their salience to the main themes discussed in the IPCC's fourth assessment report (IPCC AR4), for example, it is remarkable that human rights are scarcely signalled in almost 3,000 pages of analysis.[4] This would appear to indicate a near complete disciplinary disconnect, an impression borne out by a glance at the 10,000-strong participants' list for the thirteenth Conference of the Parties of December 2007, among whom no more than a tiny handful hailed from human rights backgrounds. Scanning for human rights 'language' is, of course, a poor analytical tool. Similar concerns may be addressed using different terms – and this appears to be at least partly true in this instance. Nevertheless, the choice of language and disciplinary lens will determine to some extent the relevance of certain kinds of information, orientation and response. Since the IPCC reports are

[3] The situation is now changing. At its seventh session, in March 2008, the United Nations Human Rights Council passed a resolution on human rights and climate change. See UN Doc. A/HRC/7/L.21/Rev.1 (26 March 2008). The Office of the High Commissioner of Human Rights subsequently undertook 'a detailed analytical study of the relationship between climate change and human rights' for consideration by the Council. A series of projects investigating the link have been initiated at universities and non-governmental organisations and elsewhere.

[4] Human rights are mentioned on a handful of occasions in the fourth assessment report (hereafter IPCC AR4, with each volume named after its relevant working group (WG)). The discussion of legal instruments for mitigation in ch. 13 (IPCC AR4, WGIII, 793–4) notes the existence of human rights litigation, without commentary. Passing references also appear, again without analysis, in IPCC AR4, WGII, ch. 15, 661; ch. 17, 736; and ch. 20, 818.

essentially literature reviews, the paucity of rights references no doubt indicates a mere vacuum rather than any conclusion, bias or failing on the part of the IPCC authors. That vacuum says as much about an absence of interest in climate change among human rights professionals to date as vice versa.

Why the silence on human rights?

What explains this mutual disinterest? The primary cause appears to be a kind of disciplinary path-dependence. The study of climate change began among meteorologists, became firmly entrenched in the physical sciences and has only gradually – if inevitably – reached into the social sciences, where the basic orientation has remained pre-eminently, though not solely, economic. Climate change negotiations have centred on consensus-driven welfare-based solutions, approaches that have historically thrived independently of, and in parallel with, the human rights register. Human rights organisations, for their part, are unlikely, as a matter of professional orientation, to take up issues framed as 'hypothetical' or scenario-based, quite aside from the disciplinary boundaries that have long existed between environmental and human rights law. It may be that consideration of new and additional *future* harms simply escapes the ordinary purview of human rights analysis. The confluence has consequently been marginal: on the few occasions that human rights are mentioned in the IPCC reports, it is almost exclusively in connection with harms that have already taken place.[5]

On reflection, scholars and practitioners in either discipline might identify plausible reasons for doubting that a 'human rights approach' would assist the formation of effective policies to address climate change. Five such reasons are set out below.[6]

The rights at issue are difficult to enforce. Climate change generally (if not exclusively) affects categories of human rights that have notoriously weak enforcement mechanisms under international law: social and

[5] The 'Inuit case' is the primary example. See Petition to the Inter-American Commission on Human Rights Seeking Relief from Violations Resulting from Global Warming Caused by Acts and Omissions of the United States, Submitted by Sheila Watt-Cloutier, with the Support of the Inuit Circumpolar Conference, on Behalf of All Inuit of the Arctic Regions of the United States and Canada (7 December 2005), 70 and the short discussion in IPCC AR4, WGIII, ch. 13.

[6] These schematic points are not intended as expressions of legal doctrine or political fact.

economic rights; the rights of migrants; rights protections during conflicts.[7] Even those rights that have strong protections, such as rights to life and to property, are not subject to their normal enforcement procedures, because the harms caused by climate change can be attributed only indirectly. In the absence of strong enforcement institutions, either at national or international level, it is not immediately obvious what human rights can add to a policy discussion that is already notably welfare-conscious, even if focused on the general good rather than on individual complaints.

Extraterritorial responsibility is hard to establish. Under human rights law, a person's government ordinarily has the primary duty to act when rights are violated. In the context of climate change, however, responsibility for impacts in the most vulnerable countries often lies not with the government nearest to hand, but with diffuse actors, both public and private, many of whom are located far away. Human rights law does not easily reach across international borders to impose obligations in matters such as these.[8]

Local accountability is hard to establish. Although countries that lack economic resources and infrastructure are least likely to be major emitters of greenhouse gases, they are most likely to suffer devastating effects of climate change – effects whose human consequences will be worsened by their low capacity to adapt. Resource constraints inevitably impair a state's ability to provide quality public goods to its population. This

[7] Nevertheless, some human rights bodies, notably the European Court of Human Rights, have found rights violations due to environmental impacts, including of the right to health. See Shelton (2001), 225–31; Robb (2001). In a recent case, *Öneryıldız* v. *Turkey* (App. No. 48939/99, decision of 30 November 2004), the Court found against Turkey for failing to act on an environmental impact assessment, thereby contributing to deaths caused by a methane explosion at a rubbish tip.

[8] Existing case law suggests that states have responsibility for: (i) state actions taken in other countries; (ii) human rights protections in countries where they exercise 'effective control'; and (iii) some violations committed abroad by private actors who fall under their jurisdiction. See, for example, *Lopez Burgos* v. *Uruguay*, HRC Comm. No. R12/52 (1979), Views of 29 July 1981; *Legal Consequences of the Construction of a Wall in the Occupied Palestinian Territory*, ICJ Advisory Opinion of 9 July 2004; *Coard et al.* v. *United States*, IACHR Case No. 10.951, Reports No. 109/99, 29 September 1999; *Banković* v. *Belgium* (App. No. 52207/99, Decision of 12 December 2001). However, the case law is sparse and its applicability to climate-related harms is unclear. Alternative mechanisms involving 'long-arm' domestic jurisdiction – such as the US Alien Tort Claims Act – may be of limited value. Although state responsibility for extraterritorial violations of social and economic rights has not been widely discussed, the particular harms caused by global warming may generate plausible claims of this kind.

problem, which underpins the inadequate fulfilment of social and economic rights in some countries, has led to the notion of 'progressive realisation' of those rights under international law. Under existing circumstances, however, climate change is likely to lead to a *progressive deterioration* of those same rights. If a government cannot be held accountable for failing to fully protect those rights in the ordinary course, it will surely be even harder to hold it responsible for circumstances it did not create.[9]

Emergency conditions limit the application of human rights law. The most severe climate change impacts will be catastrophic – drought, floods, famines, mass migration, wars – and will affect large numbers of people. In such circumstances, a common response is to declare an emergency. International human rights treaties and most national constitutions typically allow for the suspension ('derogation') of many human rights in times of emergency.[10] Emergency regimes are habitually critical or dismissive of human rights constraints, tending instead to adopt an ends-oriented and charity-centred language of humanitarian relief. Governments are empowered to act expediently, with less regard to individual rights and interests that might act as a brake on achieving the greater good. Human rights, traditionally conceived as a bulwark against expansive state discretion, become less relevant as *legal* tools at such times (although their rhetorical force may increase). Indeed, some human rights traditionalists might be expected to seek limits on climate action on precisely the grounds that it will empower government, both nationally and internationally, at the expense of individuals.[11]

Rights may conflict.[12] Human rights protect others besides those who are potentially harmed by climate change. Economic actors are also rights-holders and it is foreseeable that some of them will invoke the human right to property or peaceful enjoyment of their possessions to

[9] Some of these vulnerable countries are themselves becoming significant GHG emitters, notably China and to a lesser extent India and Brazil. In these cases, the relevance of human rights law will depend increasingly on the legal expression and enforcement capacity of human rights norms in the countries in question, which varies dramatically from place to place.

[10] For accounts of the applicability of human rights during emergencies see IASC (2006) and OHCHR (2003), ch. 16.

[11] It has become increasingly common to adopt the language of emergency when referring, not only to climate change effects, but to the phenomenon in its entirety. Even if this language is intended to be emotive rather than literal, it tends to remove climate change impacts from the ordinary reach of human rights law, at least rhetorically.

[12] Thanks to Dinah Shelton for much of the substance of this paragraph.

prevent or reduce action on climate change. The right to property has been given a broad interpretation by international tribunals and could be asserted by those who have been licensed to act in ways that harm the environment. Other human rights claims too – such as to culture, or freedom of religion, or family reunion – may bring individuals into conflict with climate change policies. All of these rights, like other rights, may be limited for the public good, and struggles can be expected over exactly where the line should be drawn in such cases. Adversarialism is, of course, part of the ordinary human rights landscape. As climate change policies will necessarily generate choices about the distribution of costs and benefits, the invocation of human rights can be expected to produce struggles, pitting interest groups against one another in a way that is markedly different from the consensus-building and compromise that has traditionally guided climate negotiations.

The above objections are not negligible. Legal scholars in particular will quickly recognise a long-standing dichotomy between formal and substantive justice: the hard rule of law formalism of international human rights law, on the one hand, versus the softer, substantivist, policy orientation of the UN Framework Convention on Climate Change (UNFCCC), on the other hand. The ethical language of 'equity' and 'common but differentiated responsibilities' (CBDR) of the UNFCCC has a quite different texture from the moral certainty and universalism of statements like the Universal Declaration of Human Rights (UDHR) and the international human rights covenants. Indeed, 'equity', as it appears in the UNFCCC, might be thought difficult to reconcile with the formal equality that underpins human rights law, much as the UNFCCC's distinction between 'Annex I' (wealthy or 'developed') and 'non-Annex I' ('developing') countries seemingly runs counter to the universal obligations held by all countries under human rights law. Climate change law and policy have striven to avoid absolute or universalist claims of a kind that pepper human rights law and writing, in favour of a flexible and discretionary 'framework' language better suited to guiding compromise and consensus.

Yet these distinctions need not necessitate a sharp divide between the disciplines; indeed, as these two areas of law and practice are forced into contact by circumstance, the distinction between them is likely to narrow. A first response to the concerns outlined above might thus be assertive: human rights law is relevant to climate change for the simple reason that climate change affects and will increasingly impinge upon human rights. A second might be predictive. As harms due to climate change are felt, it is likely that those affected will turn to the hard language of human rights enforcement mechanisms for protection. Indeed, this is already happening.[13] At the same time, while neither of these factors comes with a ready-made account of the appropriate posture to take at the interface of the two regimes, the unavoidability of negotiation between them is likely to bring cross-fertilisation. There is plenty of scope for exchange and evolution.

The present book is a first attempt to examine this interface from an interdisciplinary perspective, by picking out some areas where interaction between these two disciplines is to be expected, examining where it is already taking place and forecasting the sort of techniques and strategies it may engender or adopt. Before summarising the book's contributions, this Introduction provides some further background on the extent to which rights language has already featured in the climate change debate and the legal framework within which human rights and climate change must negotiate – before turning to the human rights relevance of the evolving climate change adaptation and mitigation frameworks.

Rights, needs, development and the state

Human rights and climate change draw on quite different vocabularies, each with their own referential history and associations: terms familiar from one register may jar in the other, or mean different things to different audiences. A quick review of the key terms as they appear in this and subsequent articles may, therefore, be useful.

'Human rights', as used here, refer to a specific set of claims about the entitlements of all human beings regardless of 'race, colour, sex, language, religion, political or other opinion, national or social origin, property, birth or other status'.[14] These claims, initially laid out in the 1948 Universal Declaration of Human Rights (UDHR), are understood to carry both a widespread moral authority, on the one hand, and a (somewhat more circumscribed) legal authority, on the other hand. As the UDHR is not legally binding, the primary source texts under international law are the 1966 International Covenants on Civil and Political Rights (ICCPR) and on Economic, Social and Cultural Rights (ICESCR). The two Covenants are legally binding on all states that have ratified

[13] See, for example, the Inuit case (footnote 5 above).

[14] Common Article 2(1) of the International Covenant on Civil and Political Rights and the International Covenant on Social, Cultural and Economic Rights.

them – that is, the vast majority of the world's countries – and are supplemented by further binding treaties that protect the rights of children, migrant workers and people with disabilities, and that prohibit torture as well as racial and gender discrimination. Regional binding human rights treaties also exist within Africa, the Americas and Europe. All these texts are supported by the case law of international, regional and national tribunals, by a body of 'soft law' (that is, non-binding resolutions and other texts from international bodies such as the UN General Assembly), and, to a degree, by the doctrinal analyses of international lawyers and scholars.

The human rights laid out in these documents are generally referred to as 'civil and political', on the one hand, and 'social, economic and cultural', on the other hand. The former include rights to life, liberty, property, freedom of expression and assembly, political participation, a fair trial, privacy and home life and protection from torture. The latter include rights to work, education, social security, as well as 'enjoyment of the highest attainable standard of physical and mental health' and 'adequate food, clothing and housing, and ... the continuous improvement of living conditions'. Whereas the former rights are typically guaranteed through judicial mechanisms, including at international level, the latter have generally been achieved through domestic welfare mechanisms rather than courts.[15] Social security has typically been more available in wealthy than poorer countries; the latter are exhorted, under the ICESCR to achieve the 'progressive realization' of these rights within the bounds of the means available to them.

Human rights, therefore, capture a range of concerns that are evidently relevant to climate change, including many that have elsewhere been framed as 'basic needs'. For example, the assertion in the first Article of both Covenants that '[i]n no case may a people be deprived of its own means of subsistence' is clearly relevant where a changing climate is having precisely this effect.[16] To speak of basic subsistence needs (water, food, healthcare, shelter and so on) in terms of rights does not merely mean adopting a legal vocabulary in place of a charitable one. In principle at least, it also implies referral to a body of internationally agreed norms that have raised those needs to the level of entitlements for

[15] Social rights have increasing traction in some national and regional judiciaries, however, and a new Optional Protocol to the ICESCR would create an international forum for individual complaints. See footnote 7 above.

[16] My thanks to Kate Raworth for this observation.

all. Nevertheless, these entitlements do not translate unproblematically into corresponding obligations, much less into fulfilled demands. Under human rights treaty law, duties lie with states toward citizens – they are not straightforwardly attributable to other corporate (non-state) actors or to the 'international community' at large. Each state that has ratified the ICESCR has a duty to 'respect, protect and fulfil' the rights laid down in the treaty for those within their jurisdiction.[17] The obligation to *respect* a right is understood to mean that the state must take no steps that would violate that right; the obligation to *protect* requires that states act to ensure that other actors, including private and international, are not permitted to violate the right; the obligation to *fulfil* requires that states take steps over time to 'progressively realize' rights to food, shelter, health, education and so on.[18] The Committee on Social, Economic and Cultural Rights, the UN body that oversees the ICESCR, commonly requests that states demonstrate steady progress in the fulfilment of these rights.

The ICESCR is not entirely silent on the role that wealthier countries might play in securing the social and economic rights of those living in poorer countries, where protection of these rights is often weak. Article 2 of the ICESCR requires states to 'undertake steps individually and through international assistance and cooperation' to fulfil these rights and to use 'the maximum available resources' to that end. But while the treaty, reinforced by the Committee's commentaries, thus encourages wealthier states to assist other states to fulfil social and economic rights, the extent to which this exhortation comprises an obligation remains deeply contested. Although social and economic rights are clearly relevant to economic development in 'developing countries', the language of rights has been only partially integrated into development discourse. (The Committee provides guidelines on the integration of human rights assessments into development planning.[19]) In practice, however, international financial institutions, multilateral development banks and

[17] There are 149 states parties to the ICESCR. The United States is not among them, having signed but not ratified it.

[18] See, for example, UN Docs, E/C.12/1999/5, CESCR General Comment No. 12, The right to adequate food (Article 11) (12 May 1999); E/C.12/2002/11, CESCR General Comment No. 15, the right to water (Articles 11 and 12) (2002); E/C.12/2000/4, CESCR General Comment No. 14, the right to the highest attainable standard of health (Article 12) (11 August 2000).

[19] UN Doc. E/C.12/1991/1, Revised general guidelines regarding the form and contents of reports to be submitted by states parties under Articles 16 and 17 of the International Covenant on Economic, Social and Cultural Rights (17 June 2001).

private foreign investors have largely refused to treat international human rights law as legally binding upon their activities, and there is little recourse under international law to require them to do so. Indeed, the very applicability of international human rights law to these actors has often appeared uncertain, given that they are neither states nor, so it is argued in some cases, subject to specific territorial jurisdictions.

More than any previous issue, climate change places the question of human rights fulfilment firmly within the context of development policy.[20] This is because tackling climate change will require revisiting development models and making far-reaching decisions about access to and use of resources, questions which in turn have direct human rights consequences. But international law does not provide a clear means by which to evaluate development activities for their impacts upon human rights nor to hold the principal development actors to account on this basis.[21] This partly explains, no doubt, the relative neglect of human rights in climate change discussions. However, it also alerts us to the importance, as we examine climate change, of the first, ethical, deployment of the language of 'human rights' – for it is frequently used in situations where hard legal obligations are unavailable or disputed. Indeed, the assertion of universal human rights is not, at base, a legal assertion at all; it is first a moral or political assertion, and as such frequently carries greater weight and authority than its narrower legal cousin. In the context of climate change, the fact that it is precisely this moral or ethical force that is most frequently invoked does not, of course, indicate that hard human rights law is inapplicable; rather, it draws a focus to the potential for a significant gap between human rights as proclaimed and discussed, and human rights as practised in law.

From this perspective, state obligations under the human rights and climate change regimes – though they differ markedly – may turn out to be complementary. Under the UNFCCC (as with most international treaties) states' primary obligations are held toward one another. Whereas human rights also carry formal interstate obligations, their duties are primarily held toward citizens (and, in some cases, other inhabitants or entrants), and so are generally kept, broken, or challenged at national level. States' human rights duties toward their citizens do, however, carry into the international arena. This is apparent not only in ICESCR Article 2, but also

[20] This argument is followed in more detail in Chapter 1 below.

[21] The literature on the human rights obligations of the main development actors is voluminous. For a good recent overview, see Tan (2008).

in the Aarhus Convention, which guarantees human rights to information and public participation in environmental matters; Article 3(7) requires that its parties 'promote the application' of its principles in 'international decision-making processes' and international organisations. State responsibility for protecting human rights thus extends, in principle, into the negotiation of other regimes, particularly where these will have direct human rights consequences, such as in finding a solution to climate change. Meanwhile, as we shall see in what follows, wealthy states have concrete obligations in the climate change regime to assist poorer states in achieving developmental goals – which turn out on inspection to have much in common with basic human rights.

A final note on language: here and elsewhere in this book, the text follows the UNFCCC in speaking of 'developed' (or Annex I) and 'developing' (non-Annex I) countries even though these categories are clearly simplistic. Neither category is monolithic: each contains countries that have very different characteristics in terms of those who need most protection from climate change harms and those who bear most responsibility. Similar differences exist within individual countries, both rich and poor. Elite groups in poor countries occupy a disproportionate share of the environmental space as they do in rich countries, and these groups are often allied. Powerful political and economic links exist between 'North' and 'South'; and the major companies in large developing countries are increasingly significant global producers in their own right. Finally, the responsibility and negotiating stances of outlier countries, particularly those, 'developed' and 'developing' (or 'emerging') alike, that have been acting with least apparent regard for the shared environment need to be viewed in a distinct and nuanced manner.

Rights language in the climate change debate to date

Several attempts have been made to place rights at the centre of the climate change debate. These have not, however, generally been *human rights*-focused: that is, they have not been based upon or referred to human rights law, jurisprudence, policy experience or practice. When human rights have been invoked, it has been in a schematic fashion, as a set of background ethical assumptions that, for example, everyone has an equal entitlement to 'fair treatment' in a 'just' climate change regime, particularly in the context of mitigation options.

A general premise underlying many rights-based approaches to climate change mitigation is the distinction between 'luxury emissions' and 'subsistence' or 'survival emissions' first put forward in 1991 by the India-based Centre for Science and the Environment and further consolidated by the political philosopher Henry Shue.[22] Rather than assuming that everyone has an equal right to emit greenhouse gases in a world where overall emissions must be limited, this approach distinguishes the usage of carbon fuels (and other greenhouse gas (GHG) sources) to fulfil basic human needs from those used to perpetuate luxurious lifestyles. Whereas the former might be regarded as a fundamental (or human) right, the latter cannot be. This view has proved helpful by contrasting excess GHG use among some populations with continued need for future GHG use in others. The problem then becomes one of redressing an imbalance, which in turn involves inter-state obligations. This case might arguably be strengthened by linking 'subsistence emissions' to the satisfaction of basic human rights, such as to food, health, water and so on – on the grounds that these rights are already widely accepted and governments are already bound by them. However, there have been curiously few attempts to explore this connection. One reason for caution in reading human (that is, social and economic) rights into any right to 'subsistence emissions' might be a concern that obligations would then be deflected from the governments of countries producing excess luxury emissions onto those in low-emission countries, who are less responsible for climate change.

The best known rights-based approach to climate change mitigation is the 'contraction-and-convergence' (C&C) framework presented by the Global Commons Institute (GCI) at the second Conference of the Parties to the UNFCCC in 1996. The idea, very briefly, was to articulate a long-term mitigation strategy that, while reducing the overall amount of GHG in use over time, would *also* tend toward equalising GHG emissions per person on a global scale. In such a regime, as overall global emissions dropped, the fall would be more precipitate in wealthy countries, while usage in poorer countries would continue to rise for a period in line with their greater development needs – toward convergence between rich and poor countries at some point in the future. Initially, GCI abjured the term 'rights' in reference to C&C, because they regarded the atmosphere as a global commons that 'cannot be appropriated by any state or person'.[23]

[22] Agarwal and Narain (1991); Shue (1993).

[23] AGBM/1.9.96/14, 'Draft Proposals for a Climate Change Protocol based on Contraction and Convergence: A Contribution to Framework Convention on Climate Change', Ad Hoc Group on the Berlin Mandate, 1996, available at: www.gci.org.uk/contconv/protweb.html. The authors suggest using 'quotas' rather than rights.

Today, however, GCI claims that C&C 'establishes a constitutional, global-equal-rights-based framework for the arrest of greenhouse gas emissions'.[24] This new formulation appears to be in line with a general shift toward the language of rights in the climate change arena.

Whereas the 'rights' at issue in models such as C&C amount to speculative universal 'rights to emit' GHGs, with no obvious basis in human rights law, they might be framed as deriving from the 'right to development', which is mentioned somewhat obliquely in the UNFCCC.[25] Such a derivation would depend on demonstrating that 'subsistence emissions' were in fact required to achieve basic human rights, a claim that is at least plausible. The right to development is a difficult and somewhat confusing notion. In international law, it has had, since 1986, declaratory (non-binding) status, and has been a subject of protracted and sometimes polarising discussion within the United Nations.[26] But whatever its doctrinal status, discussion of the right to development has evolved with time, albeit rather as a space for negotiating the differing interests of different parties in the international system rather than as law in the ordinary sense. For many, particularly in countries most vulnerable to climate change, it still provides a natural hook for assessing the rights implications of climate change and the policy premises that should underlie solutions.

One recent model for GHG mitigation is explicitly based upon the right to development: the 'greenhouse development rights' (GDR) framework put forward by Tom Athanasiou, Paul Baer and Sivan Kartha in 2007.[27] They suggest that any climate change regime should, while curbing GHG emissions, give priority to assuring the long-term fulfilment of human rights (to food, water, health and shelter) associated with current low levels of development. In terms of allocating rights and duties, the GDR framework is less concerned with convergence toward equivalent emissions than with ensuring that all countries are permitted (and aided, where necessary) to reach a comparable 'development

[24] See www.gci.org.uk.

[25] UNFCCC, Article 3(4): 'The Parties have a right to, and should, promote sustainable development.' In this ambiguous wording, however, the guaranteed right appears to be the state's 'right to promote' development.

[26] See Salomon (2005); contributions to Andreassen and Marks (2006); also Alston (2001), 283.

[27] Baer *et al.* (2007). The report was co-produced by the Stockholm Environmental Institute, EcoEquity and Christian Aid.

threshold' at which basic rights might be fulfilled.[28] The GDR framework offers pointers for determining the level at which different countries should cap their GHG emissions and emphasises the importance of technology transfer, swift and substantial adaptation funding and other forms of assistance. These require levies on wealthy countries, which the authors calculate on the basis of excess GHG usage. In common with C&C and the luxury/survival emissions frameworks, the GDR authors do not examine vulnerability beyond state level; the 'development threshold' is based on national gross domestic product (GDP) per capita and does not account for distribution within states. From this perspective, GDR is not truly human rights centric: it works with aggregate rather than individual effects and harms.

Finally, a rights-based approach has, in fact, been adopted at the heart of the climate change regime through the construction of emissions markets, as introduced in the Kyoto Protocol. The capacity to buy or sell emission reductions amounts in effect to a right to emit GHGs for those who obtain emission credits. As noted above, when rights to the atmosphere were put forward in the early climate change debates, they were consistently treated as fundamental, universal and inalienable. Yet, the legal incarnation of use-rights to the atmosphere has instead taken the very different form of exclusive tradable commodities. These rights are not human rights – they are alienable, as opposed to inalienable, and they are not conceived of as universal, but bestowed upon only a comparatively tiny section of the global population. (Nevertheless, in practical terms, such rights amount to quite concrete 'rights to develop' as it is access to GHGs that currently, and for the foreseeable future, drives development). Moreover, since rights to emit are themselves a source of income, the creation of these rights appears to bestow rewards upon the perpetrators of climate change, who have so far been the overwhelming beneficiaries of this innovation. The ease with which exclusive alienable rights to emit have passed into international law (through the Kyoto Protocol) arguably demonstrates the comparative facility of establishing new property rights under international law as compared with new human rights.

In summary, although human rights appear to play a more prominent role in each successive rights-sensitive proposal on climate change, the relevant accounts have remained generally utilitarian, relying on cost–benefit and other welfare analyses. They have drawn on human rights

[28] The 'threshold' is schematically set at US$9,000 per capita at purchasing power parity.

primarily for their normative value, to underpin models involving various kinds of distributional justice, but have given little weight or effect to their achieved status as positive international law. Existing approaches mobilise human rights rhetoric to underpin a just global climate change regime; they do not examine specific human rights violations resulting from climate change or seek to inject human rights principles into climate change law. At bottom, they invoke human rights in order to spur action on climate change rather than advocating climate change action in order to prevent human rights consequences.[29]

Human rights and climate change adaptation

In thinking through the human rights implications of climate change, it appears sensible to begin with the scientific and policy terminology that has already evolved within the climate change arena to describe the phenomenon and responses to it, and to examine where human rights considerations might fit within these realms, even if they have been absent to date. This and the following section look, respectively, at two key policy areas of climate change discussion: adaptation and mitigation.

Climate change 'adaptation' refers to actions taken to adjust lives and livelihoods to the new conditions brought about by warming temperatures and other physical and weather-related events associated with climate change.[30] It is commonly used in three distinct ways. It refers, first, to actions that individuals take on their own initiative. Confronted by warmer weather or more severe storms, for example, people may choose to use new materials in home construction or switch crops or livelihoods. Second, to government measures designed to achieve the same or similar ends (the Netherlands plans to build sea-walls to protect against rising tides, for example). Third, adaptation has a more technical meaning derived from the UNFCCC and subsequent negotiations. Because the resource imbalance between the perpetrators of climate change and its victims was recognised from the outset, the UNFCCC included a requirement that wealthier countries should provide 'new and additional funding' to poorer countries

[29] Some have called for adaptation funding as 'compensation' for harms inflicted by the actions of the rich world. This model, too, invokes human rights as an ethical rather than legal imperative. See, for example, Oxfam International (2007).

[30] The third IPCC Assessment Report defined adaptation as 'adjustments in ecological, social, or economic systems in response to actual or expected climatic stimuli and their effects or impacts. [Adaptation] refers to changes in processes, practices, and structures to moderate potential damages or to benefit from opportunities associated with climate change'. Smit and Pilifosova (2001), 877–912.

to enable them to address climate change.[31] This funding was to be 'additional' to official development assistance (ODA). The practical content of 'additionality', as it is called, has remained elusive, however. This is partly because there is no clear baseline, since few wealthy countries have reached the agreed international ODA target of 0.7 per cent of GDP, and partly because very little adaptation funding has ever materialised. In what follows, adaptation is used in this third sense, to refer to the elaboration of an international policy that will deliver adaptation funding to the countries that most need it, and to programmes that such funding might support.

Extrapolating from existing 'climate sensitive' ODA, the World Bank estimates that adaptation is likely to cost anywhere from US$4 billion to $37 billion each year.[32] Yet, at present adaptation funding has not approached even the lower end of this scale; and what has been pledged has not been committed or spent. Four adaptation funds exist, all managed by the Global Environmental Facility (GEF), a World Bank-hosted entity that works through three implementing agencies (the Bank, United Nations Development Programme (UNDP) and United Nations Environment Programme (UNEP)) to channel multilateral funding for projects related to the principal multilateral environmental treaties.[33] Climate change is one of six GEF focal areas, but adaptation has consistently been a much lower priority for the GEF than mitigation.[34] Expenditure has been, and remains, excruciatingly slow, application procedures are complex and many eligible countries are not aware of what is on offer or how to access these funds.[35]

31 UNFCCC Article 4(3). This paragraph, and much of the section, relies on Mace (2005); Müller (2006) and (2007).

32 Cited in the *Stern Review* (2006), Part V, Chapter 20, 442.

33 See for a good overview, *Stern Review*, Part VI, 557. Known as the Rio Conventions because they were all signed in Rio in 1992, these are the UNFCCC, the UN Convention on Biodiversity and the UN Convention to Combat Desertification.

34 Partly to address criticism of its lack of an effective adaptation policy, the GEF introduced a Special Priority on Adapation (SPA) fund in 2005. The SPA (which never graduated beyond a 'pilot' phase), was available to developing countries on application, subject to a complex assessment of their capacity. An original allocation of US$50 million to the SPA had still not been spent by the end of the initial pilot period in 2006, but no further funds were added for the next 'replenishment' period (2007–2010). See FCCC/CP/2007/3, Report of the Global Environment Facility to the Conference of the Parties, 13th session Bali, 3–14 December 2007 (27 November 2007), para. 8.

35 For example, only one of ten GEF-supported climate change projects in FY 2006–2007 concerned adaptation through the SPA, amounting to just US$1 million of a total US$81 million spent on climate change projects. The rest was geared toward mitigation (developing countries do not have mitigation obligations). Ibid., paras. 16–17. On the other funds – the Special Climate Change Fund (SCCF) and Least Developed Country Fund (LDCF) both created under the UNFCCC – see ibid., paras. 19–27 and Mace (2005).

The GEF has also provided US$200,000 to individual countries for the preparation of National Adaptation Programmes of Action (NAPAs), designed in-country to address urgent and priority adaptation needs (thirty-two have been finished to date).[36] On the basis of NAPAs existing at the time, the *Stern Review* projected that US$1.3 billion would be required for the 'immediate' adaptation needs of the forty-seven least developed countries (LDCs).[37] Again this has not been forthcoming.[38]

It is widely recognised that adaptation funding cannot be delivered effectively until it is known where assistance will bring the most benefit. Unfortunately, it is just this information that is generally lacking. The reason, as with so much in the climate change debate, is resource related. Because expertise and financing are concentrated in wealthy countries, the latter have much more complete information about the likely impacts of climate change and suitable responses to it, compared with sub-Saharan Africa, for example. The IPCC reports cite countless practical examples of adaptation in rich countries, many of which are already under way; forecasts for poorer countries, by contrast, remain vague and sweeping. The *Stern Review* makes the point as follows:

> Adaptation will depend on comprehensive climate monitoring networks, and reliable scientific information and forecasts on climate change – a key global public good ... [D]eveloping-country governments should provide information to their own citizens but currently lack the capacity to do this, demonstrated by the shortage of weather watch stations. The international community should therefore support global, regional and national research and information systems on risk, including helping developing-country governments build adequate monitoring and dissemination programs at the national level. Priorities include measuring and forecasting climatic variability, regional and national floods, and geophysical hazards.[39]

[36] Another Adaptation Fund was created through the Kyoto Protocol, to be replenished from a 2 per cent levy on clean development mechanism (CDM) projects. The GEF acts as the Fund's Secretariat, subject to a Board with strong developing country representation, a compromise reached at Bali. To these funds might be added the World Bank's Pilot Program for Climate Resilience, one of its Climate Investment Funds introduced in 2008.

[37] *Stern Review*, 442.

[38] By late 2007, US$0.6 million (of a pledged total of US$163 million) had been allocated to preparing NAPA projects in four countries. The GEF notes that 'approximately US $150m remains to be programmed to meet the urgent and immediate adaptation needs of the LDCs under the LDCF'. FCCC/CP/2007/3, para. 27.

[39] *Stern Review*, Part VI, 563.

The list of priority areas identified in the *Stern Review* demonstrates the scale of the challenge. Physical science data must necessarily precede, and provide a base for, research on social and rights impacts. But the latter, too, are critically important, since the primary purpose of policy in this area is to reshape the human, social and economic environment. In this context, human rights may provide a compass for policy orientation, helping to decide where research should be directed and how to prioritise policy. So while it is vital to know at what temperature increase we might expect severe droughts to occur or sea levels to rise, for example, it is no less important to learn who these events will affect and where precisely, what institutional or other support is available and where further support will be most useful.

These considerations fit naturally within the agenda outlined in the Bali Action Plan of December 2007, which calls for:

> Enhanced action on adaptation, including ... International cooperation to support urgent implementation of adaptation actions, including through vulnerability assessments, prioritization of actions, financial needs assessments, capacity-building and response strategies, integration of adaptation actions into sectoral and national planning, specific projects and programmes, means to incentivize the implementation of adaptation actions, and other ways to enable climate-resilient development and reduce vulnerability of all Parties, taking into account the urgent and immediate needs of developing countries that are particularly vulnerable to the adverse effects of climate change, especially the least developed countries and small island developing States, and ... countries in Africa affected by drought, desertification and floods.[40]

Each of the priority areas identified here arguably touches upon human rights concerns; but this not only indicates the likely fruitfulness for human rights scholars and organisations of attending to climate change, it also points to the potential usefulness to the climate change policy arena of attention to the phenomenon's human rights implications.

The human rights dimensions of mitigation policies

Perhaps inevitably, the greater part of climate change negotiation has been devoted to 'mitigation'. This term refers to the actions and policies that seek to prevent global warming from causing 'dangerous

[40] Decision -/CP.13, Bali Action Plan (Advance Unedited Version), Article 1(c)(i).

anthropogenic interference' with the climate, as required by the UNFCCC.[41] Although no 'dangerous' threshold is mentioned in the treaty, a rise of average global temperatures by no more than 2°C above pre-industrial levels was until recently cited in most policy documents (it seems increasingly unlikely that this target will be achieved). Before investigating the human rights dimensions of mitigation policies, the scientific and policy context is briefly set down in the following two paragraphs.[42]

In the IPCC's 2007 Fourth Assessment Report (AR4), GHG emission levels in the atmosphere were estimated at 455 parts per million of carbon dioxide equivalent (ppm CO_2e),[43] almost double pre-industrial levels and rising fast. Current concentrations of GHGs have already warmed the globe and will lead to further warming even if emissions were stopped immediately. However, high levels of emissions are certain to continue in the short to medium term, and discussion has, therefore, centred on identifying a point at which emissions concentrations might be stabilised in future to keep warming to a minimum. There is little agreement on the appropriate stabilisation level: different studies reach different conclusions, and all are couched in the language of probability. Recent estimates suggest that if emissions levels are stabilised at 445–490 ppm CO_2e there will be an even chance (50 per cent) that the average global temperature rise will still exceed 2–2.4°C.[44] At 550 ppm CO_2e, the probability of temperatures exceeding 2°C is closer to 80 per cent, and there is an even chance that average global temperatures will rise by 3°C over pre-industrial levels.

Keeping emissions to 450 ppm CO_2e presents an immense political challenge and few rich country governments are currently aiming at national emissions targets consistent with a global peak of 2°C. The consequences of overshooting will be much worse for some, however, than for others, and is likely to destroy life and livelihoods on some small islands and certain Arctic regions (none of those affected can take the needed policy steps alone). According to IPCC AR4, even a loose target

[41] For a discussion, see the *Stern Review*, Part III, ch. 13, 289.

[42] This account relies on IPCC AR4, WGIII, Technical Summary, and on the *Stern Review*, Part III, especially ch. 7–10. More detailed information is provided in IPCC AR4, WGIII, ch. 1–3.

[43] The figure of 455 ppm CO_2e accounts for the intensity of *all* GHGs in the atmosphere, measured as equivalents of carbon dioxide. The amount of carbon dioxide itself is estimated at 379 ppm. IPCC AR4, WGIII, Technical Summary, 27.

[44] See Table TS.2 in IPCC AR4, WGIII, Technical Summary, 39. Also UNDP (2007), 46.

of 490–535 ppm CO_2e is formidably daunting. For that, total global emissions must still peak by 2020, and then fall sharply by 2050, by between 50 and 85 per cent from 2000 levels.[45] Over that same period, the world's population is expected to increase by about 50 per cent, to 9 billion or so, while economic growth, particularly in fast growing economies such as China's, will drive energy demand ever higher. Viewed in this light, the mitigation task is truly gargantuan. Despite multiple upward pressures – population, economic growth and development – emissions will need to fall dramatically between 2020 and 2050, by at least 85 per cent from 2000 levels in rich countries, given that elsewhere they must initially rise. By about 2030 it is unlikely that emissions levels can increase anywhere: in developing countries, too, they will need to have peaked.[46]

It is a widely accepted principle, entrenched in the UNFCCC, that developed countries – which historically contributed most to the problem – have greater obligations to mitigate than developing countries. Greenhouse gas emissions can be reduced in several ways. At present, negotiations seek to establish emissions caps. Though these have yet to be agreed, binding national targets were accepted by those developed countries that ratified the Kyoto Protocol.[47] Having accepted commitments, individual countries can meet their obligations in a variety of ways. Mitigation strategies may include fuel switching (to biofuels, renewable energy sources or possibly nuclear), carbon taxes and forestry growth or preservation. But while there is general consensus that developing countries should not have to compromise their future economic growth, there is little agreement on how sharp global cuts are to be achieved while growth, especially in poorer countries, continues.

[45] Even these figures may be optimistic. Jim Hansen, a leading climate commentator, now claims that current CO_2 levels are already unsustainable: 'If humanity wishes to preserve a planet similar to that on which civilization developed and to which life on Earth is adapted, paleoclimate evidence and ongoing climate change suggest that CO_2 will need to be reduced from its current 385 ppm to at most 350 ppm [CO_2 (not CO_2e)] … If the present overshoot of this target CO_2 is not brief, there is a possibility of seeding irreversible catastrophic effects'. See Hansen *et al.* (2008).

[46] IPCC AR4, WGIII, Technical Summary, 90.

[47] Developed country parties to the Kyoto Protocol agreed to reduce their emissions by varying amounts from 1990 levels by 2012. Not all will reach their targets. At time of writing, no framework has been agreed for the post-2012 period.

What does the choice of mitigation policies imply for human rights? Human rights fulfilment in any given state depends upon a basic level of economic wherewithal and stable access to resources. However, a mitigation regime – or mix of regimes – will work only if it succeeds in reorienting productive capacities and access to resources on a massive scale. Whatever the mix of mitigation strategies arrived at, if effective it will have two broad effects. First, it will drastically reduce access to and dependence upon fossil fuels – currently the most reliable and cost-effective fuel source available (measured in terms of energy yield against cost of extraction/generation). Second, it will curtail the development policy options available to governments everywhere – an implication that matters especially in those countries that have not yet reached a level of economic growth sufficient to guarantee basic needs. Not only will climate change mitigation policies profoundly influence the allocation and use of scarce resources, they will do so far into the foreseeable future. In short, climate change mitigation efforts will reorient and fix national development paths over the long term, and these in turn will tend to set limits on the capacity of countries to fulfil basic human rights, albeit to different degrees.

This linkage between climate change mitigation, development paths and human rights fulfilment is recognised explicitly in IPCC AR4:[48]

> Development paths underpin the baseline and stabilization emissions scenarios discussed [elsewhere in the report] and are used to estimate emissions, climate change and associated climate change impacts. For a development path to be sustainable over a long period, wealth, resources, and opportunity must be shared so that all citizens have access to minimum standards of security, human rights, and social benefits, such as food, health, education, shelter, and opportunity for self-development.

Ultimately, as the IPCC report acknowledges here (without elaboration), the ability to orient and implement any mitigation policy depends upon identifying and prioritising acceptable social outcomes in advance, human rights among them. Human rights fulfilment depends upon development capacity, and that consideration must in turn guide the choice of paths toward carbon stabilisation. Latent within this view is the understanding that human rights protection is costly.[49] It is not so much a question of a right *to* development but a more basic concern: without development there can be only limited

[48] IPCC AR4, WGIII, 696.

[49] For a good account, see Holmes and Sunstein (1999).

fulfilment of human rights (indeed, this is the principle underlying 'progressive realisation').[50]

Moreover, although there is consensus that any mitigation strategy will have distributional consequences, to date these have remained largely underexplored. The fourth IPCC report is explicit on this point too. It suggests that distributional outcomes should be one of four criteria for evaluating mitigation policies, but admits that comparison in terms of this criterion 'has proved difficult – and ranking impossible' because, according to the report's authors, assessment is inevitably subjective.[51] This is no doubt true, although it is also true that there is a degree of subjectivity in evaluating *any* of the criteria. Even so, the charge of subjectivity takes little account of the relation between resource distribution and human rights fulfilment, on the one hand, and the fact that the parties to any agreement also (for the most part) already *have* binding human rights obligations to which they must attend, on the other hand. Indeed, human rights standards may offer a way to manage the dilemma of subjectivity – in principle, they provide benchmarks of acceptable outcomes based on widely-agreed principles and, moreover, on legal stricture. If a global regime proceeds without integrating human rights, it might be argued, it will not only miss an opportunity to promote and fulfil human rights but will also be blind to countless possible harms that might otherwise be foreseen and averted.

Those with human rights expertise, therefore, have good reason to think through the human rights consequences of different mitigation strategies – at national and local, but perhaps especially at international level – given that the effects will be profound, of long duration and probably irreversible. At national level, for example, what will be the consequences in human rights terms of large forest conservation efforts, extensive biofuel cultivation for export markets, or nuclear power

[50] See Baer *et al.* (2007), 23:

> [T]here is no road to development, however conceived, that does not greatly improve access to energy services. Yet, as economies are now structured, as development is now envisioned, and as long as we rely on today's energy technologies, this will imply increases in CO_2 emissions that are entirely incompatible with a precautionary climate policy. And thus our dilemma: There is simply not enough 'environmental space' for the still-poor to develop in the same way – or in anything like the same way – as that which was taken by the already-rich.

[51] IPCC AR4, WGIII, 752. The other three criteria are environmental effectiveness, cost efficiency and 'political acceptability', each of which has a better established role in mitigation choices. All, of course, are 'subjective' to some degree.

dependence? Who will be affected and how? Are institutional forms of redress available in cases of rights violations? Can long-term development be maintained if carbon use is restricted? How will hard choices be decided? At international level, how will differential access to the 'global carbon dump' affect local development paths?[52] Where the effect is harmful, are compensatory mechanisms available, and are they effective and appropriate? In principle, the likely human rights and developmental consequences of different mitigation strategies should be built into forecast scenarios for comparative purposes, something that has not been done systematically to date.[53]

Any such analysis will need to take account of the particular role that developing countries are likely to play in any global mitigation regime. It is generally agreed that it is cheaper to cut emissions in poorer than richer countries (as transitions to new energies, for example, involve fewer infrastructural shifts). As the *Stern Review* states, '[s]preading the mitigation effort widely across sectors and countries will help to ensure that emissions are reduced where it is cheapest to do so, making policy cost-effective'.[54] Of course, this provides an incentive for wealthy countries (and their companies) to try to meet their targets through actions undertaken in developing countries. The *Stern Review* is quick to point out that social and other factors must be taken into account in making decisions about where and how to make cuts. Yet, the absence of such data to date has not stopped a surge in efforts to achieve cuts in developing countries – efforts which may, in consequence, have deleterious human rights outcomes in those very countries, either in the immediate or in the longer term. Deforestation, biofuel cultivation and emissions trading will in different ways each operate to alter the economic stakes and capacities of persons who already, in many cases, lack secure access to basic needs. Assessing the possible human rights impacts of strategic decisions in these areas, though urgent, nevertheless requires considerably more knowledge than is currently available.

[52] The term 'global carbon dump' captures the notion that the atmosphere can support only a limited amount of GHGs – and so there can be no unrestricted right to send carbon into it. See Lohmann (2006).

[53] Climate change narratives have traditionally focused on *impacts* in developing countries and *mitigation* in developed countries. While this seems sensible, because carbon emissions are concentrated in rich countries while poorer countries suffer the brunt, it leaves one vital issue undiscussed – the future development of poor countries under global emissions constraints.

[54] *Stern Review*, 239. See also 245–6: 'some countries can cut emissions more cheaply than other countries, so 'what' flexibility is important'.

By extension, it would be useful to analyse the likely impact of given mitigation strategies on the potential for *alternative* development paths for poorer countries. Is clean technology transfer facilitated? If so, is this done in a sustainable and equitable manner, geared to a country's development needs rather than to the economic interests of the exporting country alone? Does the policy mix shift development paths, stimulate wealth creation and *also* consolidate basic threshold rights for all? Clearly such questions go beyond the ordinary scope of human rights inquiry. Clearly, too, they imply a need for significant new research.

Human rights and climate change at the confluence of law, science, ethics and policy

The various chapters in this book are a first attempt to look systematically at the relevance of human rights for climate change and vice versa. They begin from a broad angle, opening up difficult framing questions: what is the ethical case for introducing human rights to climate change? How adequate is the dominant human rights framework to climate change? How might inter-state claims draw upon intra-state human rights norms? How can the moral and legal obligations that climate change raises be aligned and addressed? What about the obligations of private companies? Other chapters then turn to a specific set of rights and policy areas that climate change raises: the nascent forestry regime, emissions trading under the Kyoto Protocol, the right to health, the contribution of human rights fulfilment to climate change vulnerability.

Chapter 1 raises some background justice questions presented by climate change, and assesses the availability and adequacy of human rights instruments to address them. I contend that four divergent justice questions arise in climate change, each emphasising different interests and underpinning different solutions. The nascent climate regime incorporates elements from each of these perspectives and uses flexible conciliatory language – 'equity' and 'common but differentiated responsibilities' – to mediate between them. Nevertheless, these terms do not in themselves determine the shape of a final regime; in practice different elements of the regime have moved ahead at different speeds and, in the process, certain justice claims have been prioritised, often following the perspectives of better resourced and positioned actors. The chapter examines whether and how consideration of human rights law and priorities, which have hitherto been largely absent, might rebalance or reorient these initial questions. Much as human rights principles do not provide clear-cut answers – and might, indeed, be used to support most or all justice claims – there are clear areas where the human rights and climate regimes complement and corroborate certain pictures of an 'equitable' regime over others.

Dinah Shelton draws upon a similar intuition in her contribution, arguing that, when it comes to climate change, the formal sovereign autonomy of states in the international arena dovetails with their human rights obligations under international law. Shelton examines the recent US Supreme Court case, *Massachusetts* v. *EPA*, in which the Court recognised Massachusetts' standing on the basis of its sovereign right, as '*parens patriae*', to protect the health and welfare of its citizens from the harmful actions of others in the absence of federal regulation. Such a principle might also apply at international level, Shelton suggests, for states whose inhabitants experience the harms of climate change due to the actions and failures of other states to regulate their domestic pollution. The corollary of such a sovereign right to protect citizens, in the international order, is a duty on states not to abuse their sovereign right to pollute – a duty that Shelton identifies in the famous *Trail Smelter* case, in which an international arbitrator required the Canadian province of British Columbia to stop cross-border pollution into the United States. Similar principles apply in the case of global climate change: where human rights are at risk due to climate change, affected states arguably have a right under international law to challenge the pollution of other states. The global stakes of climate change thus alters a familiar context in which sovereignty and human rights are generally perceived as being at loggerheads (perhaps most strikingly illustrated in the recent debates over the 'responsibility to protect').

In his chapter, Simon Caney provides the ethical case for attending to the human rights implications of climate change. His argument has three phases: first, he shows that climate change does in fact affect the full enjoyment of certain key human rights, such as to life and health and what Caney calls the 'human right to subsistence'. Second, Caney suggests that an approach to climate change that attends to its human rights implications carries significant advantages over other, currently more widespread, approaches – such as those that prioritise cost–benefit analyses or threats to security – in determining what particular impacts should have priority attention. Third, Caney proposes that 'a "human rights" centred analysis of the impacts of climate change has far-reaching implications for our understanding of the kind of action that should be taken and who should bear the costs' of action. It is clear, Caney concludes, that the burdens should fall largely on those actors most responsible for creating the problem, including the costs of treating the impacts of climate change on actors who are not so responsible. Viewing human rights as 'minimum moral thresholds to which all individuals are entitled, simply in virtue of their humanity, and which override all other moral values' provides a means to assess the distribution of the burdens of treating climate change.

A different approach is taken by Sam Adelman. Like Shelton, Adelman regards unrestrained sovereignty as the principal underlying obstacle to addressing the human rights violations caused by climate change; unlike her, however, he regards the doggedly state-centric dominant discourse of human rights as itself part of the problem. For Adelman, the currently leading solution to climate change (the creation of a market in carbon emission reductions) reproduces and reinforces the most undesirable traits of state sovereignty – self-interest and excess – rather than providing a basis for a more just global governance. The underlying rationality of each is nevertheless quite different: paradoxically, although the carbon market is the outcome of a state-centric decision-making process, it is also an abdication of sovereign responsibility. Adelman suggests, however, that human rights may yet help to mobilise the regenerative evolution within international law that climate change must ultimately require – perhaps through the eventual establishment of a new meta-right or gründnorm that would recognise the close interlinkage between human rights and the environment, and finally curb the excesses of the present global economic order, in the mutual interests of greater justice and environmental stewardship.

Peter Newell examines the key role of the private sector in relation to responses to climate change and asks whether human rights norms and law might provide a useful means of ensuring the accountability of such actors for their contribution to the problem. Historically, human rights instruments have not proved effective in holding corporations to account for actions undertaken abroad. While Newell notes the ongoing progress in defining the nature of corporate responsibilities, notably in the work of the UN Secretary General Special Representative, John Ruggie, it is unclear, he argues, that even were legal mechanisms stronger, they would be available in anything other than a limited number of cases. Moreover, private actors have played an active part in shaping the international climate change regime, which consequently places few direct regulative burdens on the private sector. The obligations of corporations tend to depend on the jurisdiction of operation, but this may mean little in a context where operations can easily be moved between jurisdictions. Beyond this, various forms of private regulation have emerged, but while useful, these do not have effective means of sanctioning non-compliance. Under these conditions, the scope for private accountability in cases where activities which contribute to climate change result in actual human rights violations appears weak.

Frances Seymour provides a concise yet informed picture of the human rights implications of forest governance, given in particular the rise of 'reduced emissions from deforestation and [forest] degradation' (REDD), a set of policy directions endorsed at the Bali COP in late 2007. Forests provide important 'sinks' for GHGs, with the result that their conservation has become a priority for climate change mitigation. As Seymour points out, forest preservation is also critical to adapting to climate changes in many parts of the world. Although the details are still to be decided, REDD schemes will reward the preservation of forests either through direct monetary incentives or emissions credits. In both cases, forests stand to become an even more significant and desirable resource than they already are, which may in turn exacerbate the often fraught relationship that frequently obtains between forest dwellers and forest-dependent peoples, on the one hand, and well-resourced, sometimes state-backed, loggers or other large corporate concerns, on the other hand. Seymour outlines several human rights risks that arise in a context of complex governance and property rights arrangements that exist in many of the world's remaining large forests, and in particular notes the potential trade-offs that may arise where forest protection measures must also incorporate the costs of rights protections. All else being equal, where private ordering is preferred over public, resources will flow to forests with maximum carbon-storage potential and minimal rights implications rather than to those forest communities most in need. Such trade-offs might be avoided by building strong safeguards into REDD regimes from the outset and ensuring that a clear view of the public interest – both global and local – remains uppermost in forest management systems as they evolve under REDD.

Philippe Cullet's thoughtful contribution to this volume reflects upon the notions of vulnerability and equity, two significant planks of the nascent climate regime, and examines their human rights dimensions. An association between human rights and vulnerability has long been established in international instruments, notably the ICESCR, even if there is little agreement on the precise legal obligations of governments toward vulnerable populations. Vulnerability to the effects of global warming, by comparison, has long been front and centre of the climate change debate, in part because predicting and assessing vulnerability is the key to any successful adaptation policy. Equity is a more difficult principle, as human rights instruments rarely encourage differential treatment, tending rather to a formalist equalitarian approach to law. Cullet, however, notes that climate change solutions that exacerbate or worsen basic human rights protections for any given group, particularly those who are already vulnerable, cannot be viewed as 'equitable'; attending to 'equity' thus involves accounting for human rights, at a minimum to establish legitimacy. However, Cullet observes that the Kyoto Protocol's 'flexible' mechanisms – emissions trading and the clean development mechanism – have not (so far) been constructed with a view to prioritising the rights and needs of vulnerable persons or the development needs of vulnerable countries. Using India as a case study, Cullet suggests how the regime might look if vulnerability resided at its heart. Pursuing the argument further, he suggests the need for a radical rethink of rights over the use of the air – as a common heritage of mankind – in place of the default rights to emit GHGs assumed by the Kyoto mechanisms and distributed narrowly among existing polluters.

Paul Hunt and Rajat Khosla give an overview of the international right to the highest attainable standard of health ('right to health') in the context of climate change, and show that this fundamental human right not only encompasses access to timely and appropriate medical care, but also to the underlying determinants of health, including a safe environment. Observing that climate change represents an extremely grave risk to the health of individuals, communities and populations, especially those living in poverty in developing countries, they argue that states have an obligation, arising from the right to health, to take reasonable steps to slow down and reverse climate change. They give particular attention to four elements of the right to health that are especially important in the climate change context: attention to the vulnerable and disadvantaged; active and informed participation; international assistance and cooperation; and monitoring and accountability. Regarding international assistance and cooperation, Hunt and Khosla argue that high income countries have a human rights responsibility to help developing countries establish healthy environmental conditions. They also argue that the right to health requires that monitoring and accountability mechanisms be strengthened in relation to climate change, including measures to check whether high income countries are fulfilling their responsibility of international assistance and cooperation in health matters.

Jon Barnett also identifies vulnerability to climate change as a central human rights concern. Using three case studies, Barnett examines how poor human rights fulfilment in a country can itself exacerbate vulnerability to climate change impacts. In East Timor and, in quite different ways, in China, inadequate access to human rights protections has left ordinary citizens poorly equipped to prepare for the expected ravages of a changing climate, Barnett contends. By contrast, the extreme vulnerability of populations in the Pacific atolls, Barnett's third case study, cannot be attributed to human rights weaknesses in those countries, essentially because their extremity – the possible disappearance of the territories themselves – poses existential problems that transcend the political or legal terrain.

John Mutter and Kye Mesa Barnard examine the effects of economic and social vulnerability in the context of natural disasters. While neither of their two case studies, Hurricane Katrina in New Orleans nor Cyclone Nargis in Myanmar, can be ascribed with certainty, much less solely, to climate change, they provide good studies of our current preparedness to deal with events of a kind that will increase in frequency and intensity as climate change takes hold. Mutter and Barnard describe the conditions and phases that transform 'natural extreme events' into 'human disasters': the evolution of vulnerability before an event; the event itself; and the recovery that follows. The first and last phases depend upon human agency: both are exacerbated where human rights protections are poor or absent, as they were in the case of both Nargis and Katrina. In New Orleans in particular, mortality rates were higher in poorer areas – the worst effects of the hurricane were exacerbated by poor rights protections – low rates of healthcare, poor housing and low levels of education which combined with poor access to information and inadequate transport, to produce far higher levels of risk for a section of the population who were, in any case, disproportionally exposed to discrimination. A plausible link can be drawn, although the authors do not do so, between the inaction of the United States government on climate change, as sanctioned in *Massachusetts* v. *EPA*, on the one hand (at time of writing in late 2008), and the absence of strong protections of basic social rights (rights to health, shelter, food and water, for example), on the other hand. The dearth of legal responsibility at any government level, for either cause or effect, allows for policies that do not merely neglect, but actively harm, vulnerable populations.

Finally, in conclusion, on the basis of these and other sources, I provide an overview of the potential fit between human rights and climate change, as two mediating languages of justice and two differing arenas of international law. I begin with a glance at Thomas Pogge's analysis of the structural exacerbation of global poverty in international law and its potential application to the climate change phenomenon. I then turn to two areas of overlap between climate change and human rights – one where common themes are easily neglected (emissions trading) and one where they are readily exploited (procedural rights), before finally following up the theme of human rights as thresholds, suggested by Simon Caney, for its potential policy applications. While the scale of the climate change challenge is recognised within, and indeed drives, environmental scholarship and negotiation, its transformational power has largely passed the world of human rights law by. This book is intended as a contribution to opening discussion about the challenge climate change holds out for human rights, a challenge which, if inevitable, does not appear, at this juncture, easily manageable.

References

Agarwal, Anil and Sunita Narain, *Global Warming in an Unequal World: A Case of Environmental Colonialism* (New Delhi: Centre for Science and Environment, 1991).

Alston, Philip, 'People's Rights: Their Rise and Fall' in Philip Alston (ed.), *People's Rights* (Oxford University Press, 2001).

Andreassen, Bård A. and Stephen P. Marks (eds.), *Development as a Human Right: Legal, Political, and Economic Dimensions* (Cambridge, MA: Harvard University Press, 2006).

Baer, Paul, Tom Athanasiou and Sivan Kartha, *The Right to Development in a Climate Constrained World* (Washington, DC: Heinrich Böll Foundation, 2007).

Brooks, Nick, W. Neil Adger and P. Mick Kelly, 'The Determinants of Vulnerability and Adaptive Capacity at the National Level and the Implications for Adaptation', *Global Environmental Change*, 15 (2005), 151.

Craven, Matthew, *The International Covenant on Economic, Social and Cultural Rights: A Perspective on Its Development* (Oxford: Clarendon Press, 1998).

German Advisory Council on Global Change, *Climate Change as a Security Risk* (London: Earthscan, 2008).

Guèye, Cheikh, Abdou Salam Fall and Serigne Mansour Tall, 'Climatic Perturbation and Urbanization in Senegal', *The Geographical Journal*, 173 (2007), 88.

Hansen, James, Makiko Sato, Pushker Kharecha, David Beerling, Valerie Masson-Delmotte, Mark Pagani, Maureen Raymo, Dana L. Royer and James C. Zachos,

'Target Atmospheric CO_2: Where Should Humanity Aim?' (2008). Available at: www.columbia.edu/~jeh1/2008/TargetCO2_20080407.pdf.

Holmes, Stephen and Cass Sunstein, *The Cost of Rights: Why Liberty Depends on Taxes* (New York: Norton, 1999).

Inter-Agency Standing Committee, *Protecting Persons Affected by Natural Disasters: IASC Operational Guidelines on Human Rights and Natural Disasters* (Brookings–Bern Project on Internal Displacement, 2006).

International Council on Human Rights Policy, *Beyond Voluntarism: Human Rights and the Developing International Legal Obligations of Companies* (ICHRP, 2002).

Climate Change and Human Rights: A Rough Guide (ICHRP, 2008).

IOM, *Migration and Climate Change*, IOM Migration Research Series, 31 (2008).

IPCC, *Climate Change 2007: Impacts, Adaptation and Vulnerability, Contribution of Working Group II to the Fourth Assessment Report of the IPCC* (Cambridge University Press, 2007).

Climate Change 2007: Mitigation of Climate Change, Contribution of Working Group III to the Fourth Assessment Report of the IPCC (Cambridge University Press, 2007).

Technical Summary: *Climate Change 2007: Impacts, Adaptation and Vulnerability, Contribution of Working Group II to the Fourth Assessment Report of the IPCC* (Cambridge University Press, 2007).

Technical Summary: *Climate Change 2007: Mitigation of Climate Change, Contribution of Working Group III to the Fourth Assessment Report of the IPCC* (Cambridge University Press, 2007).

IUCN, 'Gender Aspects of Climate Change' (2007). Available at: www.iucn.org/en/news/archive/2007/03/7_gender_climate_change.pdf.

Indigenous and Traditional Peoples and Climate Change, Issue Paper, March 2008.

Lohmann, Larry, *Carbon Trading: A Critical Conversation on Climate Change, Privatization and Power* (Uppsala: Dag Hammarskjöld Foundation, 2006).

Mace, M. J., 'Funding for Adaptation to Climate Change: UNFCCC and GEF Developments Since COP-7', *Review of European Community and International Environmental Law* (*RECIEL*), 14 (2005), 225.

McCarthy, James J., Osvaldo F. Canziani, Neil A. Leary, David J. Dokken and Kasey S. White (eds.), *Climate Change 2001. Impacts, Adaptation and Vulnerability. Contribution of Working Group II to the Third Assessment Report of the Intergovernmental Panel on Climate Change* (Cambridge University Press, 2001).

McGray, Heather, Anne Hammill, Rob Bradley with Lisa Schipper and Jo-Ellen Parry, *Weathering the Storm: Options for Framing Adaptation and Development* (Washington, DC: World Resources Institute, 2007).

Müller, Benito, 'Climate of Distrust: The 2006 Bonn Climate Change Adaptation Fund Negotiations' (Oxford Institute for Energy Studies, 2006).

'The Nairobi Climate Change Conference: A Breakthrough for Adaptation Funding', OIES Working Paper (2007).

OHCHR, *Human Rights in the Administration of Justice: A Manual on Human Rights for Judges, Prosecutors and Lawyers* (United Nations, 2003). Available at: www.ohchr.org/Documents/Publications/training9chapter16en.pdf.

Oxfam International, 'Adapting to Climate Change: What's Needed in Poor Countries and Who Should Pay', Oxfam, Briefing Paper, 104 (2007).

Ribot, Jesse, 'The Causal Structure of Vulnerability: Its Application to Climate Impact Analysis', *GeoJournal*, 35 (1995), 119.

Robb, Cairo A. R. (ed.), *International Environmental Law Reports Vol. 3, Human Rights and Environment* (Cambridge University Press, 2001).

Salomon, Margot E, 'Towards a Just Institutional Order: A Commentary on the First Session of the UN Task Force on the Right to Development', *Netherlands Quarterly of Human Rights*, 23 (2005), 409.

Sands, Philippe, *Principles of International Environmental Law* (Cambridge University Press, 2003).

Shelton, Dinah, 'Environmental Rights' in Philip Alston (ed.), *People's Rights* (Oxford University Press, 2001).

Shue, Henry, 'Subsistence Emissions and Luxury Emissions', *Law & Policy*, 15 (1993), 39.

Smit, B. and O. Pilifosova, 'Adaptation to Climate Change in the Context of Sustainable Development and Equity', *Climate Change 2001, Impacts, Adaptation and Vulnerability, Contribution of Working Group II to the Third Assessment Report of the Intergovernmental Panel on Climate Change*, James J. McCarthy, Osvaldo F. Canziani, Neil A. Leary, David J. Dokken and Kasey S. White (eds.) (Cambridge University Press, 2001).

Stern, Nicholas, *Stern Review on the Economics of Climate Change*, HM Treasury, Office of Climate Change (Prepublication) (2006).

Tan, Celine, 'Mandating Rights and Limiting Mission Creep: Holding the World Bank and the International Monetary Fund Accountable for Human Rights Violations', *Human Rights and International Legal Discourse*, 2 (2008), 79.

UNDP, *Human Development Report 2007/2008: Fighting Climate Change: Human Solidarity in a Divided World* (UNDP, 2007).

PART I

Rights perspectives on global warming

1

Competing claims: human rights and climate harms

STEPHEN HUMPHREYS*

To speak of climate change and human rights in the same breath is not merely to draw a connection between the activities that generate global warming and the subsequent deterioration of many human rights – a connection that is self-evident even if relatively unexplored (see my Introduction to this volume). It is also to juxtapose two very different bodies of professional expertise and spheres of international law, which comprise two disciplines or discourses, or ways of thinking and speaking about society and government and the place of the individual, about public and private obligations, responsibilities and solutions – and, indeed, about international cooperation and obligation. In the following overview of the relation between human rights and climate change, I begin by taking a wide-angle view of the justice questions that climate change raises, and then look at some of the ways in which the existing human rights regime and nascent climate regime treat them, in order to identify common themes and compatibilities between two regimes whose mutual disregard to date offers a good example of a phenomenon that has been called the 'fragmentation of international law'.[1]

That contact between these two disciplines has been largely absent to date is on its face surprising. Quite aside from the human rights implications of climate change, these are two areas of activity, whose recent evolution is contemporaneous, each accelerating around the end of the

* Research Director, International Council on Human Rights Policy. See Acknowledgements above, for the background to this article and a list of those who have contributed comments to some sections of it. Special thanks to Robert Archer for valuable editorial and substantive suggestions.

1 See UN Doc. A/CN.4/L.682, Fragmentation of International Law: Difficulties Arising from the Diversification and Expansion of International Law, Report of the Study Group of the International Law Commission Finalized by Martti Koskenniemi, 13 April 2006. This issue is further discussed in the Conclusion to the present book.

1980s. The creation of the Intergovernmental Panel on Climate Change (IPCC) in 1988 and the subsequent swift progress to a climate treaty in 1992 were coincident with the collapse of communism in Eastern Europe, which provided the context for reinvigorated interest in human rights principles, leading to their reaffirmation in the Vienna Declaration of 1993. Both regimes were supposedly free of the ideological paralysis of the Cold War years.

The disciplinary disconnect between human rights and climate change is all the more peculiar as both are primarily concerned with questions of justice, albeit in different guises. The differences between the two approaches – of style, tone and objective – are worth considering. From the moment climate change arose as an international concern, it raised justice issues that were essentially new to international law. These derived from its specifically global nature – the fact that actions taken in one part of the world had consequences in other parts – and raised largely unprecedented inter-state problems: of cause and effect; of perpetration and victimhood; of the relative burdens of impacts; and of the distributional consequences of any actions taken to address it. More than any previous international concern, the very nature of climate change drew a direct line between the wealth and lifestyles of some and the suffering of others. In this, the emerging phenomenon of climate change revived (indeed almost coincidentally) themes that had first arisen (unsuccessfully) in the immediate post-colonial period, when Third World states sought a legal basis for resource redistribution between states, based on redressing the injustices of the colonial past, on the one hand, and on notions of global solidarity in a post-war and post-colonial environment, on the other hand.

Climate change further problematised the language of 'development' itself: continued economic expansion in poor countries along the lines rich countries had followed in the past was no longer sustainable. The resolution of these various issues in the United Nations Framework Convention on Climate Change (UNFCCC) revitalised the notion of global solidarity and revived a diluted subset of earlier efforts in that direction, but now delinked from critique of the colonial past. That fundamental framework remains in place today, even if it is still incomplete, lacks certain key participants and has assumed some innovative tendencies, in particular through the Kyoto Protocol. The emphasis in the nascent climate change regime has remained on pragmatism, problem-solving and compromise, with the law viewed primarily as the expression of agreement between the parties based on contingent needs and capacities.

The authority of human rights law, by contrast, has always been premised on absolutes and universals rather than on contingencies and compromise. Human rights discourse not only favours, but actively promotes court or court-like resolution and interpretation over negotiation or mediation. Over the period of the emergence of a climate change regime, moreover, the general trend in human rights appears to have been to sharpen precisely those tendencies that seem least useful or appropriate to the problems posed by climate change. It is not just that the rights most affected by climate change are those usually called 'social, economic and cultural' (which still struggle to appear robust in the rule of law context presupposed by human rights[2]), it is also that a certain performative spirit prevails – freedom from government, the prioritisation of individual entitlement over 'solidarity', problem resolution through competition and/or litigation and a general preference for adversarial contestation over collective action and for private over public ordering – that sits uncomfortably next to the language and spirit of environmental negotiation, which is primarily concerned with *management*. Furthermore, human rights have evolved as a 'political' language that characteristically avoids overt economic detail or dispute; it is not that human rights issues do not have economic dimensions, but rather that recourse to human rights law and language has tended to allow or facilitate the suppression or neglect of economic inquiry.

This distinction should not be overdrawn. Environmental conflicts are frequently resolved through litigation, including in human rights courts; in such cases courts typically ratify arguments that associate human rights and environmental harms, drawing on and contributing to human rights case law.[3] Furthermore, human rights have come to provide a primary language for the expression and contestation of justice claims, including distributional claims (albeit generally framed in terms of discrimination), on which environmental issues frequently turn. As such, it was always likely that, sooner or later, the implicit relevance of human rights to climate change would become explicit, a transition that appears to be taking place today. And yet, since the specificity of climate change as an environmental problem lies in its global nature, the scope for activating human rights *law* is probably limited, as I will explore further below.

[2] On this, see Humphreys (2006).

[3] On the various incarnations of a 'right to a clean environment', see Adelman in this volume. On environmental litigation, see Shelton, Chapter 3 of this volume; Cairo A. R. Robb (ed.), *International Environmental Law Reports Vol. 3: Human Rights and Environment* (Cambridge University Press, 2001).

The present chapter first explores four different justice demands that have informed climate change negotiations and that provide the contested ethical backdrop against which human rights concerns must be placed. Second, it examines whether flexible provisions of the UNFCCC, such as 'equity' and 'common but differentiated responsibilities' (CBDR) can provide a framework for addressing human rights violations attributable to climate change. Third, it discusses the prospects for legal redress for climate change-related harms, drawing on human rights experience in relation to states, on the one hand, and to private actors, on the other hand. It concludes by assessing the gaps in a global system that appears, at first sight, poorly equipped to manage the human rights implications of climate change.

Four justice claims about climate change

Human rights concerns reside, albeit somewhat uncomfortably, within the ethical conundrums posed by the existence and impacts of climate change and the measures taken to treat it. Although the ethical stakes are familiar – they have been raised repeatedly from the earliest days of the climate negotiations – they are nevertheless complex, because they involve justice claims that are different in kind and are not all obviously compatible.

At least four types of justice claim have been raised in the context of climate change. The first and most straightforward arises because the activities of one group of persons – those who overuse the carbon dump – have caused and continue to cause injuries affecting a different (much larger) group, who live in parts of the world likely to be hardest hit by climate change. This claim has the familiar contours of *corrective justice*. A is engaging in activities that are wrongfully injuring B, so A should (i) desist from these harmful actions and (ii) compensate B for any injuries experienced. Initially, this looks like a human rights problem or at least a tort problem: there are actors and injured parties, perpetrators and victims; the question is what mechanisms will serve to stop the perpetrators from acting in ways that are injurious to the victims and will compensate the latter for the harms they have experienced? For reasons discussed further below, tort-like litigation is likely to be more fruitful in the national than the international context, and with regard to past rather than expected harms. Nevertheless, in part precisely because recourse to litigation is likely to be limited, this background justice claim will continue to influence the evolution of climate change responses.

A second justice claim concerns the loss of future capacity and potential. The solution to climate change is generally acknowledged to require a steep reduction of greenhouse gas (GHG) emissions globally. But since the path to economic growth and prosperity (as generally understood) has relied until now on fuels and technologies that produce these emissions, a global freeze on their usage will tend to lock-in vast wealth disparities between groups in different regions, without offering any obvious or reliable means of reducing the gap in future. This is a *substantive justice* claim, in that it recognises that an injustice has taken place even though there was no relevant law to ward against it, and the relevant actors were probably acting in good faith (at least, with regard to the atmosphere, until relatively recently), at the time they generated the problem. As such it amounts to something like the following question: how can a solution be found that will effectively reduce global dependence upon GHG emissions without in the process permanently disadvantaging a global majority who were not responsible but who may forfeit their future prosperity? This claim, too, has been front and centre of much climate change debate.

Third, if climate change is viewed, as it often is, as a 'global problem requiring a global solution' (that is, assuming some form of global solidarity) the justice issues again look different. Everyone, after all, is affected by climate change, not just those living in poor countries. The appropriate question is to ask who should pay how much of the cost of dealing with it? How should the burden of solving the problem be distributed? The justice stakes are well described by Henry Shue:

> [F]our questions ... are deeply involved in every choice of a plan for action. (1) What is a fair allocation of the costs of preventing the global warming that is still avoidable?; (2) What is a fair allocation of the costs of coping with the social costs that will not in fact be avoided?; (3) What background allocation of wealth would allow international bargaining (about issues like (1) and (2)) to be a fair process?; and (4) What is a fair allocation of emissions of greenhouse gases (over the long-term and during the transition to the long-term allocation)?[4]

It is in the context of a 'global community' affected by a common problem that the 'polluter pays principle' – which Shue recommends – is applicable.[5] There are two elements to the justice claim presented here.

[4] Shue (1993), 40.

[5] Simon Caney draws a distinction between a 'beneficiary pays' and a 'polluter pays' principle, noting that Shue's formulation tends rather to the former, at least for the current and future effects of past GHG emissions. See Caney (2005).

The claim involves distributional justice, because the emphasis is on allocation of costs and benefits, but it also focuses on *procedural justice* – on constructing mechanisms that will ensure that a just solution can be reached, that the concerns and interests of different stakeholders are heard fairly and that steps are taken as a result. Since *each* of the four claims reviewed in this section has distributive (and redistributive) assumptions and consequences,[6] for present purposes I will characterise this third claim as procedural.

A fourth way to think about climate change is to view it in terms of entitlements derived from prior usage, that is, 'legitimate expectations'. This account, like the second claim above, begins by noting that carbon-intense economies have become hazardous to the global environment through no obvious wrongdoing. When carbon-intensive economies emerged, it was not realised that they posed a profound threat to the environment. Given that many livelihoods (indeed hundreds of millions) now depend upon such economies, a legitimate entitlement has been generated among carbon users that cannot be rescinded arbitrarily in favour of a larger policy goal. A persuasive argument might even be made that compensation should be due *to the polluters* if they are to give up their acquired entitlement to the global carbon dump.[7] At the least, they might expect to have a decisive say, or veto, over the form that any solution takes. Paradoxically, the greater the scale of pollution, the stronger is a given polluter's claim to shape the regime. This might be regarded as a *formal justice* or *rule of law* perspective on climate change, in that it relies upon a strict reading of existing legal norms even though they may seem ill-suited to the problem at hand. It warns against the elimination of private rights in the public interest except under the strictest necessity, and against retroactive penalisation of actions that were legal at the time they were taken. Its strength lies not only in the claim of strict legal rectitude but also in the fact that any GHG abatement regime will be likely to be of immense interest to these actors, who are generally politically powerful.

Each of these four discourses of justice has been present within the climate change debate from the outset – although, unsurprisingly,

[6] Caney (2005), contends that it is precisely because distributional justice saturates the climate change issue that it must be viewed primarily through the lens of ethics and rights.

[7] Robert Nozick argued that any distributional outcome must be just if it results from lawful transactions following from an original just allocation or acquisition. Nozick (1974), 151.

different perspectives have been favoured by different actors.[8] No outcome can satisfy every claim, and in some cases the solutions suggested by each will conflict. This is in part because these various visions of justice conceive of the relevant rights- and duty-bearers in different ways. The first two claims clearly affirm that the primary relationship exists between states; individuals and other private actors are second-order bearers of rights and duties. The richer states, in both pictures, are the primary duty-bearers and the poorer states are (potential) rights-bearers. The third claim need not assume that states are the primary actors, but most versions (Shue's paper, for example) do so in practice. Negotiated regimes that allocate burdens and benefits will inevitably impact on individuals' rights, but the third scheme assumes that these decisions are best made and regulated by means of inter-state negotiation, in which states *represent* (and therefore manage) individual rights within the context of an overriding public interest. In principle, any inter-state agreement will impose duties on private actors, but in practice, the scope of these obligations (regulations) has generally shied away from responsibilisation of the private sector. By contrast, the fourth vision assumes that the primary *rights-bearers* are private, though states remain the primary duty-bearers. If states are to mandate emission cuts in the public interest, they must do so while respecting the rights of individuals. All states might, in principle, be duty-bearers, required to agree a scheme globally that will respect private rights locally.

To some extent, each of these justice claims has generated its own climate change solution. First, international funding for the adaptation needs of vulnerable countries appears as a proposed solution to the problem of corrective justice. It may be conceived in terms of compensation owed by those responsible for global warming to parties who are injured by it (even though there has been no acknowledgement of liability).[9] Technology transfer appears intended to help overcome the second, substantive, justice problem of prohibited carbon-intensive growth. Those in poorer countries agree not to compound a problem

[8] A fifth commonly raised justice claim is not examined here: that is, the claim of future generations to environmental and developmental resources equivalent to, or not significantly worse than, those available to present generations. On 'intergenerational equity' (as this claim is generally known in international law) see Brown-Weiss (1989). The claim is not pursued here under the assumption that, in terms of fundamental human rights, the claims of future generations, viewed locally, are not fundamentally different from those of present generations, viewed globally.

[9] For a refutation of this view, see Caney, Chapter 2 of this volume.

they have not caused on condition that income and development inequalities are not locked-in as a result. The transfer of adaptive and clean technologies (developed on the basis of prosperity derived largely from dirty technologies) is not merely a condition of their engagement in the process, but a condition of their future prosperity. The injustice of locked-in inequality is thus potentially avoided. Third, the claim for procedural justice is partially met through fidelity to the arduous processes of negotiation itself. Indeed, the difficulty of reaching agreement on the appropriate allocation of burdens among states demonstrates the complexity of this task in the light of the presence of each of the other claims outlined above. Arguably, however, other elements of procedural justice – rights to information and participation of the most affected – are poorly treated within the current regime.[10] Finally, the desire for formal justice is presumably met by the emissions trading regime established under the Kyoto Protocol, which grants emissions rights on the basis of prior usage, that are passed on to the primary (mainly private) polluters. Indeed, the scheme has been elaborated in close consultation with affected private actors.[11] The trading regime is sensitive to claims that emissions entitlements were legitimately acquired by these actors. It provides them with a voice in the regime and flexibility in deciding how to alter their behaviour. Private actors have the potential to make a profit while making amends, and effectively may even be compensated by doing so.

A number of observations leap out from the above description. First, it is not clear whether these different discourses of justice can coexist without generating contradictions or inconsistencies. However, the impact of any one claim, if adopted, on the others is not evident, in part because they are discussed by different parties in different venues with relatively little overlap, and in part because the substantive impacts will mainly be felt in the future and cannot easily be predicted. Second, some justice claims have had more practical traction than others. Emissions trading (as exemplified in the EU's ETS), including the clean development mechanism (CDM), is at an arguably more advanced stage than international adaptation funding or technology transfer. It thus appears that the 'entitlements' claim, although it is much less widely advertised than the corrective and substantive justice claims and by no

[10] This theme is further pursued in the concluding chapter to the present volume.

[11] On private shaping of the climate regime see Newell, Chapter 4 of this volume; see also Newell (2000).

means enjoys universal support, has nevertheless been more effective than those other, better known and more widely agreed claims. This counter-intuitive observation is perhaps less surprising when it is remembered that the main actors asserting prior entitlements are also leaders in energy production and distribution and so have immense power to shape debates affecting energy futures. Lavanya Rajamani writes 'It is indeed curious that international law can be read to endorse claims based on historical entitlements, yet deny claims for rectification of historical wrongs'.[12]

While none of the justice claims outlined in this section translates unproblematically into human rights language, they are nevertheless relevant to the present investigation for two related reasons. First, because any human rights claim, particularly where it is innovative, must rely for its force on the recognition of a breach of background normative propriety – the sense that an injustice has been committed or is under way. Second, human rights today occupy much of the space of justice discourse, to the extent that injustices that cannot easily be articulated in human rights terms may appear exotic or abstruse. Climate change, however, arguably presents a challenge to the authority of human rights as the dominant language of justice. If human rights law cannot accommodate these important claims, it risks becoming less relevant in much of the world, particularly those places where the effects of climate change will be increasingly suffered. The injustice of climate change effects is such that the failure of human rights to provide effective remedy can only work against their current hegemonic status (or aspiration).[13] Conversely, if some or all of the justice claims already acknowledged within the climate regime can be refined and successfully channelled through the law, or even just the language, of human rights – or if human rights can provide a basis for choosing between them – both disciplines presumably stand to gain in legitimacy and strength. But neither scenario, on its own, tells us much about the substantive justice outcomes that might result from the importation of a human rights lens into the climate change debate.

Furthermore, neither scenario has been developed as yet. Human rights have not featured, except peripherally, in any of the four justice claims presented above. In principle, they might be applied to further each of the four claims, albeit with degrees of difficulty. Social, economic and cultural rights would appear cogent to the first two justice claims, participatory rights (to public participation and information) seem

[12] Rajamani (2006), 143. [13] See Adelman, Chapter 5 of this volume.

relevant to the third, and classical property rights protections (which also and increasingly enjoy the status of human rights) appear to suit the fourth. There is no particular reason to assume that recourse to human rights law or principles *tout court* would necessarily help to choose between these different claims however. And if hard choices have to be made, there seems little reason to assume that social, economic and cultural rights claims, for example, might 'trump' property rights – in recent decades, the reverse has more often been the case. It is perhaps for this reason that some commentators have opted to focus on the third claim. An embrace of 'process rights' tells us little about which justice claim will win out, but, regardless of the substantive outcome, it prepares those who do not win to accept the outcome.

In this volume, Simon Caney puts forward a view of human rights that might help mediate between the different justice claims outlined here. In the Conclusion I pick up on this idea in more detail and apply it in the context of other concerns taken from the present collection as a whole. In the following sections I first interrogate some of the mediating language already present in the climate change regime for its compatibility with the human rights claims raised by the issue; I then turn to the capacity of standard human rights law notions – liability and accountability – to treat these claims.

Equity and 'common but differentiated responsibilities'

The UNFCCC includes language designed to mediate the various justice claims outlined above. Developing countries argued for a treaty that would recognise three fundamental distinctions between wealthier and poorer countries: different historical (and present day) responsibility for climate change; differing likely impacts of climate change, predicted to be far greater in poorer than richer countries; and different capacity to deal with the problems resulting from climate change and to develop non-carbon intensive energy technologies.

These distinctions are central to the 'principles' laid out in Article 3 of the UNFCCC:

> In their actions to achieve the objective of the Convention and to implement its provisions, the Parties shall be guided, *inter alia*, by the following:
>
> 1. The Parties should protect the climate system for the benefit of present and future generations of humankind, on the basis of equity and in accordance with their common but differentiated responsibilities and

respective capabilities. Accordingly, the developed country Parties should take the lead in combating climate change and the adverse effects thereof.

2. The specific needs and special circumstances of developing country Parties, especially those that are particularly vulnerable to the adverse effects of climate change, and of those Parties, especially developing country Parties, that would have to bear a disproportionate or abnormal burden under the Convention, should be given full consideration.

In principle, as attested by the record of negotiations, the language of 'equity' and of CBDR was introduced precisely to acknowledge the justice claims of developing countries, and in particular to aim at a balance between the differentials of *contribution* to the problem and *capacity* to treat it.[14] The clauses above appear to promise that responsibility would be attributed fairly: that those living carbon-intensive lives in richer countries would be held accountable, to a degree, while those in poorer countries would be compensated for the resulting harms. If that was the promise, however, there is little sign to date that these principles are being fulfilled, or even that they are adequate to translate ethical imperatives into legal obligations.

Equity appears comparatively rarely in international law, and plays an unusual and ill-defined role. The introduction of 'equity' recognises that the law is not always ideally formulated to treat every case; that occasions arise where proper application of formal law may lead to unjust or discriminatory outcomes. The search for 'equitable' solutions under conditions of CBDR seems intended to compensate for the shortcomings of formal equality under law by acknowledging the reality that substantive differences exist between equally sovereign states. Equity might be thought to provide a means to reach a decision, given its association with deliberation and fairness (procedural and substantive justice). According to Dinah Shelton, '[t]he procedural and substantive dimensions of equity are often perceived as inter-related, based on the assumption that fairer proceedings lead to fairer outcomes'.[15] But the two may also exist in tension, insofar as 'substantive' justice outcomes are often expected to be redistributive, whereas procedures are often designed to entrench formal equality regardless of distributional inequality.[16] In other words, even if it is reasonably clear that equity is *not* the application of general rules

[14] The present account relies heavily on Rajamani (2006); on 'contribution and capacity', see ibid., 129–33.

[15] Shelton (2007a), 640. This section relies heavily on this text and also on Shelton (2007b).

[16] Shelton (2007a), 640, citing Franck (1995), 7–9.

uniformly in all contexts, it is much less clear what it *is*.[17] '[D]ebate exists on the appropriate principles to determine equitable allocation, e.g. whether decisions should be based on need, capacity, prior entitlement, 'just deserts', the greatest good for the greatest number, or strict equality of treatment', says Shelton.[18] Philippe Sands explains further:

> In the absence of detailed rules, equity can provide a conveniently flexible means of leaving the extent of rights and obligations to be decided at a subsequent date ... In many respects, UNCED [the 1992 United Nations Conference on Environment and Development at which the UNFCCC was signed] was about equity: how to allocate future responsibilities for environmental protection between states which are at different levels of economic development, which have contributed in different degrees to particular problems, and which have different environmental needs and priorities.[19]

In short, equity appears in the UNFCCC in part because while there was agreement at the time of its signature that action must be taken, there was much less agreement about who should pay the costs of taking action, and how. The insertion of equity acknowledges in principle the validity of different justice claims, while postponing any decision on their relative merit.[20] This has been especially true in the climate regime, where the various different justice claims are unusually knotty and interdependent. While not an empty gesture, then, equity does not amount to a redeemable promise in favour of any particular outcome.

[17] Equity is, therefore, an uneasy subject of judicial pronouncement. In *North Sea Continental Shelf Cases* (1982), the ICJ proclaimed that 'the justice of which equity is an emanation is ... justice according to the rule of law: which is to say that its application should display consistency and a degree of predictability; even though it looks with particularity to the more peculiar circumstances in an instant case, it also looks beyond it to principles of more general application', *North Sea Continental Shelf Cases* (1982) ICJ Rep. 18. Shelton (2007a, 647) notes that in this passage 'the court seeks a degree of legal certainty in its choice and application of norms, but it must take into account the facts, the situations, and the specific interests or claims of the parties. Equitable norms themselves provide no guidance in selecting among the various facts or factors that could weigh in the decision. Thus, an element of subjectivity is probably present in all efforts to achieve an equitable result'. Commenting on the same case, Sands (2003, 153) describes the court's view negatively, as follows: 'equity was not an exercise of discretion or conciliation or the operation of distributive justice'.

[18] Shelton (2007a), 653. [19] Sands (2003), 262.

[20] In short, to say that a solution must be 'equitable' says little about what that solution should look like. Sands (2003), 152: 'in applying equity in [environmental] treaties, it will be proper to establish its meaning in the context of its use in a particular treaty. [H]owever, treaties rarely provide a working definition of equity ...'.

CBDR, the second and closely related Article 3 principle, might appear to hold out greater hope of a resolution. It has a pedigree in international environmental law beginning with the Stockholm Declaration of 1972 and continuing through to the declaration of the World Summit on Sustainable Development in 2002. Its definitive expression occurs in the (non-binding) 1992 Rio Declaration on Environment and Development, Principle 7:

> States shall cooperate in a spirit of global partnership to conserve, protect and restore the health and integrity of the Earth's ecosystem. In view of the different contributions to global environmental degradation, States have common but differentiated responsibilities. The developed countries acknowledge the responsibility that they bear in the international pursuit of sustainable development in view of the pressures their societies place on the global environment and of the technologies and financial resources they command.

Here is a clear recognition that richer countries are more at fault for 'global environmental degradation' and should, therefore, play a greater role in mitigating the damage elsewhere while also contributing to 'sustainable development'. The principle would thus seem to support the first two justice claims outlined above. Yet, in the transition from the (non-binding) Rio Declaration to the (more binding) UNFCCC Article 3 much of this strong language was lost or removed at the insistence of the wealthier states.[21]

As Lavanya Rajamani makes clear, CBDR in the UNFCCC nevertheless entails a positive obligation on wealthier countries to 'assist' poorer countries. But the terms are narrow. Industrial countries agree to provide 'new and additional financial resources' to developing countries to meet the 'agreed full incremental costs' of complying with their commitments and to cover the 'agreed full costs' of their reporting obligations.[22] The wording skews contributions toward funding mitigation in developing countries, rather than adaptation (where human rights needs are arguably most urgent). Its reference to 'incremental costs', for example, makes sense for mitigation activities, where the GHG emissions of a given activity can ordinarily be reduced at an additional cost – but this seems inappropriate for adaptation activities, where costs are likely to be wholly new, like the causes they address.[23]

In sum, whereas the treaty references to equity and CBRD make clear that climate change responsibilities are relative and differentiated, and

[21] For a full account, see Rajamani (2006), 137 and 196–7. [22] Rajamani (2006), 108–9.
[23] McGray *et al.* (2007), 33; Mace (2005), 226–8, 244, 335–6.

underpin some binding requirements on rich parties to fund adaptation by poor parties, these principles in themselves appear inadequate to creating a binding obligation on the perpetrators of climate change to compensate its victims. So whereas they may be invoked in support of the first two justice claims outlined above, they appear to establish only weak obligations in that regard.[24] In this context, might human rights law help?

International human rights law presupposes a world of formally equal states. At first glance there is little scope for any arrangement (such as 'equity') that would disturb the supposed universality of human rights. Nevertheless, a principle similar to CBDR operates in the International Covenant on Economic, Social and Cultural Rights (ICESCR), which implicitly acknowledges differences in capacity (if not responsibility), when it speaks of 'progressive realisation' of the Covenant rights.[25] Indeed, the ICESCR stipulation (in Article 2(1)) that developing countries should use international assistance first and foremost to attend to social and economic needs at home receives indirect support in the UNFCCC, in a further application of the CBRD principle found in Article 4(7):

> The extent to which developing country Parties will effectively implement their commitments under the Convention will depend on the effective implementation by developed country Parties of their commitments under the Convention related to financial resources and transfer of technology and will take fully into account that economic and social development and poverty eradication are the first and overriding priorities of the developing country Parties.

This important proviso introduces a key condition that must kick in before poorer countries undertake caps of their own. Not only must rich countries fulfil their obligation to provide financial resources and relevant technology, but also, in language that echoes the ICESCR, 'economic and social development' and 'poverty eradication' are recognised

[24] The most significant application of the principles of equity and CBDR in the climate regime is the central role they play in the Kyoto Protocol, in setting emissions caps for developed but not developing countries. Kyoto is unusual in that the key commitments are taken on by only a subset of parties. It is precisely this aspect of the Kyoto Protocol that has fuelled US opposition to ratification. See in this regard, Biniaz (2002); Weisslitz (2002). But Kyoto's narrow interpretation of CBRD nevertheless leaves other climate change consequences and demands unaddressed.

[25] See Craven (1998), 144–52; Rajamani (2006), 20–4, who also points out that the legality and frequency of reservations to human rights treaties involves a *de facto* licensing of differential treatment under human rights law.

as the 'first and overriding priorities' for poor countries. The UNFCCC does not, therefore, restrict adaptation assistance and technology transfer to a global mitigation effort, it also places them in the context of adaptation and the fulfilment of social and economic rights in those countries. This requirement fits easily with the corresponding obligation in the ICESCR on developing countries to use international assistance to those same ends. Together, the UNFCCC and the ICESCR appear to require that international assistance be made available, and that efforts to address climate give priority to social and economic rights fulfilment.

Overall, nevertheless, the picture is hardly edifying. Even though (i) actions by rich countries resulting in global warming lead to the non-fulfilment or violation of human rights in poor countries of an increasingly severe and extreme nature; that (ii) the great majority of states parties to the UNFCCC also have obligations to respect and protect human rights; that (iii) rich countries are obliged by the UNFCCC to assist poor countries to tackle climate change; and that (iv) poor countries are obliged by the ICESCR and by UNFCCC, Article 4(7) to channel resources made available by rich countries toward economic and social development first and foremost – despite all these circumstances, the legal obligation on rich countries to provide redress for harms caused by their actions in developing countries still appears to be extraordinarily elusive. At best, the increasingly clear evidence of harm, including human rights violations, might contribute to the 'pressure' on wealthier countries expressed in Rio Principle 7, to make amends in ways that go beyond the mere encouragement of mitigation measures (in developing countries as well as at home). It presumably requires robust support for adaptation at a minimum, and substantial transfers of relevant technologies. But translating these implications into practice evidently requires further work.

State responsibility and private liability

If my neighbour converted her bungalow into a palace, and in the process directed a channel of toxic sludge through my garden, killing off the sheep and herb garden upon which I depended for food and income and leaving me and my family destitute, all else being equal, I would have a good case against her in a court of law.[26] The case would remain good even if my neighbour could show the splendour of her new life and the

[26] This example borrows from Caney (2005).

difficulty of returning to the cramped space of her original bungalow. It would probably be good enough to ensure that my neighbour took steps to rectify or avoid the situation rather than let it go to court. The questions of justice in this hypothetical example look fairly clear-cut.

Viewed from the perspective of the injured, climate change, too, looks fairly clear-cut. Yet the international legal system is simply not constructed to deliver justice in situations of this kind. More than most global issues, climate change throws into relief the shortcomings of transnational justice arrangements, given the scale and intimacy of global interdependence that drives the problem and must also drive its solutions. The sheer difficulty of locating a judicial venue or attaching responsibility in relation to climate change highlights the inadequacies of the world's institutions to that end. Human rights litigation (and tort litigation generally) ordinarily works by addressing specific injuries caused by specific perpetrators and experienced by specific victims, who must have standing to bring the case before a competent tribunal. Litigation of harms resulting from climate change is troubled on almost every count. No-one doubts that climate change has victims – specific individuals who undergo suffering due to climate change, such as, for example, contracting a tropical disease in northern Italy, or losing a season's (or a decade's) crop to drought in Sahelian Africa. But the events that create such victims result from numerous diffuse acts performed by countless individuals in scores of locations, generally unrelated to one another. And the actual harms experienced are only indirectly linked, at best, to any particular act or person.[27]

Yet, for the individuals involved, things need not be so complex. A given victim of climate change-related harms can usually show a specific injury. The real dilemmas arise, then, in identifying a perpetrator, cause and form of redress. To take an extreme example, no single act caused the gradually warming temperatures in Rimini that created conditions for tiger mosquitoes to survive winter and breed, contributing to an outbreak of chikungunya there in 2007.[28] Many intermediate actors might be blamed for having allowed the outbreak to happen: the passenger on the plane that brought the disease; the airline that allowed him or her to board; the public authorities in Rimini that allowed the mosquitoes to breed. But if a key background cause is warming winters in Italy,

[27] For discussion of climate change-related litigation see generally Mank (2005); Okamatsu (2006); Posner and Sunstein (2007); Gupta (2007).

[28] Rosenthal (2007).

responsibility might be thought to lie, at least to a degree, with the countless GHG emitters around the planet whose cumulative actions have led to global warming. This group can hardly be sued.

The group can, however, be broken down. Everyone emits GHGs, but some emit more than others – only about 20 per cent of the world's population cumulatively and systematically overuse the global carbon dump. In these cases, the governments of a handful of countries would appear to be responsible, at first blush, for allowing such emissions to take place. More narrowly still, some industries – the oil, hotel, airline and automobile industries, for example – are directly or indirectly responsible for a predominant share of global emissions, enough, at least, to constitute an identifiable source of harm.[29] Are there potential openings for liability? In the following, I look, first, at states then at private parties.

Governments and public actors

The state might be thought to be responsible for harms resulting from climate change both as a direct polluter and also for failing to regulate private emissions of GHGs. In a 2007 US Supreme Court ruling, in the case of *Massachusetts* v. *EPA*, the Environmental Protection Agency (EPA) (a federal executive agency) was sued because it had failed to regulate GHG emissions, and thereby reduce the amount produced in the United States (and thus globally). As the EPA did not act and the federal government did not require it to act (as they could have done), both might be seen as responsible. Had they acted, some injuries from climate change, including in Massachusetts, might conceivably have been avoided.[30] In principle, a similar argument might be applicable globally. Governments everywhere can regulate GHGs, and where they fail to do so, resulting in harm, they might be held responsible for injuries to other states by act or omission. In principle, states or state entities might be sued either by private actors or by other states.

For countries that contribute tiny amounts to global emissions, it is implausible to suggest that any state action could meaningfully impact on

[29] Private companies are among the primary responsible actors in GHG emission – for example, according to the Pew Centre on Global Climate Change, Royal Dutch Shell alone emits more GHGs than the United Kingdom if downstream emissions are included. See Pew Center (2006).

[30] *Massachusetts* v. *Environmental Protection Agency*, 549 U.S. 497 (2007). The fact that the plaintiff too was a state entity weighed in the Court's ruling, since the state may act as *parens patriae* on behalf of its citizens' health and welfare. See Shelton, Chapter 3 of this volume.

global warming. Even large emitters, however, can perhaps claim that they *are* acting merely by dint of engaging in prolonged negotiations to regulate emissions. Furthermore, rich governments could reasonably claim that it would be foolhardy to take steps to reduce emissions without first having the agreement of other states: a country that did so would lose the benefits of a carbon economy while nevertheless suffering the consequences of climate change.[31] Ultimately, governments must generally act to benefit a parochial 'national' interest, and this need not always coincide precisely with the global interest, at least not in the short to medium term.[32] Add to this the fact that some rich countries may actually stand to benefit (in terms of local climate and food produce) from warming of 2°C or thereabouts, providing a perverse incentive for them not to act precipitously.

Assuming that two conditions are met – that appropriate legal and institutional tools are available and that a plausible case can be made against state actors – a plaintiff would still generally have to have been harmed on the territory of the relevant state in order to raise a complaint. It is not impossible to sue foreign governments or officials for acts or omissions that have resulted in harms to individual non-citizens in other territories, but the barriers to doing so are high.[33] Individuals have standing to sue states before certain international institutions, such as the European Court of Human Rights (ECtHR). Human rights lawyers (and others) may also approach international human rights fora, such as the UN's Human Rights Committee.[34] Once again, however, a victim would

[31] This argument was put forward by the US Government in the case *Massachusetts* v. *EPA*. The fact that major emitting countries (such as China) were not bound by Kyoto targets meant, so it was claimed, that any US actions would be ineffective as well as painful.

[32] Strong arguments can be made that acting to stop or slow climate change might *not* be in the 'public interest' – or not, at least, in the narrow national interest of states that have much to lose from economic restructuring but little to fear from limited global warming ('the American way of life is not up for negotiation').

[33] For such cases in US courts, see Dellapenna (1988). Where evidence of egregious harms is forthcoming, the relevant legal framework is long-arm jurisdiction in certain national courts or, at international level, the International Criminal Court. These systems, which may involve trying public officials in a private capacity, are discussed below.

[34] The Committee on the Elimination of Racial Discrimination has already spoken out against violations of the rights of indigenous persons in Indonesia, in the context of large state-driven biofuel plantations. CERD Concluding Observations on Indonesia, 71st session, CERD/C/IDN/CO/3 (15 August 2007), para. 17. Further background is available in the NGO submission to CERD entitled 'Request for Consideration of the Situation of Indigenous Peoples in Kalimantan, Indonesia, under the United Nations Committee on the Elimination of Racial Discrimination's Urgent Action and Early Warning Procedures' (6 July 2007).

ordinarily need to have been injured on the territory of the state in question in order to bring suit. As a recent treatise on the ECtHR notes:

> The case law of the Court demonstrates that its recognition of the exercise of extra-territorial jurisdiction is exceptional: it has done so when the respondent State, through the effective control of the relevant territory and its inhabitants as a consequence of military occupation, or through the consent, invitation or acquiescence of the authorities of that territory, exercises all or some of the public powers normally exercised by the latter.[35]

It is unlikely that climate change harms will meet this narrow test, although the law may yet evolve, given the scale of the unfolding catastrophe that climate change could represent. Essentially, however, petitionable tribunals such as the ECtHR operate rather as courts of final appeal for harms at national level rather than as fora for the arbitration of transnational harms.

Once climate change really gets under way, and produces victims *within* countries with strong legal redress, such as the United States and many European states, cases by nationals against their own governments are likely to proliferate. So far, the United States leads the way – no doubt as much because little real federal action to treat climate change has been taken so far as because of the generally litigious environment.[36] Outside the United States, climate change-related litigation is also underway in Germany (where export-credit agency support for fossil fuel-related commercial activities have been challenged); Nigeria (where Shell and the national oil company are being sued for damages associated with gas-flaring, although the suit is not framed as climate change-driven); and Australia (challenging the use of coal and claims to use 'clean coal' in power stations).[37] Even if such cases are successful, however, they will still have a very limited capacity to address the broader human impacts of climate change. For one thing, measures that target individual governments will have only a limited overall impact on global emissions (for reasons that will be further elaborated below). For another, the main victims of climate change will not be resident in the wealthy polluting

[35] Van Dijk *et al.* (2006), 21.

[36] Results so far have been mixed. See *Green Mountain Chrysler-Plymouth-Dodge et al.* v. *Crombie et al.* (ruling of 12 September 2007); *People of the State of California* v. *General Motors et al.* (ruling of 17 September 2007); *Center for Biological Diversity* v. *National Highway Traffic Safety Administration*, 508 F.3d 508, 547 (9th Cir. 2007). For discussion, see ICHRP (2008), 41–18.

[37] Documentation on each of these cases can be found on the Climate Law website at www.climatelaw.org.

countries, but in countries responsible for negligible emissions (such as in sub-Saharan Africa), or where the legal system are unlikely to support legal action of this kind (such as, for example, China currently).

In principle, then, these larger justice claims are perhaps better addressed through the inter-state framework. There are many options but few clear paths to success. Cases alleging state responsibility for transboundary harms, originally recognised in the pathbreaking *Trail Smelter* case in the 1930s, will be hard pressed to succeed in the face of the complexities of climate change and the ongoing international quest for a treaty solution.[38] Given the immense economic considerations involved in climate change action, there are many levers available to large states to dissuade smaller vulnerable states from litigation. Tuvalu – a small island state likely to sink due to climate change-related rising sea levels – sought legal advice on who might be held responsible for the imminent loss of homes and lifestyles, but chose not to pursue litigation.[39] Indeed, the legal options available to vulnerable island states facing disappearance at low warming thresholds have been investigated in some detail.[40] Relevant questions concern the rights of 'environmental refugees', the status of states whose territory disappears and the obligations upon other states to receive climate migrants and provide conditions for the continued survival of their cultures. Should negotiations continue to stall, the effects of climate change on societies and public policy is likely to trigger renewed examination of legal options of this kind.[41]

There may be scope for inter-state litigation elsewhere,[42] for instance, under the WTO dispute settlement process, if, for example, states find

[38] *Trail Smelter Arbitration (United States* v. *Canada)*, 3 R.I.A.A. 1907 (1941); *Gabčikovo-Nagymaros Project* (*Hungary* v. *Slovakia*), Judgment, ICJ Rep. [1997], 7. See Sands (2003), 291–307; Gupta (2007), 78. See Shelton, Chapter 3 of this volume.

[39] Price (2002); Ralston *et al.* (2004); Okamatsu, (2006); Rinnerberger (2006).

[40] See especially E/CN.4/Sub.2/2005/28 (16 June 2005), Expanded working paper by Françoise Hampson on the human rights situation of indigenous peoples in States and other territories threatened with extinction for environmental reasons (16 June 2005). A total of just under half a million individuals are likely to be affected, from the islands of Tuvalu, Nauru, Kiribati, Maldives and the Bahamas (ibid., para. 25).

[41] If so, related ongoing work in the UN's International Law Commission on the governance of 'shared natural resources' (recently moving beyond transboundary groundwater to examine oil and gas deposits) may further buttress the doctrinal weight of such cases. See UN Doc. A/CN.4/580, Fourth Report on Shared Natural Resources: Transboundary Groundwaters, by Mr. Chusei Yamada, Special Rapporteur (March 2007).

[42] Other fora include the international tribunal under the United Nations Convention on the Law of the Sea (UNCLOS), which may provide possible recourse for states that can demonstrate harms resulting from pollution of the marine environment (such as to coral

themselves disadvantaged by the unwillingness of other states to control the emissions of private actors within their jurisdictions.[43] The failure of some states to honour obligations to prevent climate change under the UNFCCC might be viewed as an effective subsidy.[44] Continued subsidies to national fossil fuel industries are also emerging as a possible target of trade action. At a minimum, global warming may easily reignite the long-running disputes between environmentalists and trade fundamentalists, played out inconclusively in well-known cases such as *Shrimp/Turtles*.[45]

Nevertheless, the principle that each state has 'responsibility to ensure that activities within their jurisdiction or control do not cause damage to the environment of other States or of areas beyond the limits of national jurisdiction', although well established in international law, has rarely been invoked in cases of unintentional transboundary injuries.[46] After Chernobyl, for example, as Dinah Shelton and Alexander Kiss point out, other affected states did not sue, and chose not to support the construction of a remedial framework against future harms of this kind: 'The emphatic preference remains measures of prevention rather than cure.'[47] In fact, states have, to date, accepted little liability for the damaging effects of pollution, at least when caused by actions themselves sanctioned under international law.[48] Over time, however, climate change may yet transform the relevant context in this area also. Having accepted emissions commitments under the Kyoto Protocol, for example, states that have missed their targets might presumably have committed

reefs). See Burns (2006). A series of petitions have been filed with UNESCO's World Heritage Committee, which oversees the World Heritage Convention, seeking recognition of the threat of climate change to world heritage sites and requesting special protection. Available at: www.climatelaw.org/cases/country/intl/media/2004Nov17.

43 See IPCC AR4, WGIII, 793; New Economics Foundation (2003); Stiglitz (2006); Cosbey (2007).

44 Stiglitz (2006).

45 WTO Appellate Body Report on US – Import Prohibition of Certain Shrimp and Shrimp Products, WT/DS58/AB/R (12 October 1998). See also GATT Dispute Panel Report on US Restrictions on Imports of Tuna, 3 September 1991, GATT B.I.S.D. (39th Supp.), 155 (1993). For discussion see Howse (2002).

46 UNFCCC preamble. See also Principle 21 of the 1972 Stockholm Declaration; Principle 2 of the 1992 Rio Declaration; Convention on Biological Diversity, Article 3. See Kiss and Shelton (2007), 2.

47 Kiss and Shelton (2007), 4.

48 Pollution in relevant international agreements to date is defined so as to effectively preclude the damage to the atmosphere caused by burning fossil fuels ('loss or damage caused outside the [vehicle] carrying oil by contamination resulting from the escape or discharge of oil from the [vehicle], wherever such escape or discharge may occur'.) See Kiss and Shelton, (2007), 10.

'internationally wrongful' acts.[49] The 'wrong' to other states represented by such breaches will only increase with future agreements and stricter commitments. If the status of international law remains unclear today on many of these issues, it may be that the urgency of climate change threats will force clarification.

Transnational private actors

For a variety of reasons, even if suits against states or state officials were to be successful for particular plaintiffs, they would be unlikely on their own to cause policy changes that would reduce emissions sufficiently to end further harms. One reason for this is that settlements in such cases will usually reflect national rather than global priorities. As wealthy countries can withstand greater climate pressures – since they are better equipped to adapt, and are mostly, in any case, less vulnerable to the most severe climate effects – their *national* thresholds for tolerating climate change are likely to be higher than those elsewhere and than the global threshold. This means that generating sufficient pressure to reduce the number of likely climate victims in rich countries like the United States and in Europe might not in itself contribute to lowering global emissions to a point at which victim rates in highly vulnerable areas, such as sub-Saharan Africa, for example, might drop substantially.

A second, more significant reason that national-level litigation may have only marginal impact is the fundamentally transnational basis of much GHG *production*. Many of the biggest emitters do not operate in only one state: they act globally. The biggest American and European emitters (oil and gas and logging companies) generate many of their emissions abroad, in countries that do not have emissions caps or robust regulation or judicial enforcement. American and European car producers build and sell cars in multiple locations: even if fuel efficiency regulations are introduced in their home countries, they can still be avoided elsewhere. Many of the poorest countries, for example, rely for transport on discarded fuel-inefficient vehicles from the West. Airlines and shipping companies escape global emissions accounting altogether

[49] The Draft Articles on the Responsibility of States for Internationally Wrongful Acts, which codifies state practice in this area, were adopted by the International Law Commission in July 2001. The General Assembly 'took note' of them in December of that year (UN GA Resolution 56/83), but they do not (yet) have treaty status. For an overview, see the contributions to Provost (2002).

(although this is likely to change). Furthermore, if emission levels are evaluated across entire production and supply chains, it is quickly apparent that many of the emissions attributed to developing countries, in fact, serve to improve the lifestyles of the wealthy.[50] In manufacturing, companies can source or outsource the most polluting phases of production to other countries, and many agricultural goods too are consumed far from the country of their production. The current global trading regime tends to ensure that a preponderant amount of global emissions are attributable to goods and processes that leave footprints in multiple countries.

For all these reasons, private actors have both the means and incentives to escape a state-centric emissions accounting regime. Indeed, a perverse effect of CBDR is that firms may seek ways to 'dump' emissions in countries that do not have caps – by moving production to those countries away from countries with strict targets. Such a move would appear to violate the spirit of CBDR as the main beneficiaries are likely to be the population of the wealthier country.

Transnational private liability for human rights, on the one hand, and for environmental harms, on the other hand, are linked both directly and by analogy. The analogical link consists in the fact that it is difficult in much of the world to hold transnational private actors to account, for *either* environmental or human rights damages committed by them, or on their behalf, or with their complicity.[51] The direct link is that acts harmful to the environment may also result in harms to human rights (and, indeed, vice versa). The link can be shown clearly when private activity results in polluted water or air, for example, in industries such as mining, or radiation in the case of the nuclear industry, or in lost livelihoods as a result of large-scale logging. In the case of climate change, this link is more nuanced: fossil fuel extraction and deforestation in poorer countries, often subject to fewer social and environmental regulations, contribute directly to global environmental damage that in turn generates human rights violations in those same countries as well as elsewhere.

In instances of both environmental and human rights harms, liability is weak because effective jurisdiction has been historically difficult to establish in cases of these kinds. The 'transnationality' of private actors is a key source of difficulty: large companies may be incorporated in multiple jurisdictions through subsidiaries or affiliates or shell companies

50 For an account of the extent to which GHG emissions are 'exported', see Simms *et al.* (2007).

51 For a full account, see Clapham (2006).

established for tax purposes or to avoid liability. Companies may also outsource key parts of their supply chains across borders while still controlling them. Often a large company operating across different jurisdictions will choose to apportion its various legal obligations among them, in each case choosing the legal regime most beneficial and cost-effective for the company's interests.

Added to this, the capacity of judicial systems to enforce environmental or human rights protections is uneven. Plaintiffs, especially the most vulnerable, often lack the means to pursue cases through the courts. Local law may be unclear or may not cover an adequate range of environmental harms; local courts may be weak or corrupt or lack independence.[52] The great importance of large foreign companies to some small developing economies may also extend their operational freedom.

For all these reasons, companies that cause harm abroad may not face effective sanctions.[53] Even if cases succeed, damages may be too small to cause a company to desist from harmful behaviour. Given that most of the companies at the centre of massive GHG pollution are based in rich countries, and that most of their products and profits return to such countries, it might seem that protection of the rights of those affected elsewhere would likewise fall to the courts and governments of rich countries. Yet this has not been the practice to date.

Large companies are often difficult to pursue in their home countries also (when one can be identified), due to a variety of legal obstacles, such as the doctrine of *forum non conveniens*.[54] The handful of legal instruments generally noted in this context typically include: long-arm domestic jurisdiction for grave breaches of international criminal law, introduced in many countries on ratification of the Statute of the International Criminal Court (ICC); that Court itself in very limited cases; and, in the United States, the long-arm jurisdiction provided by the Alien Tort Claims Act (ATCA) and Torture Victims Protection Act (TVPA).[55] Although these instruments are

[52] For an overview, see Open Society Justice Initiative (2006).

[53] For an extreme example, that of the Bhopal Union Carbide disaster, see Amnesty International (2004).

[54] The doctrine of *forum non conveniens* is often invoked by corporations in US and other courts to fend off claims based on actions in foreign territories; the essence of the claim is that a foreign court is the better location to hear such a suit. *Forum non conveniens* was famously invoked effectively in a New York court by Union Carbide in the Bhopal case.

[55] See Clapham (2006), 244–6 (on the ICC), 252–63 and 441–50 (on ATCA). On TVPA, see Fitzmaurice (2004), 205–6. On extraterritorial jurisdiction for grave breaches of international law, see Ramasastry and Thompson (2006). Climate change cases would differ

increasingly invoked, they have so far had little success in court claims against corporate entities or their representatives. From a climate change perspective, a further significant difficulty is that they apply only to egregious violations, usually international crimes, and breaches of 'the law of nations' – such as genocide, crimes against humanity, war crimes, slavery, torture and piracy. Extreme climate change harms are likely to include death or starvation resulting from drought or water salination, destruction and loss of property, shelter and livelihoods, the spread of fatal diseases and exposure to war. It is far from clear that, even in the extreme forms they are likely to take, such outcomes will amount to breaches of international criminal law or the 'law of nations', or torture (or 'cruel, inhuman and degrading treatment'), although such an eventuality is not unthinkable. Given the difficulty of attributing blame to any single actor in the case of climate change, notions of complicity, joint enterprise and aiding and abetting may be useful in such cases. All of these possibilities warrant further research, including whether liability might be attributed among multiple companies (jointly and severally), or among companies and governments.

Where caps do not exist, major polluters are free to emit as much GHG as they wish at home – but they might still be pursued in domestic courts for the human harms this behaviour causes elsewhere. Class actions might be conceivable in some contexts against major car or oil companies, for example. In United States and other courts, it may be possible to sue companies for misleading the public, using the sort of arguments successfully invoked against tobacco companies. Certain companies appear to have funded the production and dissemination of false information regarding climate change, thereby delaying public action and so worsening the overall damage caused.[56]

The very fact that such strategies would be innovative and somewhat conjectural indicates the difficulty of demonstrating liability for harms of this kind. The entrenchment of human rights norms in international law has not so far provided clear answers in such cases, in part because corporations have not generally been recognised as subjects of international law and can plausibly argue that they have few direct obligations to ensure human rights fulfilment.[57] In a perfect world where every state had the capacity and will to apply international law

from typical ATCA cases in that the acts in question would have taken place on US territory – only the victim would be located on foreign soil.

[56] For background, see Wallace (2002).

[57] See Nollkaemper (2004), 224–7; Bekker (2004), 210.

locally according to common standards, this would hardly be a problem. Failing that, however, attempts to expand international law to cover private actors directly have so far resulted in soft law 'compacts', an expanding domain of corporate social responsibility (CSR) and private law arrangements.[58] These are important developments, bringing incremental but potentially substantial improvements in behaviour and accountability, through practices that may eventually coalesce into hard law. But they do not yet provide the legal security necessary to ensure protection against human rights harms.[59]

This problem is well-known and often discussed in legal and human rights circles.[60] Its relevance has, however, so far been less discussed in the context of climate change. One reason for this may be that international climate negotiations have accommodated corporate involvement far more than human rights law has done (until recently). Companies have long been engaged in shaping the climate regime, with some success, and there has been near universal recognition that, if any regime is to work, it must eventually tie-in private actors.[61] By contrast, in the human rights domain, the question of whether human rights norms should be *legally* binding on companies acting transnationally has long been a source of dispute. Nevertheless, even on this point the issues are perhaps not as far apart as they first appear. In both cases a functional regime ultimately depends upon obligations becoming binding to some degree, and in both the true contest is rather over 'how binding': what limits should be set and how they should be monitored and enforced? In both environmental and human rights law, for a variety of systemic reasons, obligations are likely to have greater (binding) force in wealthier than in developing countries, and in both the possibility arises that a company can avoid or reduce its obligations where all or part of its operations are based in developing countries.

This last problem is arguably worsened under the climate change regime, where disparity is built explicitly into the system. As it currently

[58] See Newell, Chapter 4 of this volume.

[59] See for an overview Clapham (2006), 195–237. The principal instruments are the OECD Guidelines for Multinational Enterprises, the UN Global Compact, and the Norms on the Responsibilities of Transnational Corporations and Other Business Enterprises with regard to Human Rights adopted by the UN's Sub-Commission on the Promotion and Protection of Human Rights.

[60] See, for example, contributions to Kamminga and Zia-Zarifi (2000); International Council on Human Rights Policy (2002).

[61] See, for example, Peter Newell (2000).

operates, the CBDR principle offers transnational companies space to exploit differences of standards between states. Regardless of how emissions are reduced, companies may find ways to avoid restrictions by moving operations into non-Annex I countries. Indeed, the Kyoto Protocol's clean development mechanism (CDM), as currently constructed, creates conditions that are peculiarly beneficial for carbon-intensive companies acting transnationally. When a company starts a new project in a developing country, and can demonstrate some 'additional' GHG reductions from a notional 'business as usual' development baseline, it is entitled to use those reductions to offset the cuts it would otherwise need to have made in its home country.[62] In such a scenario, the company will: (i) be likely to produce a net *increase* in emissions (a phenomenon known as 'leakage'); (ii) avoid GHG limits on its home operations; and (iii) stand to make an additional profit on any excess 'reductions' (from 'business as usual') it may be able to sell afterwards.

The CDM has not yet operated on a scale large enough to produce this problem and it is likely that a post-Kyoto regime will include better safeguards against abuse. Nevertheless, the scenario described is merely an extreme illustration of a problem that will arise due to the principle of CBDR itself because, as currently conceived, it is based on a rich/poor differential between states but not between private actors. If CBDR is truly to promote development, it will need to identify *who benefits* from the looser regime in developing countries. If emissions capacity is simply handed back to rich countries, allowing companies that operate internationally to emit GHGs in developing countries but send the profits and finished products home, little will have been achieved.[63]

Conclusion

Climate change is already threatening livelihoods and food and water security across the globe. It poses an immense challenge to the development aspirations of the world's poorest countries. The scale and urgency

62 Some companies have introduced internal trading regimes that would allow different subsidiaries and national branches to trade emissions reductions with one another with a view to interchange within an international regime.

63 An increasing number of transnational companies originate in and operate out of developing countries. The relevant assessment therefore is not where a company is domiciled, but where the benefits of its emission-producing activities are ultimately consumed.

of the problem are beyond past challenges: treating it will mean destabilising and reorienting global economic flow and growth patterns. Moreover, climate change throws up questions of justice and distribution far beyond those contemplated by the current global institutional architecture, including its human rights machinery. The many areas of infirmity identified in this chapter indicate systemic inarticulacy.

International human rights law is not equipped to treat the profound justice dilemmas thrown up by climate change: indeed, many of its standard terms of art appear either outdated or overly narrow when faced with the enormity of the climate challenge. Take, for example, the 'responsibility to protect': when it comes to climate change harms, a first step towards the 'protection' of those across borders is merely to cut emissions at home. Yet, simple as this proposition may appear on its face, it is unlikely to displace the turn to 'humanitarian intervention' associated with this nascent 'responsibility'.

More fundamentally, the state-centricity of human rights obligations presents a constraint in the context of climate change, where social and economic rights in many countries are threatened primarily by actions undertaken outside their jurisdictions. Since most states so affected are under a prior obligation (under ICESCR, Article 2(1)) to 'take steps … to the maximum of its available resources' to fulfil those rights, they are presumably obliged, insofar as they can steer incoming flows of international support, to require that official development assistance contributes directly to the fulfilment of their citizens' social and economic rights, including by shielding them from the harmful effects of climate change.

In the climate change regime, the special responsibility of wealthy countries to mitigate and assist in adaptation and technology transfer is widely recognised, if barely implemented. Whereas wealthy states are exhorted, rather than obliged, by the ICESCR (Article 2(1), see also CESC General Comments Nos. 2 and 3) to underwrite the protection and fulfilment of the rights of citizens in poorer countries, they arguably *do* have such an obligation under the climate change treaty regime, insofar as obligatory international assistance towards adaptation and technology transfer must prioritise 'economic and social development and poverty eradication' as the 'first and overriding priorities of the developing country Parties', as per UNFCCC, Article 4(7).

Beyond this, rich states have a special responsibility to monitor and, where necessary, regulate the transnational behaviour of private entities within their purview. This, too, is widely recognised. Certain companies might fall through the net of a global system that (rightly) focuses on differential treatment for rich and poor countries. The companies in question are primarily large emitters who have significant presence (and political influence) in poor countries: oil, gas and other extractive and energy companies; manufacturers reliant on high-carbon production processes; loggers and industrial farmers who generate or benefit from other GHG emissions (such as methane or the elimination of carbon sinks); vehicle producers and other companies reliant on carbon-based energy distribution systems. Companies in these circumstances may find it possible to shift their emissions burdens to poor countries (with low or no caps), while redirecting the benefits of those emissions to consumers in rich countries (in the form of finished products and profits), an outcome that would appear to distort the intent of the UNFCCC's differential regime. Human rights concerns also arise where companies (including mining or private water companies) rely on or control basic resources, such as water, that will be rendered scarce by climate change. In all these areas of concern, the weaknesses of the international human rights legal framework, as well as the recent history of attempts to overcome them, may hold lessons and cautions for the climate change regime.

Over the long term, private companies have other distinctive responsibilities. Some play a key role in consolidating development paths in many countries, because they drive energy distribution and use patterns, and generate the technological innovations on which economies are built. Where poorer countries are not yet locked into carbon-intensive economies, technical innovation and transfer is mandated to make possible and to promote alternative development paths. Yet, as the patents and investment on which innovation depends are generally privately owned and protected, it is far from clear what 'technology transfer' is to mean in practice. Being at once private and multinational, companies may escape obligations here too.

In short, the disciplinary disconnect that separates the law and practice of climate change and human rights, viewed as special regimes, does not merely pose a conundrum of international law 'fragmentation'. The differing registers, interests and disciplinary biases of lawyers and professionals in each field may contribute to the mutual weakening of *each* – at a minimum the existing weaknesses of the human rights regime appear exacerbated in conditions of climate change, with little obvious sign of renewal or reinforcement in future. On the other hand, it is also possible, if improbable, that creative lawyers and practitioners may find ways to mobilise *both* regimes to address the needs of those whose human rights will be most affected by climate change, and to undo the incidental and accidental regulatory exposure that tends to produce vulnerability systematically.

References

Amnesty International, *Clouds of Injustice: Bhopal Disaster 20 Years On* (Amnesty International, 2004).

Bekker, Pieter H. F., 'Corporate Aiding and Abetting and Conspiracy Liability under International Law' in Wybo P. Heere (ed.), *From Government to Governance, 2003 Hague Joint Conference on Contemporary Issues of International Law* (The Hague: Asser Press, 2004).

Biniaz, Susan, 'Common but Differentiated Responsibilities', *American Society of International Law Proceedings*, 96 (2002), 358.

Brown-Weiss, Edith, *In Fairness to Future Generations: International Law, Common Patrimony, and Intergenerational Equity* (New York: Transnational Publishers, 1989).

Burns, William C. G., 'Potential Causes of Action for Climate Change Damages in International Fora: The Law of the Sea Convention', *International Journal of Sustainable Development Law & Policy*, 2 (2006), 27.

Caney, Simon, 'Cosmopolitan Justice, Responsibility and Global Climate Change', *Leiden Journal of International Law*, 18 (2005), 747.

Clapham, Andrew, *Human Rights Obligations of Non-State Actors* (Oxford University Press, 2006).

Cosbey, Aaron, '*The Kyoto Protocol and the WTO: Seminar Note*' (IISD, 2007).

Craven, Matthew, *The International Covenant on Economic, Social and Cultural Rights: A Perspective on its Development* (Oxford: Clarendon Press, 1998).

Dellapenna, Joseph, *Suing Foreign Governments and their Corporations* (The Bureau of National Affairs, 1988).

Fitzmaurice, Malgosia, 'Case Study: Wiwa and Royal Dutch Petroleum and Shell Transport and Trading Company before the US Court of Appeals for the Second Circuit (the Aliens Tort Claims Act and the Doctrine of *forum non conveniens*)' in Wybo P. Heere (ed.), *From Government to Governance, 2003 Hague Joint Conference on Contemporary Issues of International Law* (The Hague: Asser Press, 2004).

Franck, Thomas, *Fairness in International Law and Institutions* (Oxford: Clarendon Press, 1995).

Gupta, Joyeeta, 'Legal Steps outside the Climate Convention: Litigation as a Tool to Address Climate Change', *Review of European Community & International Environmental Law (RECIEL)*, 16 (2007), 76.

Howse, Robert, 'The Appellate Body Rulings in the Shrimp/Turtle Case: A New Legal Baseline for the Trade and Environment Debate', *Columbia Journal of Environmental Law*, 27 (2002), 491.

Humphreys, Stephen, 'Are Social Rights Compatible with the Rule of Law? A Realist Inquiry', Hauser Global Law Working Paper 10/06 (2006).

International Council on Human Rights Policy, *Beyond Voluntarism: Human Rights and the Developing International Legal Obligations of Companies* (ICHRP, 2002).

Climate Change and Human Rights: A Rough Guide (ICHRP, 2008).

IPCC, *Climate Change 2007: Mitigation of Climate Change, Contribution of Working Group III to the Fourth Assessment Report of the IPCC* (Cambridge University Press, 2007).

Kamminga, Menno T. and Samin Zia-Zarifi, *Liability of Multinational Corporations under International Law* (The Hague: Kluwer Law, 2000).

Kiss, Alexander and Dinah L. Shelton, 'Strict Liability in International Environmental Law', copy on file with ICHRP, 2007.

Mace, M. J., 'Funding for Adaptation to Climate Change: UNFCCC and GEF Developments Since COP-7', *RECIEL*, 14 (2005), 225.

Mank, Bradford, 'Standing and Global Warming: Is Injury to All Injury to None?', *Environmental Law*, 35 (2005), 1.

McGray, Heather, Anne Hammill, Rob Bradley with Lisa Schipper and Jo-Ellen Parry, *Weathering the Storm: Options for Framing Adaptation and Development* (Washington, DC: World Resources Institute, 2007).

New Economics Foundation, *Free Riding on the Climate: The Possibility of Legal, Economic and Trade Restrictive Measures to Tackle Inaction on Global Warming* (London: New Economics Foundation, 2003).

Newell, Peter, *Climate for Change: Non-State Actors and the Global Politics of the Greenhouse* (Cambridge University Press, 2000).

Nollkaemper, Andre, 'Translating Public International Law into Corporate Liability' in Wybo P. Heere (ed.), *From Government to Governance, 2003 Hague Joint Conference on Contemporary Issues of International Law* (The Hague: Asser Press, 2004).

Nozick, Robert, *Anarchy State and Utopia* (New York: Basic Books, 1974).

Okamatsu, Akiko, 'Problems and Prospects of International Legal Disputes on Climate Change' (2006). Available at: web.fu-berlin.de/ffu/akumwelt/bc2005/papers/okamatsu_bc2005.pdf.

Open Society Justice Initiative, *Legal Remedies for the Resource Curse* (New York: Open Society Institute, 2006).

Pew Center on Global Climate Change, *Getting Ahead of the Curve: Corporate Strategies that Address Climate Change* (October 2006).

Posner, Eric A. and Cass R. Sunstein, 'Climate Change Justice', *University of Chicago Law & Economics, Olin Working Paper*, 354 (2007).

Price, Tom, 'The Canary is Drowning: Tiny Tuvalu Fights Back against Climate Change', *Global Policy Forum*, 3 December 2002.

Provost, René, *State Responsibility in International Law* (Farnham: Ashgate, 2002).

Rajamani, Lavanya, *Differential Treatment in International Environmental Law* (Oxford University Press, 2006).

Ralston, Holley, Britta Horstmann and Carina Holl, *Climate Change Challenges Tuvalu* (Berlin: Germanwatch, 2004).

Ramasastry, Anita and Robert C. Thompson, *Commerce, Crime and Conflict: Legal Remedies for Private Sector Liability for Grave Breaches of International Law* (Oslo: Fafo, 2006).

Rinnerberger, Nina, 'Drowning Islands: Social Justice through Litigation?' (2006). Available at: www.clocksandclouds.org/issue_one_articles/Drowning_ Islands.pdf.

Robb, Cairo A. R. (ed.), *International Environmental Law Reports Vol. 3, Human Rights and Environment* (Cambridge University Press, 2001).

Rosenthal, Elisabeth, 'As Earth Warms Up, Tropical Virus Moves to Italy', *The New York Times*, 23 December 2007.

Sands, Philippe, *Principles of International Environmental Law* (Cambridge University Press, 2003).

Shelton, Dinah, 'Environmental Rights,' in Philip Alston (ed.), *Peoples' Rights* (Oxford University Press, 2001).

'Describing the Elephant, International Justice and Environmental Law', copy on file with ICHRP, 2007a.

'Equity' in Dan Bodansky, Jutta Brunnée and Ellen Hey (eds.), *The Oxford Handbook of International Environmental Law* (Oxford University Press, 2007b).

Shue, Henry, 'Subsistence Emissions and Luxury Emissions', *Law & Policy*, 15 (1993), 39.

Simms, Andrew, Victoria Johnson and Joe Smith, *Chinadependence: The Second UK Interdependence Report* (London: New Economics Foundation, 2007).

Stern, Nicholas, *Stern Review on the Economics of Climate Change* (HM Treasury, Office of Climate Change (prepublication, 2006).

Stiglitz, Joseph, 'A New Agenda for Global Warming', *The Economists' Voice*, 3: 7, article 3 (2006).

Van Dijk, Pieter, Fried van Hoof, Arjen, van Rijn and Leo Zwaak (eds.), *Theory and Practice of the European Convention on Human Rights* (Mortsel: Intersentia, 2006).

Wallace, Perry, 'Global Climate Change and the Challenge to Modern American Corporate Governance', *SMU Law Review*, 55 (2002), 493.

Weisslitz, Michael, 'Rethinking the Equitable Principle of Common but Differentiated Responsibility,' *Colorado Journal of International Environmental Law & Policy*, 13 (2002), 473.

2

Climate change, human rights and moral thresholds[1]

SIMON CANEY*

It is widely recognized that anthropogenic climate change will have harmful effects on many human beings and, in particular, on the most disadvantaged. Specifically, it is projected to result in flooding, heat stress, food insecurity, drought and increased exposure to water-borne and vector-borne diseases. Various different normative frameworks have been employed to think about climate change. Some, for example, apply cost–benefit analysis to climate change. The *Stern Review* provides a good example of this approach.[2] It proceeds by comparing the costs (and any benefits) associated with anthropogenic climate change with the costs and any benefits of a programme for combating climate change. On this basis it argues that an aggressive policy of mitigation and adaptation is justified. Whereas the costs of combating climate change, according to Stern, are quite low, the costs of 'business of usual' would be considerable. Other analysts adopt a second perspective and conceive of climate change in terms of its impact on security.[3] For

* Simon Caney is Professor in Political Theory, Oxford University; Fellow and Tutor in Politics, Magdalen College.

[1] The research for this paper was undertaken while I held a Leverhulme Research Fellowship and the paper was completed while I held an ESRC Climate Change Leadership Fellowship. I am grateful to both the Leverhulme Trust and the ESRC for their support. I am also grateful to Stephen Humphreys for his comments on an earlier draft.

[2] Sir Nicholas Stern, *The Economics of Climate Change: the Stern Review* (Cambridge University Press, 2007).

[3] It is very important to distinguish this traditional type of security-based argument, with its emphasis on violent conflict, from other conceptions of security. It should, for example, be contrasted to the concept of 'human security'. The latter breaks with notions of security that define it wholly in terms of the extent of violent conflict and defines it more broadly. A canonical characterization of human security can be found in the UNDP's 1994 *Human Development Report*. It is argued there that human security comprises 'economic security', 'food security', 'health security', 'environmental security',

example, the High Representative and the European Commission to the European Council issued a statement on *Climate Change and International Security* which argues that climate change is 'a threat multiplier which exacerbates existing trends, tensions and instability'.[4] It argues that climate change will contribute to insecurities, such as tensions over scarce resources, land loss and border disputes, conflicts over energy sources, conflict prompted by migration and tensions between those whose emissions caused climate change and those who will suffer the consequences.[5] In addition to the 'economic' approach and 'security'-based approach, some adopt a different third perspective, according to which the natural world has intrinsic value. This ecological approach condemns human-induced climate change because it is an instance of humanity's domination and destruction of the natural world.

For all of their merits these three perspectives omit an important consideration – the impact of climate change on the fundamental human rights of people. In this chapter I shall argue that a human rights approach provides an appropriate way in which to evaluate the effects of climate change. There are historical precedents for applying human rights to evaluate environmental change. Principle 1 of the 1972 Stockholm Declaration of the United Nations Conference on the Human Environment declares that '[m]an has the fundamental right to freedom, equality and adequate conditions of life, in an environment of a quality that permits a life of dignity and well-being, and he bears a solemn responsibility to protect and improve the environment for present and future generations'.[6] More recently, on 14 November 2007, a conference of AOSIS members adopted the Malé Declaration on the Human Dimension of Global Climate Change.[7] This invoked 'the

'personal security', 'community security' and 'political security', United Nations Development Programme, *Human Development Report 1994: New Dimensions of Human Security* (Oxford University Press, 1994), ch. 2, especially 24–5. My concern here is with traditional conceptions of security. For a good application of the concept of human security to climate change see Karen O'Brien 'Are we Missing the Point? Global Environmental Change as an Issue of Human Security', *Global Environmental Change*, 16:1 (2006), 1–3.

4 The paper can be found at: www.consilium.europa.eu/ueDocs/cms_Data/docs/pressData/en/reports/99387.pdf. The quotation is from p. 2.

5 High Representative and the European Commission to the European Council, *Climate Change and International Security*, Section II.

6 United Nations Environment Programme, *Declaration of the United Nations Conference on the Human Environment* (Stockholm, 1972). Available at: www.unep.org/Documents.multilingual/Default.asp?DocumentID=97&ArticleID=1503.

7 *Malé Declaration on the Human Dimension of Global Climate Change*, available at: www.ciel.org/Publications/Male_Declaration_Nov07.pdf.

fundamental right to an environment capable of supporting human society and the full enjoyment of human rights' and it expressed concern:

> that climate change has clear and immediate implications for the full enjoyment of human rights including *inter alia* the right to life, the right to take part in cultural life, the right to use and enjoy property, the right to an adequate standard of living, the right to food, and the right to the highest attainable standard of physical and mental health.[8]

The Human Rights Council of the United Nations has since passed a resolution which found that 'climate change poses an immediate and far-reaching threat to people and communities around the world and has implications for the full enjoyment of human rights'.[9]

I believe that this is a promising approach. In what follows I shall argue that:

(1) climate change jeopardizes some key human rights;
(2) a 'human rights' centred analysis of the impacts of climate change enjoys several fundamental advantages over other dominant ways of thinking about climate change; and
(3) a 'human rights' centred analysis of the impacts of climate change has far-reaching implications for our understanding of the kind of action that should be taken and who should bear the costs of combating climate change.

The nature of human rights

It is useful to begin with an analysis of 'human rights'. The concept of 'human rights' has several components. I shall highlight four. Human rights: (1) are grounded in a person's 'humanity'; (2) represent moral thresholds; (3) respect each and every individual; and (4) take general priority over other values. Let us consider each of these in turn.

(1) *Humanity.* First, human rights refer to those rights that persons have *qua* human beings. There are a number of different kinds of rights. H. L. A. Hart, for example, distinguishes between 'special rights' and 'general rights'. Special rights, on his account, are rights that persons have by virtue of some action that they and some other party have performed (for example, they have signed a contract or one has

8 Malé Declaration.

9 This was agreed at the seventh session of the Human Rights Council on 26 March 2008 (A/HRC/7/L.21/Rev.1).

authorized the other to do something), or by virtue of a special relationship (for example, they have been born into one state and, therefore, have the rights of citizenship).[10] These special rights can be contrasted to what Hart terms 'general rights'. These are the rights that people have by virtue of their humanity, and not because of the nation or state into which they were born or any actions that they have performed. Hart's concept of 'general rights' captures well the traditional understanding of 'human rights'. They are the rights that people possess independently of any social convention or social practice. They are grounded in respect for a person's humanity.

(2) *Moral thresholds.* Second, human rights represent moral 'thresholds' below which people should not fall. They designate the most basic moral standards to which persons are entitled. This point is nicely conveyed by Henry Shue who writes that '[b]asic rights are the morality of the depths. They specify the line beneath which no one is to be allowed to sink'.[11] As such they are only part of a complete political morality. They leave room for other moral ideals and values. To reiterate, they simply designate the most fundamental moral requirements which individuals can claim of others.

(3) *Universal protection.* Third, and related to this, human rights represent the entitlements of *each and every individual* to certain minimum standards of treatment, and they generate obligations on all persons to respect these basic minimum standards. Article 1 of the Universal Declaration of Human Rights (UDHR) (1948) captures this well. As it states, '[a]ll human beings are born free and equal in dignity and rights'. A human rights approach thus stands opposed to aggregative political moralities that simply sum the interests of all with a view to increasing the total social good. A human rights approach insists on the protection of the entitlements of *all* individuals and condemns

[10] H. L. A. Hart, 'Are There Any Natural Rights?', *The Philosophical Review*, 64:2 (1955), 183–8. I dissent from one aspect of Hart's characterization of general rights: he ascribes general rights to all humans capable of choice, whereas I ascribe general/human rights to all humans whether or not they can exercise choice. Hart's position here follows from his commitment to the 'choice' theory of rights which he pioneered and defended in 'Are There Any Natural Rights?'. (I endorse the alternative theory of rights, that is, what has come to termed the 'interest' theory of rights. For a canonical statement of this approach see Joseph Raz, *The Morality of Freedom* (Oxford: Clarendon Press, 1986), ch. 7. Evaluating the debate between the choice theory and the interest theory would take us too far afield.)

[11] Henry Shue, *Basic Rights: Subsistence, Affluence, and U.S. Foreign Policy*, 2nd edn., with a new afterword (Princeton University Press, 1996), 18.

any trade-offs which would leave some below the minimum moral threshold.

(4) *Lexical priority.*[12] Fourth, human rights generally take priority over moral values, such as increasing efficiency or promoting happiness.[13] They constrain the pursuit of other moral and political ideals, and if there is a clash between not violating human rights, on the one hand, and promoting welfare, on the other hand, then the former should take priority.

In short, then, and combining each of the four properties above, we may say that human rights specify minimum moral thresholds to which all individuals are entitled, simply by virtue of their humanity, and which override all other moral values.[14]

[12] The concept of 'lexical priority' comes from John Rawls, *A Theory of Justice*, rev edn (Oxford University Press, 1999), 37–8. As Rawls employs this term, to say that A enjoys lexical priority over other values is to say that it is morally more urgent and may not be sacrificed to pursue any of these other values.

[13] This priority may not be absolute in all circumstances. One can, of course, envisage situations where sacrificing the rights of one person will save very many people. Some might then condone the sacrifice of one right in such scenarios. Three points should be made here. First, these refer to exceptional cases and so one might say (as I do in the text above) that human rights generally take priority. Second, even if one thinks that an individual human right may be violated one may hold that such a violation is permissible only to honour other human rights. So even if an individual human right may be overridden this does not entail that human rights as a category can be overridden to further some other goal. Indeed, the standard cases presented to show that human rights might be overridden always present examples in which the case for violating one human right (e.g., torturing a terrorist suspect) is that it would uphold other human rights (e.g., the right to life) of many others. Finally, though I cannot argue the point here, I agree with those who argue that even if one could conceive of a case where, in principle, violating one human right would protect more human rights, institutionalizing it in practice would be wrong because it would in all likelihood lead to unjustified human rights violations. Accepting that in a hypothetical situation a right might be violated does not show that in practice institutions should be given the power to do so, simply because one might think that the relevant decision-makers are fallible or might abuse the power. See Peter Jones, *Rights* (Basingstoke: Macmillan, 1994), 203–4.

[14] The account I have sketched conforms to what Charles Beitz terms an 'orthodox' conception of human rights. See Beitz, 'Human Rights and the Law of Peoples' in Deen Chatterjee (ed.), *The Ethics of Assistance: Morality and the Distant Needy* (Cambridge University Press, 2004), 193–214. He contrasts the 'orthodox' account to what he terms the 'practical' account. The latter maintains that human rights should be defined in terms of the role that they play in political practice. More precisely, human rights, on this view, specify the conditions under which some kind of intervention in another society is justified. Beitz raises a number of objections to the orthodox conception and proposes the practical conception as a superior alternative. For Beitz's description of the practical account see 'Human Rights and the Law of Peoples', especially

Two further points about the concept of human rights bear noting. First, it is conventional to distinguish between positive and negative rights, where positive rights require others to perform certain actions and where negative rights require others simply to abstain from certain actions. To illustrate the difference: one might affirm that there is a negative right not to be tortured. This generates duties on all not to perform this kind of action. Alternatively, one might affirm a positive right, say, to education. This requires not simply that others do not deprive persons of education but also that others perform positive actions to ensure that all have access to education.[15]

Finally, it bears noting that there are a variety of different justifications of human rights. Following Thomas Nagel, I shall distinguish between 'intrinsic' and 'instrumental' justifications of human rights.[16] An 'intrinsic', or deontological, approach is grounded in the idea of respect for persons. It holds that to violate a person's human rights is to fail to show them the respect that they are owed. It does not, in Kant's phrase, treat people as ends in themselves. Nagel himself adopts an intrinsic approach, where he defends human rights on the grounds that they reflect the 'value of inviolability'.[17] Each person, on this view, has a certain 'moral status' or standing and should not be treated as a potential means to an end.[18] To view them as potentially usable in this way is to fail to recognize their inviolability. This intrinsic rationale for human rights can be contrasted to instrumental or teleological approaches. The latter justify human rights on the grounds that they enable each person to enjoy certain fundamental goods. Unlike deontological accounts, they justify human rights in terms of their consequences for people's lives and the state of affairs produced. Human rights, on this second account, are valuable because they enable people to be autonomous or to achieve a decent standard of living.[19] To give one recent example, in his important work, *Justice, Legitimacy,*

201–5, and also Charles Beitz, 'Human Rights as a Common Concern', *American Political Science Review*, 95:2 (2001), 269–82, especially 276ff.

[15] This is a necessarily abbreviated discussion of this distinction. For a fuller analysis see Shue, *Basic Rights*, ch. 2 and Caney, 'Global Poverty and Human Rights: the Case for Positive Duties' in Thomas Pogge (ed.), *Freedom from Poverty as a Human Right: Who Owes What to the Very Poor*? (Oxford University Press, 2007), 275–302.

[16] Nagel, 'Personal Rights and Public Space', *Philosophy and Public Affairs*, 24:2 (1995), 86.

[17] Nagel, 'Personal Rights and Public Space', 89, see also 89–93.

[18] Nagel, 'Personal Rights and Public Space', 89. Nagel here is developing ideas defended by Frances Kamm and Warren Quinn (89, note 3).

[19] This does not exhaust the different approaches to grounding human rights. For a contrasting view see that expressed by John Rawls in *The Law of Peoples with 'The*

and Self-Determination: Moral Foundations for International Law, Allen Buchanan argues that human rights have value because they protect interests that 'are constitutive of a decent life; they are necessary conditions for human flourishing'.[20] A similar position is taken by Martha Nussbaum, who argues that human rights are valuable because they protect vital 'capabilities' that are necessary to lead a decent life.[21] The teleological position is also defended by James Griffin in his recent work, *On Human Rights*.[22] In what follows, I shall be neutral between the intrinsic and instrumental accounts.[23] Both, I suggest, will endorse the human rights I propose.

Climate change and human rights

Having clarified the concept of human rights, I now want to turn to the linkages between anthropogenic climate change and human rights. Climate change, so I shall argue, jeopardizes three key human rights: the human right to life; the human right to health; and the human right to subsistence. Each of these will be examined in turn.

Prior to discussing each of these human rights, it is worth drawing attention to one aspect of the arguments that follow. In the case of each of the human rights that I will identify, I will present what I take to be the least contentious and most modest formulation of the human right in question and show that even using such minimal conceptions of human rights, anthropogenic climate change violates human rights. In doing so, I am *not* rejecting other more expansive interpretations of

Idea of Public Reason Revisited' (Cambridge, MA: Harvard University Press, 1999). Rawls approaches human rights in a different way. He argues that human rights perform three roles: (i) they specify an essential condition for any 'decent' society; (ii) if they are honoured then any kind of intervention is illegitimate; and (iii) they constrain the extent of permissible diversity among different societies (*The Law of Peoples*, 80, compare further 79–81). Rawls proposes a set of human rights that both 'liberal' and 'decent' non-liberal peoples can embrace and he rejects an account of human rights that is predicated on a commitment to liberalism, *The Law of Peoples*, 37 and 65.

20 Buchanan *Justice, Legitimacy, and Self-Determination: Moral Foundations for International Law* (Oxford University Press, 2004), 127. See more generally Buchanan's excellent analysis of the nature of, and case for, human rights, ibid., ch. 3.

21 Martha Nussbaum, 'Capabilities and Human Rights' in Pablo de Grieff and Ciaran Cronin (eds.), *Global Justice and Transnational Politics: Essays on the Moral and Political Challenges of Globalization* (Cambridge, MA: MIT Press, 2002), 117–49.

22 James Griffin, *On Human Rights* (Oxford University Press, 2008), especially 33–7 and 57–82.

23 I have defended an instrumental approach in Simon Caney, *Justice Beyond Borders: A Global Political Theory* (Oxford University Press, 2005), ch. 3.

each of these human rights. My point is that one does not need to rely on more controversial or ambitious conceptions of human rights in order to see how climate change jeopardizes human rights.[24]

The human right to life

The right to life has been conceptualized in various ways. Controversies surround what entities hold this right (do fetuses have a right to life?), and what exceptions apply to it (consider, for example, debates concerning the justifiability of capital punishment and killing during warfare). The claim that I wish to defend does not require us, however, to take a stand on either of these controversial issues. It states that:

> HR1 – the human right to life: all persons have a human right not to be 'arbitrarily deprived of his life'. (International Covenant on Civil and Political Rights (ICCPR) (1976), Article 6.1)

Two comments are in order here. First, note that this formulation of the right to life conceives it simply as a negative right. As such, it does not make the more contentious claim that each person has a positive right to have their life saved from all kinds of threats. Second, HR1 makes reference to 'arbitrarily' depriving people of life. The point of this wording is to allow the possibility that it might, in principle, be justifiable to deprive people of their life. Such a loss of life would not be 'arbitrary'. As noted above, some might hold that capital punishment is justified and hence would reject HR1 if it claimed that all loss of life counts as human rights violation. By insisting that only 'arbitrary' loss of life counts as a rights violation (and by allowing the possibility that capital punishment can be a non-arbitrary loss of life) one avoids this controversy. This addition does not have any further implications, but it is important to present as compelling a conception of the human right to life as possible.

Now once we interpret the human right to life along the lines suggested by HR1, and thereby avoid the controversies mentioned above, it is clear that it would be endorsed by both deontological and teleological

[24] My approach here is indebted to that advanced by Thomas Pogge in his pioneering work on global poverty. See his important work, *World Poverty and Human Rights: Cosmopolitan Responsibilities and Reforms*, 2nd edn. (Cambridge: Polity, 2008). I do disagree with some aspects of Pogge's methodology. See on this Caney, 'Global Poverty and Human Rights' and 'Global Justice, Humanity, and the Eradication of Global Poverty' in Alison Jaggar (ed.) (Cambridge: Polity, forthcoming 2009).

approaches to human rights. If recognizing the value of inviolability entails anything it surely entails that one does not act so as to arbitrarily deprive people of their lives. It is similarly clear (obvious even) that from a teleological point of view, each person has a right that others do not arbitrarily deprive them of their own life. This is a necessary condition of leading a minimally decent life.

Having identified a plausible conception of the human right to life, it is clear that anthropogenic climate change violates this right. It does so in at least two ways. First, climate change is projected to result in an increasing frequency of severe weather events, such as tornadoes, hurricanes, storm surges and floods, and these can lead to a direct loss of life. Storm surges, for example, can have a devastating effect. R. F. Mclean and Alla Tsyban write that:

> Storm-surge flooding in Bangladesh has caused very high mortality in the coastal population (e.g., at least 225,000 in November 1970 and 138,000 in April 1991), with the highest mortality among the old and weak ... Land that is subject to flooding – at least 15% of the Bangladesh land area – is disproportionately occupied by people living a marginal existence with few options or resources for adaptation (references omitted).[25]

Climate change will also produce flooding and landslides and these can be devastating. The Fourth Assessment Report of the IPCC reports that '[i]n 1999, 30,000 died from storms followed by floods and landslides in Venezuela. In 2000/2001, 1,813 died in floods in Mozambique' (references omitted).[26] In addition to severe weather events, climate change will also involve heat waves and these, too, will lead to loss of life. For example, studies have found that a five-day heat wave in Chicago in 1995 led to at least 700 extra deaths.[27] Furthermore, in 2003 the heat wave in Western Europe also resulted in a considerable increase in

[25] R. F. Mclean and Alla Tsyban, 'Coastal Zones and Marine Ecosystems,' in James J. McCarthy, Osvaldo F. Canziani, Neil A. Leary, David J. Dokken and Kasey S. White (eds.), *Climate Change 2001: Impacts, Adaptation, and Vulnerability. Contribution of Working Group II to the Third Assessment Report of the Intergovernmental Panel on Climate Change* (Cambridge University Press, 2001), 366–7.

[26] Ulisses Confalonieri and Bettina Menne, 'Human Health,' in Martin Parry, Osvaldo Canziani, Jean Palutikof, Paul van der Linden and Clair Hanson (eds.), *Climate Change 2007: Impacts, Adaptation and Vulnerability. Contribution of Working Group II to the Fourth Assessment Report of the Intergovernmental Panel on Climate Change* (Cambridge University Press, 2007), p. 398.

[27] Jonathan Patz *et al.*, 'The Potential Health Impacts of Climate Variability and Change for the United States: Executive Summary of the Report of the Health Sector of the U.S. National Assessment', *Environmental Health Perspectives*, 108:4 (2000), 370.

deaths from respiratory, cardiovascular and cerebrovascular problems. Haines *et al.* report, for example, that:

> More than 2000 excess deaths were reported in England and Wales during the major heat wave that affected most of western Europe in 2003 ... The greatest impact on mortality occurred in France, where it was estimated that 14800 excess deaths occurred during the first 3 weeks of August 2003 than would be expected for that time of year. Deaths in Paris increased by 140% (references omitted).[28]

By virtue of both of these mechanisms, we may conclude that the current anthropogenic climate change violates the human right to life.[29]

The human right to health

The effects of climate change will not be restricted to its impact on the human right to life; they will also undermine the human right to health. Again, though, we need to be careful in framing this right. A canonical statement of the right to health can be found in the International Covenant on Economic, Social and Cultural Rights (ICESCR) (1976), which affirms 'the right of everyone to the enjoyment of the highest attainable standard of physical and mental health' (Article 12.1). In a similar vein the Convention on the Rights of the Child (1990) asserts 'the right of the child to the enjoyment of the highest attainable standard of health' (Article 24.1).

These maximalist conceptions of the right to health will be challenged by some. A critic might baulk at the claim that all are entitled to 'the highest attainable standard of physical and mental health'. He or she might contend that to attain the highest possible standard of health would require diverting all resources to this single objective, and this would be implausible given the need to resource other important rights

[28] A. Haines, R. S. Kovats, D. Campbell-Lendrum and C. Corvalan, 'Climate Change and Human Health: Impacts, Vulnerability, and Mitigation', *The Lancet*, 367, June 24 (2006), 2103.

[29] Of course, one cannot specify in advance which particular individuals will suffer, but this does not undermine the moral point that the actions in question undermine human rights. If a saboteur weakens a viaduct on which people drive to work so that after a while it will collapse under the weight of traffic, he or she violates the human rights of those who subsequently plunge to their deaths even if no one can predict in advance who will suffer from this fate. For instructive remarks see Joel Feinberg, 'The Rights of Animals and Unborn Generations,' in *Rights, Justice, and the Bounds of Liberty: Essays in Social Philosophy* (Princeton University Press, 1980), 181–2.

or moral objectives.[30] In the light of these possible concerns, I shall propose a less ambitious conception of the human right to health. This affirms the following:

> HR2 – the human right to health: all persons have a human right that other people do not act so as to create serious threats to their health.

This differs from the ICESCR and CRC conceptions in two related ways. First, it does not require people to maximize the health of all. Second, it does not affirm a positive right to be (maximally) healthy. It affirms only a negative right that persons do not harm the health of others. Note, however, that HR2 is, of course, presupposed by the interpretation of the human right to health found in the ICESCR. For the latter also holds that persons should not act in such a way as to create an unhealthy environment: it is just that it goes much further, calling for positive action to ensure the highest attainable standard of health.[31]

Again, it is, I hope, clear that both deontological and teleological approaches would vindicate HR2. Judged from a deontological point of view, the argument for HR2 is that acting to expose others to dangerous diseases manifests a lack of respect for their status as free and equal persons. To engage in activities which create serious health hazards for others constitutes a severe failure to recognize their moral standing and their inherent dignity as persons. The teleological approach would similarly endorse HR2. The capacity to lead a decent life requires that persons are not exposed to serious threats to their health. Their capacity for agency, their ability to pursue their conception of the good, will be undermined, if not thwarted altogether, by disease and injury.

With this in mind, let us now turn our attention to the health effects of climate change. There is by now an extensive literature chronicling the severe health effects of anthropogenic climate change. The

[30] Such a critic should take into account General Comment No. 14 (2000) on Article 12 of the ICESCR, which elaborates how this concept is to be interpreted. General Comment No. 14 can be found in Sofia Gruskin, Michael A. Grodin, George J. Annas and Stephen P. Marks (eds.), *Perspectives on Health and Human Rights* (New York: Routledge, 2005), 473–95.

[31] Note in this context that General Comment No. 14 on Article 12 of the ICESCR makes clear that the human right to health 'extends to the underlying determinants of health, such as food and nutrition, housing, access to safe and potable water and adequate sanitation, safe and healthy working conditions, and a healthy environment'. See para. 4 of General Comment No. 14. This point is also reiterated in General Comment No. 14, 'I. Normative Content of Article 12', paras. 11 and 12 (as well as in para. 15).

Fourth Assessment Report of the IPCC notes, for example, that anthropogenic climate change will:

> increase the number of people suffering from ... disease and injury from heatwaves, floods, storms, fires and droughts;
> increase the range of malaria in some places but decrease it in others;
> increase the burden of diarrhoeal diseases;
> increase cardio-respiratory morbidity ... associated with ground-level ozone'; and
> increase the number of people at risk of dengue.[32]

To develop some of these points in more detail: the IPCC reports that '[c]limate change is projected to increase the burden of diarrhoeal diseases in low-income regions by approximately 2 to 5% in 2020'.[33] It adds that dengue, too, will increase dramatically and it reports research that estimates that: 'in the 2080s, 5–6 billion people would be at risk of dengue as a result of climate change and population increase, compared with 3.5 billion people if the climate remained unchanged'.[34] Human-induced climate change thus clearly results in a variety of different threats to the human right to health.

The human right to subsistence

Thus far we have seen how anthropogenic climate change undermines two fundamental human rights. Let us turn now to the third human right which I claim is harmed by anthropogenic climate change. This third human right makes the following claim:

> HR3 – the human right to subsistence: all persons have a human right that other people do not act so as to deprive them of the means of subsistence.

Note that HR3 is more minimal than the human right to food affirmed in human rights documents. Both the International Covenant on Economic, Social and Cultural Rights (ICESCR) and the Universal Declaration of Human Rights (UDHR) (1948) appear to affirm a positive right to food. For instance, Article 11 of the ICESCR asserts 'the right of everyone to an adequate standard of living for himself and his family, including adequate food' (Article 11.1), and Article 25.1 of the UDHR uses similar wording. Furthermore, the ICESCR also simply

[32] Confalonieri and Menne, 'Human Health', 393.
[33] Ibid., 'Human Health', 407. [34] Ibid., 'Human Health', 408.

asserts 'the fundamental right of everyone to be free from hunger' (Article 11.2). These formulations, thus, presuppose HR3 but go further, insisting that there is also a positive right to receive aid to ensure that no one suffers from hunger no matter what the cause of that hunger.[35]

Note, further, that HR3 enjoys support from both deontological and teleological perspectives. From a deontological perspective, the claim is that to deprive others of the possibility of meeting their basic needs is to treat them without due respect. To deny others of the ability to satisfy their subsistence needs fails to acknowledge their moral standing and their dignity as persons. This is especially so when, as is the case with climate change, the majority of emissions come from the advantaged, who do not need to engage in such health-endangering behaviour.[36] Turning now to the teleological view: again, this would endorse HR3. Food and drinkable water are necessary preconditions of the ability to act and pursue even minimal goals.

If we turn now to consider the impacts of climate change, it is clear that anthropogenic climate change violates this right. Four different mechanisms should be noted. First, temperature increases will lead to drought and thereby undermine food security. Anthony Nyong and Isabelle Niang-Diop report, for example, that '[i]n southern Africa, the area having water shortages will have increased by 29% by 2050, the countries most affected being Mozambique, Tanzania and South Africa'.[37] Second, sea level rises will involve loss of land to the sea and thus hit agriculture badly. This is especially clear in countries like Bangladesh. Third, flooding will also lead to crop failure. Fourth, freak weather events will also destroy agriculture. The upshot of these processes is that people will be deprived of the means of subsistence. Bill Hare, for instance, reports that recent research suggests that there will

[35] HR3 is closest in formulation to Article 1.2 of the International Covenant on Civil and Political Rights (1976) which states that 'In no case may a people be deprived of its own means of subsistence.' HR3, though, refers to the entitlements of individuals, not those of 'a people'.

[36] For relevant data see 'Gas Exchange: CO_2 Emissions 1990–2006', *Nature*, 447:7148 (2007), 1038, and Michael R. Raupach, Gregg Marland, Philippe Ciais, Corinne Le Quéré, Josep G. Canadell, Gernot Klepper and Christopher B. Field, 'Global and Regional Drivers of Accelerating CO_2 Emissions', *Proceedings of the National Academy of Sciences of the United States of America*, 104:24 (2007), especially 10292.

[37] Anthony Nyong and Isabelle Niang-Diop, 'Impacts of Climate Change in the Tropics: the African Experience', in Hans Joachim Schellnhuber, Wolfgang Cramer, Nebojsa Nakicenovic, Tom Wigley and Gary Yohe (eds.), *Avoiding Dangerous Climate Change* (Cambridge University Press, 2006), 237.

be '45–55 million extra people at risk of hunger by the 2080s for 2.5°C warming, which rises to 65–75 million for a 3°C warming'.[38]

Other possible human rights

Thus far we have seen that anthropogenic climate change violates three fundamental human rights. Lest this argument be misunderstood, it is important to make several additional clarificatory remarks. First, it is worth underscoring the fact that if the impacts of climate change were entirely due to natural phenomena and were not traceable to human causes then the preceding argument would not succeed. HR1 states that persons have a human right that *other people* do not deprive them of their life and so if persons lose their life because of purely natural causes then HR1 is intact. Similarly, HR2 states that persons have a human right that *other people* do not act so as to create serious threats to their health. And, as we have just seen, HR3 holds that all persons have a human right that *other people* do not act so as to deprive them of the means of subsistence. Climate scientists are unequivocal that the current and projected future climate change stems from human activities and, given this, the three preceding claims all hold. The threats to life, health and subsistence that many face, and that many more may face unless mitigation and adaptation occur, are threats that are the products of the actions of other people.[39]

[38] Bill Hare, 'Relationship between Increases in Global Mean Temperature and Impacts on Ecosystems, Food Production, Water and Socio-economic Systems' in H. J. Schellnhuber, W. Cramer, N. Nakicenovic, T. Wigley and G. Yohe (eds.), *Avoiding Dangerous Climate Change* (Cambridge University Press, 2006), 179.

[39] The IPCC states that 'It is *very likely* that anthropogenic greenhouse gas increases caused most of the observed increase in global average temperatures since the mid-20th century', Susan Solomon, Dahe Qin and Martin Manning, 'Technical Summary,' in Susan Solomon, Dahe Qin, Martin Manning, Melinda Marquis, Kristen Averyt, Melinda M. B. Tignor, Henry Leroy Miller Jr and Zhenlin Chen (eds.), *Climate Change 2007: The Physical Science Basis. Contribution of Working Group I to the Fourth Assessment Report of the Intergovernmental Panel on Climate Change* (Cambridge University Press, 2007), 60. Note that it is arguable that it would be possible for people to violate these three human rights even if climate change were not anthropogenic. Humans can violate the three human rights in two different ways. The first (and most obvious) route is for humans to emit high levels of greenhouse gases and to destroy carbon sinks, which will in turn produce high temperatures, increased precipitation and severe weather events. The second route is for humans to design social and political institutions that leave people vulnerable to the physical impacts of climate change. Suppose that climate change were non-anthropogenic (and so route 1 was inapplicable), but politicians could implement an effective programme of adaptation and design

Second, it is worth emphasizing and repeating the point that the aim of the preceding argument is to show how climate change undermines human rights, while at the same time appealing to premises that are as uncontroversial as possible. For that reason I have focused on the three rights given above and not on other more contentious candidates; I have also relied on what I take to be the most uncontroversial formulations of those rights. The aim is to identify absolutely fundamental human rights that can enjoy ecumenical support from a wide variety of different ethical perspectives. The rights not to be killed, not to have one's health jeopardized and not to be deprived of the means necessary for subsistence are all, I suggest, rights that can be adopted from within a wide variety of different conceptions of the good and ethical world views.

Third, having noted this, it is nonetheless worth mentioning that there are other possible human rights implications of climate change. For example, it is arguable that climate change jeopardizes a human right to development (HR4). Furthermore, one might argue that there is a human right not to be forcibly evicted (HR5), and that climate change violates this because people from coastal settlements and small island states will be forced to leave.

Fourth, it should be stressed that to say that climate change jeopardizes human rights is, of course, not to say that it may not also be criticized on a variety of other grounds. To take just one example, the stance defended here is, for instance, compatible with the claim that anthropogenic climate change is objectionable because it is wrong for humanity to treat the natural world in such a hubristic fashion.[40] My claim is that the human rights impacts of climate change are serious and should be addressed: it is not that they are the only morally relevant impacts of climate change.

Supplementary considerations

In the previous section I argued that climate change threatens the enjoyment of fundamental human rights. The case for a 'human rights'-centred

institutions that would safeguard the vital interests of people in life, health and subsistence, but chose not to do so. They can then be said to violate the human rights of others to life, health and subsistence for they are acting in such a way as to create threats to life, health and subsistence.

40 This view has been defended by Dale Jamieson in 'What's Wrong with Climate Change?' (unpublished paper presented at conference on 'Global Justice and Climate Change', Oxford, September 2007).

analysis of the impacts of climate change can, however, be strengthened further, and in this section I want to draw attention to the additional insights that a human rights approach brings over cost–benefit and security-based analyses.

A human rights analysis enjoys three related advantages over cost–benefit analysis (CBA). These all stem from the fact that the latter aggregates the costs and benefits felt by individuals and then selects the policy that maximizes the good. It has long been recognized that one implication of this kind of aggregative consequentialist approach is that it could call for outcomes in which some suffer greatly, but their disutility is outweighed by enormous benefits to others. Unlike a human rights approach, a CBA has only a partial and contingent commitment to the basic interests and entitlements of the most vulnerable. This problematic aspect of CBA manifests itself at several points in discussions about climate change. Consider the three illustrations of this flaw below.

(1) *Climate impacts.* One example of this kind of problem can be found in Bjørn Lomborg's book, *Cool it*. Lomborg argues that although climate change leads to loss of life from heat stress, it also leads to a much greater decrease in mortality from cold during the winter and this good outweighs the bad.[41] Anthropogenic climate change should, therefore, not be condemned. Indeed, other things being equal, it is morally required. To propose this, though, is to propose engaging in activities which one knows will directly kill some and harm the health and ability of others to subsist. This would strike many as morally unacceptable even if it has the side-effect of saving some lives. A human rights approach, however, rules out such policies.[42]

(2) *Intergenerational equity.* A second illustration of the point in hand concerns the question of whether it is appropriate to devote resources to mitigation now for the benefit of future people. It is sometimes argued that because, and to the extent that, future generations are wealthier than current generations it would be wrong to mitigate.[43]

[41] Bjørn Lomborg, *Cool it: The Skeptical Environmentalist's Guide to Global Warming* (London: Marshall Cavendish, 2007), 13–18.

[42] See also Edward A. Page, *Climate Change, Justice and Future Generations* (Cheltenham: Edward Elgar, 2006), 34.

[43] For this viewpoint see Bjørn Lomborg, *The Skeptical Environmentalist: Measuring the Real State of the World* (Cambridge University Press, 2001), 314; William Nordhaus, 'Discounting in Economics and Climate Change', *Climatic Change*, 37:2 (1997), 317; Nordhaus, 'The Question of Global Warming: An Exchange', *New York Review of Books*, 55:14, 25 September (2008), 93.

This, however, is not a compelling argument if it turns out that future generations are wealthier than current generations but that some in the future are deprived of the basic necessities of human life. By virtue of its aggregative nature, a CBA approach is concerned only with the total amount of utility and, therefore, the total wealth of current and future generations, and it is indifferent to the plight of the very severely disadvantaged if their disutility is outweighed by the utility of others. A human rights approach, however, is not vulnerable to this charge because it establishes moral thresholds below which persons should not fall.

(3) *Risk and uncertainty.* A third illustration of the point at hand arises from the risks and uncertainties associated with climate change. Climate scientists repeatedly stress that the projections of future changes to the Earth's climate are not certain and that they are characterized by both risk and uncertainty. A CBA approach will respond to risks by multiplying the probability of an event with the utility/disutility of that event, thereby arriving at the expected utility. However, by doing so it ignores a morally relevant aspect of current climate change, namely that *some* persons are imposing grave risks on *others*. It matters a great deal whether those who are taking risks are exposing just themselves to serious risks or whether they are exposing others to serious risks. In the former case, one might say that as long as the risk-takers are sufficiently well-informed and rational then their choice is permissible. The second situation is, however, quite different, for some are posing a threat to the rights of *others*. A CBA cannot capture the relevance of this distinction since its concern is simply with the aggregate level of expected utility. A human rights approach, however, captures the importance of this distinction because it disaggregates the impacts of climate change and is concerned with ensuring that none fall beneath a certain threshold. As such it would condemn as unjust a situation in which some (who are advantaged) expose others (who are vulnerable) to risks that threaten the latter's basic interests. Similarly, it would permit the first kind of risk-taking on the grounds that persons are within their rights to expose themselves to risk. A human rights perspective can thus deal better with the risk and uncertainty associated with climate change.

(4) If we turn now from CBA to the security-oriented approach presented in the introduction, we find a similar problem but for a different reason. This, too, will generate only a contingent and partial commitment to protecting the most vulnerable. It gives us reason to be

concerned about climate change only if, because, and to the extent that, it results in violent conflict.[44] It follows from this that in those cases where climate change causes death, disease, malnutrition and starvation, but in which it does not lead in turn to conflict it is silent and would devote no resources to assisting those threatened by dangerous climate change. It, therefore, fails to have an unconditional concern with the most disadvantaged. Its commitment to them is contingent on conditions that may not be met.

In short, then, a human rights approach will thus protect the vulnerable, whereas CBA fails to do so because of its aggregative character and a security-based approach fails to do so because its concern is only with climate change that causes conflict.

The implications of a human rights approach

Having argued that climate change undermines fundamental human rights and that this way of thinking about the impacts of climate change enjoys an advantage over CBA, I now want to reflect on several implications of applying a human rights approach to the impacts of climate change. First, and most obviously, a human rights approach requires us to adopt a discriminating approach to the impacts of climate change and would not, therefore, take into account all the impacts of climate change. From a purely human rights approach, only those effects that violate rights should be taken into account.[45]

A second implication of a human rights approach is that it requires us to reconceive the way in which we think about the costs involved in mitigation and adaptation. Some have argued that it would be extremely expensive to prevent dangerous climate change and hence that

[44] Note: as was stressed in note 3, I am concerned here only with traditional conceptions of security of the type expressed in the Introduction. My arguments are not directed against 'human security' and attempts to argue that climate change jeopardizes human security.

[45] In general terms this means that impacts that lead to less *preference satisfaction* or less *economic growth* do not count. In more concrete terms this means that impacts on tourism, say, or on the insurance industry are not relevant except insofar as they bear on the realization of people's human rights. (IPCC reports tend to refer to the impacts of climate change on both tourism and the insurance industry: see, for example, Tom Wilbanks and Patricia Romero Lankao, 'Industry, Settlement and Society,' in *Climate Change 2007: Impacts, Adaptation and Vulnerability*). My point is not that impacts on tourism do not matter, but that we need to distinguish between those impacts on the tourist industry that undermine human rights (for example, those whose livelihood depends on it) and those which do not.

humanity should not do this. If, however, it is true that climate change violates human rights then this kind of reasoning is inappropriate. An example can help illustrate the point. Suppose that someone builds a restaurant in their garden and makes a large profit from this. Suppose, however, that this restaurant releases fumes which threaten the lives of others nearby (thereby jeopardizing their human right to life) and it also leaks pollution into the water supply (thereby violating their human right to health). Those committed to human rights will condemn this as unjust and call for the owner of the restaurant not to engage in such rights-violating behaviour. If the owner protests that this would be very expensive the appropriate reply is that this is not germane. If a person is violating human rights then he or she should desist even if it is costly. Other examples illustrate the point: suppose that (as seems highly likely) the abolition of slavery was immensely costly to slave-owners. It does not follow from this that slave-owners should be allowed to continue in their rights-violating activity.[46] The implications for mitigation and adaptation are clear. That mitigation and adaptation would be *costly* similarly does not in itself entail that they should not be adopted. If emitting greenhouse gases (GHGs) results in rights violations it should stop, and the fact that it is expensive does not tell against that claim. A human rights approach thus requires us to reframe the issues surrounding the costs of mitigation and adaptation.

A human rights approach to climate change has a third implication. If, as argued above, climate change violates human rights then it follows that compensation is due to those whose rights have been violated. The conventional approach to climate change identifies only two kinds of response to climate change: mitigation and adaptation. The IPCC's Assessment Reports, for example, operate with this dualistic framework. The IPCC defines mitigation as '[a]n *anthropogenic* intervention to reduce the anthropogenic forcing of the *climate system*; it includes strategies to reduce *greenhouse gas sources* and emissions and enhancing *greenhouse gas sinks*'.[47] Adaptation is then defined as an '[a]djustment in natural or *human systems* in response to actual or expected climatic stimuli or their effects, which moderates harm or

[46] Part of the point here is about baselines. It is true that the slave-owners will be much worse off compared to the status quo prior to abolition but the point is that this is an illegitimate and inappropriate baseline to employ to assess what their entitlements should be.

[47] For this definition see 'Appendix I: Glossary' in *Climate Change 2007: Impacts, Adaptation and Vulnerability*, 878.

exploits beneficial opportunities'.[48] Broadly put, mitigation seeks to minimize changes to the climate system, and adaptation seeks to adjust human institutions in order to cope with the changes to the climate system. This, however, is too narrow a framework, for if there is insufficient mitigation and thus changes to the climate occur, and if, further, there is insufficient adaptation, then the fundamental human rights to life, health and subsistence will be violated. And where human rights have been violated then those who have been wronged (if they are still alive) are entitled to compensation. A human rights approach thus generates duties of mitigation and duties of adaptation, and (given the changes to the climate that are in process and given the likely lack of adequate adaptation) it also entails duties of compensation.

It is important to stress that compensation is fundamentally different from adaptation. The point of adaptation is to prevent the changes to the natural world having a malign impact on people's vital interests and human rights. If adaptation is successfully implemented then people's rights would be protected. The case for compensation, by contrast, arises when and because persons' rights were not protected. One might put it thus: the point of adaptation is to protect and uphold rights, and the point of compensation is to redress the fact that people's rights have been violated.

This third point draws our attention to a fourth implication of adopting a human rights approach to climate change: namely, that it affects the way in which one should think about inflicting harms on others and the role that compensation may play in our decision-making. On one way of thinking about harms, if one imposes a cost on people but also bestows on them a benefit then the two may cancel each other out and the affected person has no cause for complaint. This assumes that harms and benefits are commensurable and the shortfall represented by a harm is erased by the allocation of a benefit. A human rights approach adopts a different approach to the imposition of harms. For if one has a human right not to suffer a certain harm then it is wrong to violate that with a view to giving a compensatory sum to counterbalance the harm. To give an example: it is obviously impermissible for one person to assault another person with a view to then giving them a large benefit in order to somehow cancel out the harm. Similarly, one cannot destroy someone else's property and then simply write a

[48] For this definition see 'Appendix I: Glossary', 869.

cheque and think that the victim has no cause for complaint. He or she does. The point here is that if a person has a human right (and, indeed, any other kind of right) then that generates a duty to respect that right, and it is not acceptable to violate that duty with a view to then making compensation. Of course, as was argued in the previous paragraph, if people do in fact violate rights then there is a case for compensation. This, however, does not give one permission to engage in rights violations and it does not undermine the key point that a human rights approach rejects the trade-off between burdens and benefits that other approaches endorse.[49]

Let us turn now to a fifth corollary of a human rights approach to climate change. A human rights approach guides not simply our evaluation of the impacts of climate change, but also the distribution of the duties to uphold the human rights threatened by climate change. It should inform who is obligated to pay for the costs of mitigation and adaptation. The central point here is that if we accept a set of fundamental human rights then it follows that any programme to combat climate change should not itself also violate these rights. Thus, any international treaty distributing emission rights and any national level climate action plan should not jeopardize the human rights to health, life and subsistence. In practice this requires that the least advantaged – those whose human rights are most vulnerable – should not be required to bear the burden of combating climate change.

In one final point, it is worth remarking that a human rights perspective provides a useful way of conceptualizing Article 2 of the United Nations Framework Convention on Climate Change (UNFCCC). The latter states that the objective of the UNFCCC is to achieve a 'stabilization of greenhouse gas concentrations in the atmosphere at a level that would prevent *dangerous anthropogenic interference* with the climate system' (UNFCCC (1992), Article 2: my emphasis). What counts as a 'dangerous' anthropogenic interference is clearly, in part, a normative issue. It cannot be resolved by science alone for at most that can tell us the types of changes that are likely to occur. To determine when the changes are 'dangerous' we need some normative principle or principles. My proposal, in this context, is that dangerous climate

[49] For illuminating discussion see Clive L. Spash, *Greenhouse Economics: Value and Ethics* (London: Routledge, 2002), 231–6 and Henry Shue, 'Bequeathing Hazards: Security Rights and Property Rights of Future Humans' in Mohammed H. I. Dore and Timothy D. Mount (eds.), *Global Environmental Economics: Equity and the Limits to Markets* (Oxford: Blackwell Publishers, 1999), 40–3.

change should be interpreted as climate change that systematically undermines the widespread enjoyment of human rights.

Concluding remarks

The important links between climate change and human rights have been neglected. In this paper, I have sought to address this lacuna. I have defended three distinct conclusions:

(1) Climate change jeopardizes human rights and in particular the human rights to life, health and subsistence (see 'Climate change and human rights', above).
(2) Analysing the impacts of climate change in terms of its effects on human rights enjoys advantages over other ways of evaluating the impacts of climate change (see 'Supplementary considerations', above).
(3) Endorsing a human rights framework for evaluating the impacts of climate change has implications for our understanding of who should bear the burdens of climate change and what kinds of policies are appropriate (see 'The implications of a human rights approach', above).[50]

As I noted above, I am not claiming that a human rights approach captures *all* the morally relevant impacts of climate change. My argument is simply that a human rights perspective has important insights and any account of the impacts of climate change which ignores its implications for people's enjoyment of human rights is fundamentally incomplete and inadequate.

[50] In focusing on these links between climate change and human rights I am not claiming that this exhausts the relevant connections between human rights and climate change. Two other connections are worth noting. First, it is arguable that persons have a human right to have an input into any decision-making process that affects their fundamental interests. On this basis, one may argue that persons have a human right to shape the political process by which decisions about mitigation and compensation are made. One might call this the human right to procedural justice. Second, it is also arguable that the extent to which people are able to adapt to dangerous climate change is a function of the extent to which their basic human rights are respected. The more that their rights have been violated the less they are able to adapt to climate change. (This second theme is explored by Jon Barnett in 'Human rights and vulnerability to climate change', Chapter 9, below)

3

Equitable utilization of the atmosphere: a rights-based approach to climate change?

DINAH SHELTON*

Most discussions of a rights-based approach to the environmental crises facing the planet quite appropriately centre on demanding that each state take action to prevent or mitigate environmental harm that diminishes, for those within its territory and jurisdiction, the enjoyment of internationally guaranteed human rights. Environmental degradation, in particular global climate change, undeniably has a negative impact on, and will increasingly limit, civil, political, economic, social and cultural rights, especially for the world's most vulnerable populations. Nonetheless, problems of standing, justiciability, ripeness and causality have been among the prominent problems encountered when individuals have sought to vindicate their rights through human rights litigation.

Another rights-based approach is explored herein, whereby the government of a state may, and, indeed, arguably has the duty to, assert and defend the rights of its inhabitants, rather than remaining passive and ultimately defending itself for alleged rights-violating acts and omissions. The premise of the approach is that in the international community, which is organized on a territorial basis among some 192 independent, sovereign and juridically equal states, governments exist for the purpose of protecting the sovereign rights of the state and the human rights of their inhabitants, present and future, or, in constitutional language, 'to ensure domestic tranquillity, provide for the common defence, promote the general welfare and ensure liberty to its citizens now and in the future'. International human rights treaties join constitutions in demanding that governments respect and ensure proclaimed rights;

* Manatt/Ahn Professor of International Law, The George Washington University Law School.

such treaties also create subsidiary institutions and mechanisms to identify and attempt to remedy state failures in achieving these objectives. These treaties, institutions and mechanisms reflect the territorial basis of society by imposing obligations on a state primarily with respect to those within its territory and subject to its jurisdiction. The drafters of the human rights instruments no doubt assumed that in the normal course of events a state would be able to violate the rights only of those within its legal power.[1]

As structured in this way, human rights law can function to address most violations of civil and political rights, which indeed usually occur as foreseen,[2] however, harm to economic and social rights, as well as to the underlying environmental conditions necessary to the enjoyment of all rights, often originates in activities outside the jurisdiction of the state where the harm is felt. The extent of transboundary human rights obligations remains a topic of some controversy within human rights bodies, while the doctrine of sovereign immunity makes it difficult for individuals to seek a remedy in their national courts against a foreign government. Yet, it may be possible to recast the rights and duties involved when transboundary harm occurs, to achieve the goals of prevention and accountability, merging the law of state responsibility for transboundary environmental harm with international human rights law. Rather than individuals attempting to vindicate their rights, plaintiff states may represent those individuals as well as future generations in bringing claims against the responsible states, thus utilizing state sovereignty as a vehicle for implementing international human rights law and international environmental law. The potential and problems with this rights-based approach are explored in the following discussion.

[1] Violations involving the extraterritorial use of force are regulated by international humanitarian law or the law of armed conflict.

[2] This is not to say that extra-territorial violations of civil and political rights are unknown, but where they occur human rights tribunals may condemn them on the basis that the violating state had 'effective control' over the territory where the acts occurred due to the presence of state agents within the territory. Compare, e.g., *Banković and Others* v. *Belgium and Others*, E.Ct.H.R., App. No. 52207/99 (admissibility declaration, 12 December 2001) (Grand Chamber); *Issa* v. *Turkey*, E.Ct.H.R., App. No. 31821/96, 16 November 2004; *Ilaşcu and Others* v. *Moldova and Russia*, App. No. 48787/99, Judgment of 8 July 2004; and the Inter-American Commission cases: *Case 11.589, Armando Alejandre Jr., Carlos Costa, Mario De La Peña, and Pablo Morales* v. *Cuba*, Rep. 63/05, *Hossein Alikhani* v. *United States*. See generally Tarik Abdel-Monem, 'How Far Do the Lawless Areas of Europe Extend? Extraterritorial Application of the European Convention on Human Rights', *Journal of Transnational Law and Policy*, 14 (2005), 159.

A The right of permanent sovereignty over natural resources

Sovereign equality, including permanent sovereignty over natural resources,[3] is a basic constitutional principle of the international legal system.[4] It implies non-intervention or non-interference with decisions taken by each government concerning matters exclusively within that state's territory or jurisdiction. Traditional international law respects each state's exclusive jurisdiction over its territory. Yet acts which take place or originate on the territory of one state may cause damage or infringe upon the sovereignty of another state, giving rise to conflict between the sovereign rights of the two states. The conflict can be looked at from two perspectives: that of the state on whose territory the pollution originates and that of the state whose territory is affected by the pollution. The polluter state might argue the theory of absolute state sovereignty, but this approach has been repudiated in a world where states are increasingly obliged to cooperate.[5]

Doctrine and international practice are virtually unanimous in condemning claims of absolute state sovereignty, which offers no legal means to reconcile the equal rights of two opposing states, especially when the conflict is over use of a shared natural resource. Instead, equitable balancing principles and norms of transboundary conduct have been enunciated in international litigation and have emerged in state practice. The International Court of Justice (ICJ), in its judgment in the 1949 *Corfu Channel* case, referred to 'every State's obligation not to allow knowingly its territory to be used contrary to the rights of other states'.[6] The same year, the United Nations Survey of International Law concluded that there is 'general recognition of the rule that a State must

[3] See, e.g., Declaration on Permanent Sovereignty over Natural Resources, General Assembly Resolution 1803 (XVII) of 14 December 1962.

[4] See I. Brownlie, *Principles of Public International Law*, 6th edn (Oxford University Press, 2003), 287 and note 1.

[5] Absolute sovereignty is identified with the 'Harmon Doctrine,' named for the US Attorney General who, in 1885, officially claimed that the Mexican government had no right to protest about water pollution in the boundary Rio Grande River, which lowered water quality in Mexico and damaged Mexican agriculture, thereby harming Mexican farmers. Harmon contended that the rules, principles and precedents of international law imposed no obligation or responsibility on the United States to protect Mexico from pollution. Therefore, any harm to Mexico was a political rather than a legal question. 21 Op. Att Gen. 274, at 280–3 (1895).

[6] *Corfu Channel* case, ICJ Rep. 1949, 22.

not permit the use of its territory for purposes injurious to the interests of other States in a manner contrary to international law'.[7]

Principle 21 of the 1972 Stockholm Declaration on the Human Environment affirmed the general principles of state sovereignty in the environmental context. It began by proclaiming that 'States have, in accordance with the Charter of the United Nations and the principles of international law, the sovereign right to exploit their own resources pursuant to their own environmental policies.'[8] The reference to the UN Charter and principles of international law indicates that state sovereignty is exercised within international law and with respect for the rights of others. State sovereignty within international law implies, in particular, that each state's resource utilization must not harm other states. Principle 21 explicitly requires this as it balances the principle of permanent sovereignty with each state's 'responsibility to ensure that activities within their jurisdiction or control do not cause damage to the environment of other States or of areas beyond the limits of national jurisdiction'. This balance reflects the equitable principle of abuse of right or good neighbourliness among equal subjects of the international legal system.[9] It goes beyond courtesy or comity in being a normative principle regulating relations between states.[10]

The doctrine of abuse of right thus bars a state from exercising a sovereign right without an acceptable motivation or benefit when the activity undertaken causes harm to another state. Assessing whether the exercise of a right is abusive involves judging whether the harmful consequences produced outside the territory outweigh the benefits to those within. An agreement between Finland and Sweden concerning boundary waters expresses this concept:

> Where the construction would result in a substantial deterioration in the living conditions of the population or cause a permanent change in natural conditions such as might entail substantially diminished comfort

[7] UN Doc. A/CN.4/1/Rev.1 (UN Pub. 1948. V.1(1)), 34 (1949).

[8] Stockholm Declaration on the Human Environment, Principle 21, in Report of the United Nations Conference on Human Environment, UN Doc. A/CONF.48/14/Rev.1 (1972).

[9] Australia relied upon its sovereign rights to protest against acts having detrimental impacts within its territory in bringing its action against France in the *Nuclear Tests* case. See ICJ Pleadings, Nuclear Tests, I; *Australia* v. *France* (1974) 479–90. Brownlie refers to this as invoking 'the international law equivalent to trespass'. Brownlie, *Principles International Law*, note 5 at p. 275.

[10] A. C. Kiss, *Abus de Droit en Droit International*, Librairie générale de droit et de jurisprudence (thèse) (1953).

> for people living in the vicinity or a significant nature conservancy loss or where significant public interests would be otherwise prejudiced, the construction shall be permitted only if it is of particular importance for the economy or for the locality or from some other public standpoint.[11]

The language of this agreement demonstrates the importance of sovereignty to the states involved, but also the limits: each state requires that the other accept some inconveniences and deterioration in conditions as a result of utilizing the common resource for economic development, but creates a presumption that permission should be denied for an activity when it would cause 'substantial' or 'significant' harm. The presumption may be overcome by demonstrating that substantial economic benefits outweigh the significant burdens imposed. Another way of expressing this concept, utilizing human rights language, is that one state's projects aimed at the fulfilment of economic and social rights may proceed if there is no significant impairment of the human rights of individuals in another state.[12]

The principle of abuse of right thus requires balancing the interests of the acting and impacted states and respecting proportionality in conduct.[13] The OECD Principles on Transfrontier Pollution explicitly refer in their introduction to 'a fair balance of the rights and obligations among countries concerned by transfrontier pollution'.[14] The text states that 'countries should seek, as far as possible, an equitable balance of their rights and obligations as regards the zones concerned by transfrontier pollution'.[15] The factors that go into assessing equitable balance are discussed later in this chapter.

Today it is generally accepted that the principle forbidding abuse of right, whose origin lies in Roman law (*sic utere iure tuo ut alterum no laedas* – that is, use your own so as not to injure another) forms part of international law.[16] Treaties and judicial decisions apply the abuse of

[11] Art. 3(2), Agreement Concerning Frontier Rivers between Finland and Sweden (16 September 1971).

[12] A balancing of economic and social rights often occurs within a single state, as a government with scarce or limited resources must allocate the resources among, *inter alia*, education, science, housing and health.

[13] The legal principles drafted by the panel of independent experts for the Brundtland Commission reflect this concept. See Art. 12, (1987), 28.

[14] OECD, *Principles Concerning Transfrontier Pollution*, C(74)224, 14 November 1974, Introduction.

[15] OECD (1974).

[16] See, e.g., *Corfu Channel* case, ICJ Rep.; Kiss, *Abus de droit*. See also Restatement (Third) Foreign Relations Law of the United States, s. 601. (A state is obligated to take such

right principle to transfrontier pollution.[17] According to Article 5(1) of the UN Convention on the Law of the Non-Navigational Uses of International Watercourses:[18]

> Watercourse States shall in their respective territories utilize an international watercourse in an equitable and reasonable manner. In particular, an international watercourse shall be used and developed by watercourse States with a view to attaining optimal and sustainable utilization thereof and benefits therefrom, taking into account the interests of the watercourse States concerned, consistent with adequate protection of the watercourse.

The widespread adoption of the concept of equitable utilization in treaty law and state practice could be deemed to create a specific rule of international law directly forbidding significant transfrontier pollution as a *prima facie* case of abuse of right. In other words, the very fact of such pollution may be deemed *per se* a violation of the principle of equitable utilization and thus prohibited by international law. Jurisprudence supports this view, as the cases discussed in the following section illustrate.

B Interstate cases on transfrontier pollution

The well known arbitral decision between the United States and Canada[19] resulted from the activities of a Canadian smelter of zinc and lead ores, located in Trail, British Colombia. From the beginning of its operations in 1896, American farmers suffered damage due to emissions of sulphur dioxide by the plant. In 1903, the record year, these emissions exceeded 10,000 tons a month. In 1930, 300 to 350 tons of sulphur, in addition to other chemical residues, poured into the air. Initially, the smelter company paid indemnities to those suffering from the pollution,

measures as may be necessary, to the extent practicable under the circumstances to ensure that activities within its jurisdiction and control do not cause significant injury, e.g., to the environment of another state or of areas beyond the limits of national jurisdiction.)

[17] As early as 1911, in reference to international watercourses, the Institute of International Law stated that neither state bounded by a river may 'on its own territory, utilize or allow the utilization of the water in such a way as seriously to interfere with its utilization by the other State or by individuals, corporations, etc., thereof'. James Brown Scott (ed.), *Resolutions of the Institute of International Law Dealing with the Law of Nations* (New York: W.S. Hein, 2003), 169.

[18] UN Convention on the Law of the Non-Navigational Uses of International Watercourses, New York, 21 May 1997.

[19] *Trail Smelter* Arbitration, *United States* v. *Canada*, 3 UNRIAA (1905).

either following American court procedures or as a result of bilateral accords. In 1925, the case was reopened after the smelter added two 409-foot stacks to the plant to increase production, resulting in greater pollution. An association of injured persons was formed in order to obtain general damages in the place of individual recoveries. In 1927, the United States government officially took up the case and presented a claim to the government of Canada. After various efforts to settle the case by other means, the two governments submitted the matter to arbitration, signing a Convention to this effect on 15 April 1935.[20]

The arbitral commission was asked to respond to four questions:

(1) Did the Trail Smelter cause damage after 1 January 1932, and if so, what indemnity should be paid as a consequence?[21]
(2) If the first question is answered affirmatively, should the Trail Smelter be required to refrain from causing damage in the state of Washington in the future, and if so, to what extent?
(3) In light of the preceding question, what measures or regime, if any, should be adopted or maintained by the Trail Smelter?
(4) What indemnity or compensation should be paid on account of the decision of the arbitral tribunal?

The final decision of the arbitral tribunal, issued on 11 March 1941, detailed the facts and topographical, meteorological and economic conditions of the region subjected to pollution. On the question of whether, or to what extent, the Trail Smelter must refrain from causing damages on the American territory, the tribunal defined the applicable principles in referring to Article IV of the arbitration Convention. It decided that it should take into consideration the law and practice existing in federal

[20] Convention for the Settlement of Difficulties arising from Operations of Smelter at Trail, BC (Ottawa, 15 April 1935), 162 L.N.T.S. 73; 49 Stat. 3245, T.S. No. 893, 6 Bevans 60; C.T.S. 1935, No. 20, 30 A.J.I.L. 163 (Supp.).

[21] In an interim decision, dated 16 April 1938, the arbitral tribunal responded to the first question concerning damage caused by the Trail Smelter since 1 January 1932. For the period between that date and 1 October 1937, the tribunal awarded $78,000 for damage to cleared and uncleared land. The tribunal also decided that the Trail Smelter should be subject to a temporary regime to continue until 1 October 1940, including abstention from causing damage and installation of equipment to control pollution. The United States had presented claims for $1,849,156.16, including harm to: (1) cleared and uncleared land and improvements; (2) livestock; (3) property in Newport; (4) infringement of United States sovereignty; (5) unpaid interest; and (6) business losses. With interest the total came to $2,100,011.17. Only the first claim was accepted. The tribunal found that the language of the compromise precluded it from considering harm to United States sovereignty.

states, given the dearth of international precedents on this point.[22] It deemed United States and Swiss law to confer on their constituent units rights analogous to those of states under international law. The arbitrators found the air pollution law of the United States in dealing with the quasi-sovereign rights of the states of the union conformed to the general rules of international law while providing more detail. These cases are discussed in section C, below.

On the international plane, the tribunal asserted a general duty on the part of a state to protect other states from injurious acts by individuals within its jurisdiction. It also noted the difficulty of determining what constitutes an injurious act. Swiss domestic courts had concluded, and this tribunal agreed, that precautions taken by a state should be the same as those it would take to protect its own inhabitants. On these bases, the tribunal reached its conclusion:

> [t]he Tribunal, therefore, finds that the above decisions, taken as a whole, constitute an adequate basis for its conclusions, namely, that, under the principles of international law, as well as of the law of the United States, no State has the right to use or permit the use of its territory in such a manner as to cause injury by fumes in or to the territory of another or the properties or persons therein, when the case is of serious consequence and the injury is established by clear and convincing evidence.[23]

For the arbitral tribunal, Canada's liability for the Trail Smelter derived from its duty to ensure that the Smelter's activities conformed to the obligations that international law places on each state. The Trail Smelter itself should refrain from causing damage by emission of fumes on the territory of the state of Washington. The damage which did occur should be repaired by the governments, in conformity with Article XI of the arbitral Convention.

The third question posed to the tribunal concerned what measures or what regime should be adopted or maintained by Canada for the future. The Tribunal suggested a regime to eliminate further damage on US territory by air pollution from Canada. Should the Smelter fail to conform to the order given it to refrain from causing further damage, the tribunal, in response to the fourth question regarding future damages,

[22] If international courts and tribunals today should adopt a similar methodology, the *Massachusetts* v. *EPA* case that provides a focus for the latter part of this chapter could be of major importance in developing the law with respect to climate change.

[23] 3 UNRIAA 1938, 1965.

approved the principle of indemnity, leaving the extent and amount to agreement between the governments involved.

It is difficult to overestimate the importance of the *Trail Smelter* arbitration.[24] The arbitral Convention itself constitutes a noteworthy precedent, insofar as it announced two principles. First, it recognized the responsibility of a state for acts of pollution having their origin on its territory and causing damage on the territory of other states, even if the polluting acts are not imputable to the state itself or its organs. Thus, the state may be responsible for not enacting necessary legislation, for not enforcing its laws against those within its jurisdiction or control, for not preventing or terminating an activity, or for not sanctioning the person responsible for it. Second, the Convention transcended international responsibility to resolve the conflict before it, aiming toward a common regulation of the issue. The award itself affirmed the existence of a rule of international law imposing liability for failing to prevent significant transfrontier air pollution, a fact of fundamental importance to future action on climate change.[25] The Tribunal also elaborated a framework for the future, recognizing the necessity for further cooperation between the interested states, and, in particular, indicating the maximum emissions permitted under various meteorological conditions. In requiring mitigating or preventive regulation, the award indicated that polluters cannot always pay, but may be required to halt serious pollution according to the evolution of the situation and knowledge of it.

[24] The case continues to be invoked. In 1972, Canada referred to the judgment when an oil spill in Washington polluted beaches in British Colombia. *Canadian Yearbook of International Law*, 11 (1973), 333–4. It was referred to by both parties in the *Gabçikovo/Nagymaros Project* case at the ICJ, [1997] ICJ Rep. 3 (25 September). Most recently, *Pakootas* v. *Tech Cominco Metals, Ltd.*, 452 F.3d 1066 (2006) imposed liability on the owner of the Trail Smelter, which continues to operate, for unpermitted toxic waste disposal. The Ninth Circuit held that the disposal occurred within the United States, where the toxic substances leeched from the waters in which they were discharged and, therefore, there was no need to consider the case as one of transboundary pollution.

[25] There remain difficult problems in utilizing the *Trail Smelter* case. While the imposition of liability implies that a wrongful act occurred, the tribunal did not discuss the standard of care or indicate whether the case was one of strict liability, negligence or intentional wrong. Canada had acted (issuing permits for the smelter to operate) and failed to act (not regulating or taking mitigating actions). Whether the government would have escaped liability by showing 'due diligence' to reduce harmful emissions is unclear; however, it seems most likely from the opinion that the government could pay for the harm caused and allow the smelter to continue to operate – unless the harm became so severe that it would be inequitable to allow it to continue. Note, too, that the EPA argued in *Massachusetts* v. *EPA* that GHG emissions are not 'pollutants' as defined by the US Clean Air Act.

Some fifteen years after the *Trail Smelter* award, the *Lake Lanoux* arbitral decision[26] also alluded to the problem of transfrontier pollution in holding that France could use the waters of the lake for French public works, restoring the waters to the River Carol which crosses the Spanish frontier to join the Segre River. No water pollution was alleged, but the arbitral tribunal nonetheless addressed the matter:

> It could have been argued that the works would bring about a definitive pollution of the waters of the Carol or that the returned waters would have a chemical composition or a temperature or some other characteristic which could injure Spanish interests. Spain could then have claimed that her rights had been impaired in violation of the Additional Act.[27]

The Tribunal later indicated the consequences which would occur from such pollution: 'admittedly there is a rule prohibiting the upper riparian State from altering the waters of a river in circumstances calculated to do serious injury to the lower riparian State, [but] such a principle has no application to the present case, since it was agreed by the Tribunal … that the French project did not alter the waters of the Carol'.[28]

Principle 21 of the Stockholm Declaration fundamentally restated these international obligations and added duties owed to the international community as a whole. Thus, the duty not to cause damage to the environment exists not only toward other states, but also toward the 'areas beyond the limits of national jurisdiction': the high seas and the air space above them, the deep seabed, outer space, the Moon and other celestial bodies and Antarctica.

Principle 21, although part of a non-binding text, is now recognized as a rule of customary international law. It has been reaffirmed in declarations adopted by the United Nations, including the Charter of Economic Rights and Duties of States[29] and the World Charter for Nature,[30] and has been adopted by other international organizations

[26] *Lake Lanoux Arbitration (France–Spain)*, Arbitral Tribunal 12 RIAA, (1957), 281. An English translation of the award appears in *Yearbook of the International Law Commission*, vol. II (Part Two) (1974), 194–9, UN Doc. A/5409, paras. 1055–68; 53 A.J.I.L. 156–71 (1959); and *International Law Reports* (1957), 101–42.

[27] 12 UNRIAA at 303.

[28] *Yearbook of the International Legal Commission*, vol. II (Part Two) (1974) 197, UN Doc. A/5409, para. 1066.

[29] Charter of Economic Rights and Duties of States, GA Res. 3281, 12 December 1974, 29 UN GAOR Supp. (No. 31) (A/9631).

[30] World Charter for Nature, G.A. Res. 37/7, 37 UN GAOR Supp. (No. 51) at 17, UN Doc. A/37/51 (1982).

and conferences.[31] Its content is inserted in the Convention on the Law of the Sea.[32] The 1979 Geneva Convention on Long Range Transboundary Air Pollution reproduces Principle 21 stating that the Principle 'expresses the common conviction that States have' on this matter. Principle 21 as restated in the 1992 Rio Declaration also appears in the preamble of the 1992 UN Framework Convention on Climate Change and Article 3 of the Convention on Biological Diversity, to which virtually all the states of the world are contracting parties. Finally, the ICJ recognized in an advisory opinion that '[t]he existence of the general obligation of states to ensure that activities within their jurisdiction and control respect the environment of other states or of areas beyond national control is now part of the corpus of international law relating to the environment'.[33] This statement was repeated in the judgment concerning the *Gabçikovo-Nagymaros Project*, in which the Court also 'recall[ed] that it has recently had occasion to stress … the great significance that it attaches to respect for the environment, not only for states but also for the whole of mankind'.[34]

C Sovereignty as a basis for inter-state climate change litigation: from *Trail Smelter* to *Massachusetts* v. *EPA*

The landmark *Trail Smelter* arbitration[35] relied on inter-state cases in federal systems to come to its conclusion that 'no state has the right to use or permit the use of its territory in such a manner as to cause injury by fumes in or to the territory of another or the properties or persons therein, when the case is of serious consequence and the injury is established by clear and convincing evidence'. The Tribunal specifically noted that the decisions of the United States Supreme Court on which it relied were decisions in equity,[36] but also indicated that standing to sue was based on the sovereign legal interests of each state.

[31] See, e.g., Preliminary Declaration of a Programme of Action of the European Communities in respect to the Environment, O.J. C 112/1, 20 December 1973; Final Act, Conference on Security and Cooperation in Europe, Helsinki, August 1976.

[32] UNCLOS, Article 194(2).

[33] *Legality of the Threat or Use of Nuclear Weapons*, Advisory Opinion, ICJ Rep. (1996), 241–2, para 29.

[34] *Gabçikovo/Nagymaros* Case [1997] ICJ Rep. 3, para. 53.

[35] *Trail Smelter, United States* v. *Canada* (1941) RIAA iii.1905 at 1965; Ann. Digest (1938–40), No. 104.

[36] The cases relied heavily on the law of nuisance, which involves an equitable balancing of benefits and burdens to the parties.

The *Trail Smelter* panel explained that the decisions of the Supreme Court concerned controversies over the quasi-sovereign rights of states in the federal union and thus were appropriate to use by analogy in addressing inter-state disputes at the international level. While some of the early lawsuits failed for lack of proof, New Jersey succeeded in obtaining an injunction against the city of New York to prevent it from harming the coastal waters of the state by dumping sewage into the sea.[37] The leading decision on point, however, was *Georgia* v. *Tennessee Copper Company and Ducktown Sulphur, Copper and Iron Company, Ltd.*[38] This case defined the nature of the suit as one brought by the state in its capacity as quasi-sovereign, a capacity that gives it an interest 'independent of and behind the titles of its citizens, in all the earth and air within its domain'. The Supreme Court found that 'it is a fair and reasonable demand on the part of a sovereign that the air over its territory should not be polluted on a great scale by sulphurous acid gas, that the forests on its mountains, be they better or worse, and whatever domestic destruction they may have suffered, should not be further destroyed or threatened by the act of persons beyond its control, that the crops and orchards on its hills should not be endangered from the same source'. The court's reference to pollution 'on a great scale' implicitly requires that states in a community (either federal or international) accept a certain amount of pollution incidental to normal activities, a notion also inherent in the concept of abuse of right.

Air pollution is a cause of anthropogenic climate change and like other air pollution results, *inter alia*, in the destruction of forests, mountains, crops and orchards. It thereby deprives individuals of their property and may otherwise reduce the enjoyment of their human rights. If not addressed, climate change threatens the ultimate sovereign and human right, since the very existence of some states and individuals is threatened by rising sea levels. But even more than the earlier pollution cases, climate change poses complex issues of proof sufficient to impose state responsibility and demand mitigation by those responsible for the polluting activities.

The recent US Supreme Court judgment of *Massachusetts* v. *EPA*[39] considered these difficult issues, relying in part on the same cases utilized by the *Trail Smelter* arbitral panel. Aspects of the case may suggest

[37] *New York* v. *New Jersey*, 256 U.S. 296.

[38] *Georgia* v. *Tennessee Copper Company and Ducktown Sulphur, Copper and Iron Company, Ltd.*, 206 U.S. 230 (1907).

[39] *Massachusetts* v. *EPA*, 549 U.S. 497 (2007).

possible avenues for pursuing a rights-based approach to climate change in international or national tribunals: instead of human rights litigation brought by individuals, the affected states could litigate to protect their resources, and to vindicate the human rights of present and future generations of their citizens.

A dozen states in the United States[40] joined by American Samoa, the District of Columbia, the cities of New York and Baltimore,[41] and a host of non-governmental organizations[42] brought a suit in the federal court in the United States to challenge the federal government's failure to regulate greenhouse gases (GHG) under the authority of the federal Clean Air Act.[43] The petitioners alleged that the Environmental Protection Agency (EPA) had abdicated its responsibility to regulate the emissions of four GHGs, including carbon dioxide.[44] The EPA was

[40] California, Connecticut, Illinois, Maine, Massachusetts, New Jersey, New Mexico, New York, Oregon, Rhode Island, Vermont and Washington. Delaware filed an *amicus curiae* brief in support, noting that 'as a low-lying coastal state, [it] experiences daily the effects of global warming. These effects include increased flooding and coastal erosion, increased ocean temperature, and heightened damage to the environment, the property and the people of Delaware'. Five other states (Arizona, Iowa, Maryland, Minnesota and Wisconsin) also submitted a brief in support of the petitioners.

[41] The US Conference of Mayors, National Association of Counties, International Municipal Lawyers Association, American Planning Association, the City of Seattle, the City of Albuquerque, the City of Burlington, and the City and County of San Francisco filed as *amici curiae* in support of the petitioners. In their statement of interest they reported that over 18 months, mayors of 275 cities in 42 states signed the US Mayors Climate Protection Agreement (available at: www.ci.seattle.wa.us/mayor/climate). They agreed to reduce GHG emissions in their communities to 7 per cent below 1990 levels by 2012.

[42] The non-governmental organizations were the Center for Biological Diversity, Center for Food Safety, Conservation Law Foundation, Environmental Advocates, Environmental Defense, Friends of the Earth, Greenpeace, International Center for Technology Assessment, National Environmental Trust, Natural Resources Defense Council, Sierra Club, Union of Concerned Scientists and the US Public Interest Research Group. Calpine Corporation, a clean energy company, filed an *amicus curiae* brief in support of the petitioners, as did the Aspen Skiing Company and Entergy Corporation, one of the nation's largest owners and operators of electricity generating power plants. Entergy has undertaken a voluntary emissions reduction programme and argued that incentives had to be given for other companies to do the same. In addition, the company argued that regulation of GHGs would stimulate innovation in research and development of energy sources.

[43] 42 U.S.C. 7602.

[44] The states might also have brought action directly against the major emitters, the power companies and car manufacturers, but jurisdiction and proof of causation might have been more difficult and multiple lawsuits would have been required to reach all the major actors.

joined in its defence by ten states[45] and six trade associations representing the automotive and energy sectors of the economy.[46]

As in earlier cases concerning transboundary environmental harm, the states argued in part that they have a unique interest in the federal response to climate change because that response will have a significant effect on the impact of climate change on state resources. They posited that there will be more GHG emissions if states regulate individually than if there is national regulation, leading to greater harm to the states. The states' interests were asserted to include preventing loss of unique state lands and unique resources and bodies of water, preventing harm from more frequent and intense storm surges and floods and protecting shrinking water supplies. Respondents and the intervening states countered that because a large percentage of worldwide CO_2 emissions comes from outside the United States, it would be futile for the EPA to regulate such emissions and any such regulation would result in requiring states to achieve the impossible.

In the opening paragraphs of its judgment, the Supreme Court acknowledged that it had accepted to hear the case based on the 'unusual importance of the underlying issue' of global warming. As a matter of statutory analysis, the case was relatively straightforward. The Clean Air Act requires the EPA Administrator to prescribe standards applicable to the emission 'of any air pollutant from any class or classes of new motor vehicles or new motor vehicle engines, which in his judgment cause or contribute to air pollution which may reasonably be anticipated to endanger public health or welfare'.[47] The statute defines 'welfare' to include effects on weather and climate,[48] while 'pollutant' is broadly defined to include any substance or matter emitted into or entering the ambient air.[49] Based on these provisions, nineteen private organizations filed a petition in 1999 with the EPA to obtain regulation of four GHGs: carbon dioxide; methane; nitrous oxide; and hydrofluorocarbons. After fifteen months of consideration, the EPA requested public comment on

45 Alaska, Idaho, Kansas, Michigan, Nebraska, North Dakota, Ohio, South Dakota, Texas and Utah.

46 Alliance of Automobile Manufacturers, National Automobile Dealers Association, Engine Manufacturers Association, Truck Manufacturers Association, CO_2 Litigation Group and Utility Air Regulatory Group.

47 Section 202(a)(1), Clean Air Act, added by Pub. L. 89–272, section 101(8), 79 Stat. 992, and as amended by, *inter alia*, 84 Stat. 1690 and 91 Stat. 791, 42 U.S.C., section 7521(a)(1).

48 42 U.S.C. 7602(h).

49 42 U.S.C. 7602(g).

the petition in 2001. EPA received over 50,000 comments during the subsequent five months.[50]

During the comment period, the White House commissioned its own climate change study[51] which concluded that anthropogenic climate change is occurring. Despite the evidence, the EPA denied the petition on 8 September 2003, giving two reasons: (1) the Clean Air Act does not authorize EPA to issue mandatory regulations concerning greenhouse gases; and (2) even if the authority did exist, it would be 'unwise' to issue regulations, given 'residual uncertainty' about a causal link between GHGs and climate change, as well as the 'economic and political significance' of the issue.[52] Separately, the EPA stated that it 'disagrees with the regulatory approach urged by petitioners',[53] and that it would not be 'effective or appropriate for EPA to establish GHG standards for motor vehicles at this time' in part because motor vehicles are only one of many sources of air pollutants associated with climate change. Subsequently, EPA argued that no one had standing to challenge the agency's decision.

The applicants sought judicial review of the EPA's denial of their petitions. A divided Court of Appeals[54] upheld the EPA's decision, each judge in the majority doing so on a different ground. The petition to the Supreme Court followed, leading to a flurry of interventions and *amici curiae* on both sides.[55]

[50] Pet. App. A63.

[51] National Research Council, *Climate Change: An Analysis of Some Key Questions* (2001).

[52] 68 Fed. Reg. 52922.

[53] The EPA preferred instead 'near-term voluntary actions and incentives' and 'programs aimed at reducing scientific uncertainties and encouraging technological development'. Pet. App. A82.

[54] *Massachusetts* v. *EPA*, 415 F.3d 50 (DC Cir., 2005).

[55] See above, footnotes 40–42 for a partial listing of the *amici curiae*. In addition to those previously mentioned, other notable briefs were filed in the case. Four former EPA administrators supported the petitioners, showing the past practice of the EPA to protect the public from new pollutants and emerging health threats, based on available scientific information. The Alaska Inter-Tribal Council and other Alaskan indigenous groups argued on behalf of the petitioners that the impacts of climate change threaten the physical and cultural survival of Alaska natives. Impacts of global warming are already affecting their members because of thinning sea ice, increased coastal erosion, melting permafrost and changes in plant and animal distributions, thus depleting the subsistence resources of the indigenous peoples. The position of these *amici* directly opposed that of the state of Alaska, which intervened in support of the EPA's position. A brief by major religious organizations, including the National Council of the Churches of Christ, Church World Service and National Catholic Rural Life Conference, focused on the religious dimensions of combating climate change, relying on Christian tenets of stewardship for the natural world. A very large coalition of groups concerned with wildlife conservation, including sporting and conservation organizations,

Two points are of particular interest in the case. The first notable point is the Court's acknowledgement of the reality of climate change, giving a judicial imprimatur to scientific findings on the impact of GHGs. At least equally important is the Court's discussion of standing, for its holding and rationale on this issue might support future international inter-state cases similarly based on the infringement of sovereign rights, including the human rights of a state's inhabitants.[56]

The Supreme Court's judgment was close (5 to 4). The majority opinion began with a review of the emergence of concern with climate change from the enactment of the Clean Air Act to the present, including the adoption of the UN's Framework Convention on Climate Change and the Kyoto Protocol,[57] as well as the reports of the Intergovernmental Panel on Climate Change.

The petitioners, supported by numerous briefs filed by *amici curiae*, argued that global warming is not merely a future threat, but a present reality, with deadly public health consequences, storm surges and intense hurricanes. They asserted that standing does not demand waiting until there is a catastrophic level of global warming, but only requires some present actual or imminent injury. In their view 'EPA distorted two statutory terms ("air pollutant" and "judgment") and ignored a third ("welfare") in order to inject its own policy preferences into a statute that does not embody them'.[58]

The plaintiffs conceded that EPA action on motor vehicles will not stop global warming altogether but could reduce the concentration of GHGs in the atmosphere and thereby delay and moderate, to a significant extent, the impacts of global warming. A brief filed by an alliance of environmental organizations from western states[59] supported the petitioners, arguing that although the effects of global warming are

the Association of Zoos and Aquariums, religious organizations and professional associations, joined in a brief that addressed the impact of climate change on wildlife and ecosystems.

[56] The first level proceeding at the DC Circuit Court noted that only one of the plaintiffs needed to demonstrate standing and held that Massachusetts had plainly demonstrated that it had standing. Hence, the focus on that state's interest and injury in the Supreme Court judgment. *Amici* briefs nonetheless recounted 'numerous and profound, particularized and imminent' injury to the other plaintiffs, including more frequent and more damaging storms, more flooding, more erosion and an increase in summer season heat stress morbidity and mortality. Mayors brief at 24.

[57] Kyoto Protocol to the United Nations Framework Convention on Climate Change, 16 March 1998, 37 I.L.M. 32.

[58] Pet. Brief, 2.

[59] The groups joining the brief were North Coast Rivers Alliance, Desert Protection Society, Westside Association to Save Agriculture, California Sportfishing Protection

generalized, they have had specific adverse impacts on the petitioners, who have suffered particular, concrete, actual, imminent and redressable harms due to the failure to regulate carbon dioxide emissions.[60] Moreover, because certain vulnerable segments of the population suffer a disproportionate share of the harm inflicted by global warming, while others remain uninjured, the political process is unlikely to address the injuries adequately.

The EPA argued that the plaintiffs lacked standing because the widespread nature of the harm alleged from GHG emissions presented an insurmountable jurisdictional barrier.[61] The plaintiffs' injury could not be distinguished from that of the public at large and, therefore, there was no 'injury in fact'. The EPA also argued that the petitioners had failed to establish that the injuries they allege from global warming are traceable to GHG emissions from new vehicles in the United States, rather than to GHG emissions from other sources in the United States, to GHG emissions from vehicles or other sources elsewhere in the world, or to entirely different factors.[62] Third, the petitioners failed to show that a decision to require regulation of emissions of GHG from new motor vehicles in the United Sates would redress their injuries.[63] Thus, according to the EPA, the petitioners failed to show either injury in fact, causation, or that the injury would be redressed by a favourable decision.

The Court disagreed with the EPA, after testing whether the plaintiffs had 'such a personal stake in the outcome of the controversy as to assure that concrete adverseness which sharpens the presentation of issues' to be decided.[64] The judgment quoted from an earlier opinion by Justice Kennedy, which noted that 'it does not matter how many persons have been injured by the challenged action' so long as there is concrete and personal injury to the party bringing suit. This concrete and particularized injury can be actual or imminent, but must be 'fairly traceable to the

Alliance, Save Medicine Lake Coalition, Klamath Forest Alliance, San Joaquin Audubon Society and the North Cascades Conservation Council.

[60] The nature and scope of real and imminent injuries to coastal states was addressed more fully in an *amicus* brief filed by a coalition of individuals and groups concerned with ocean and coastal conservation. Those signing the brief included, *inter alia*, the Ocean Conservancy, Jean-Michel Cousteau, the Marine Conservation Biology Institute and Ocean Futures Society.

[61] Resp. Brief, 7–8, 10–20.

[62] EPA cited figures indicating that as much as 80 per cent of all GHG emissions emanate from countries other than the United States. Further, the US transportation sector is responsible only for about 7 per cent of worldwide GHG emissions. Ibid. at 13.

[63] Fed. Resp. Cert. App. at 12.

[64] *Massachusetts* v. *EPA* at 14, citing *Baker* v. *Carr*, 369 U.S. 186, 204 (1962).

defendant'. Finally, it must be shown that a favourable decision would be likely to redress the injury.[65]

The Court stressed the 'special position and interest of Massachusetts'. According to the Court, 'it is of considerable relevance that the party seeking review here is a sovereign State and not, as it was in [an earlier environmental case] a private individual'.[66] Quoting from *Georgia* v. *Tennessee Copper Co.*[67] – the major case relied upon in the *Trail Smelter* arbitration – the Court noted that the suit was being brought by the state 'in its capacity of quasi-sovereign' and that 'in that capacity the State has an interest independent of and behind the titles of its citizens, in all the earth and air within its domain. It has the last word as to whether its mountains shall be stripped of their forests and its inhabitants shall breathe pure air'. A full century after the earlier judgment, the Court thus reaffirmed that states have a sovereign interest in the environment sufficient to support standing, based on the state's 'well-founded desire to preserve its sovereign territory today'.[68] To emphasize the point, the Court quoted from another judgment affirming that states in the US federal system retain the dignity, though not the full authority, of sovereignty.[69] On this basis, Massachusetts and the other petitioners, especially those coastal states threatened with the loss of territory by global warming, were afforded standing to sue the federal government.

The Court's discussion of the states' sovereign rights and duties resonates on the international level. The Court pointed out the limited options available to a state seeking to address climate change or other transboundary environmental harm: it cannot lawfully invade another state to force reductions in GHG emissions; it cannot legislate a reduction in motor vehicle emissions outside its own jurisdiction. Yet there are

[65] The Court noted that these requirements are relaxed when Congress has granted a procedural right of action to protect the litigant's interests. In such case, the litigant has standing 'if there is some possibility that the requested relief will prompt the injury-causing party to reconsider the decision that allegedly harmed the litigant'. That procedural right to challenge agency action unlawfully withheld was in fact granted by Congress. 42 U.S.C. 7607(b)(1).

[66] *Massachusetts* v. *EPA* at 15.

[67] 206 U.S. 230, 237 (1907). See also *Nebraska* v. *Wyoming*, 515 U.S. 1, 20 (1995).

[68] *Massachusetts* v. *EPA* at 16.

[69] *Bowen* v. *Public Agencies Opposed to Soc. Sec. Entrapment*, 477 U.S. 41, 51 note 17 (1986). Arizona, Iowa, Maryland, Minnesota and Wisconsin, as *amici curiae* in support of the petitioners, had cited to the case to assert that states have standing to sue whenever they allege an interest in preserving their sovereignty and that interest has been interfered with or diminished. Brief of Arizona *et al.*, 20.

important sovereign interests at stake that are of concern to the state as a whole, including protection against any substantial impairment of the health or prosperity of the individuals within its boundaries. These persons are protected in part by entitling states to 'special solicitude' in the standing analysis. The same analysis can apply to independent states in the international system, who are obligated by human rights law to promote and protect the human rights of their inhabitants and whose recourse to force is limited by law. Where individuals may lack procedural capacity to enforce their rights against the acts or omissions of a foreign sovereign, their state can do so on their behalf and can ensure that the state's resources are protected for them and their descendants.

In finding injury to the states, the Court accepted that the harms associated with climate change are 'serious and well-recognized' and that significant harm has already been inflicted.[70] Citing petitioners' experts, the Court specifically mentioned sea level rise and increases in the spread of disease, finding that sea level rise has already begun to swallow Massachusetts' coastal land. State ownership of a substantial portion of the coastline gives the state a particularized interest in obtaining EPA action.

A key element in the litigation was whether or not EPA regulation of GHG would redress any harm to the petitioners. EPA argued that its regulation of new motor vehicles in the United States would be so insignificant that it could not provide any realistic possibility of mitigating global climate change and remedy the injury to petitioners. The agency specifically pointed to rising GHG emissions from developing nations like the People's Republic of China (PRC) and India. This 'offset' meant no effective relief could be obtained. The Court rejected the EPA's defence, accepting that incremental action could be sufficient and, therefore, required by the statute.[71] Moreover, even if the PRC and India are increasing GHG emissions, a reduction in domestic emissions would slow the pace of global increases, no matter what happens elsewhere.

[70] The harms identified include 'the global retreat of mountain glaciers, reduction in snow-cover extent, the earlier spring melting of rivers and lakes, [and] the accelerated rate of rise and sea levels during the 20th century relative to the past few thousand years'. *Massachusetts* v. *EPA* at 18, citing NRC Report, 16.

[71] The Court did not seem particularly convinced that regulating GHG from new motor vehicles would be an insignificant step. It called emissions from the transportation sector 'an enormous quantity', noting that they account for more than 6 per cent of worldwide carbon dioxide emissions. *Massachusetts* v. *EPA* at 21. Regulating these emissions would 'make a meaningful contribution to greenhouse gas concentrations'.

The final step in the standing analysis was the question of remedy: whether the Court could direct EPA to take steps to slow or reduce vehicle emissions, acknowledging that such a step would not reverse global warming. The Court again referred to the 'enormity of the potential consequences associated with man-made climate change', noting that the more drastic the injury the lesser the increment in probability necessary to support standing. It concluded that 'the rise in sea levels associated with global warming has already harmed and will continue to harm Massachusetts. The risk of catastrophic harm, though remote, is nevertheless real. That risk would be reduced to some extent if petitioners received the relief they seek. We therefore hold that petitioners have standing to challenge the EPA's denial of their rulemaking petition.'

The four dissenting justices would have found the case non-justiciable as a political question to be addressed by the Congress and Executive branches of government. In their opinion, the majority was incorrect to hold that states have a special status to litigate to assert their sovereign rights. In their view, precedent requires a particularized injury for each litigant and no authority supports the '*parens patriae*' representation of all persons within a state who have suffered no immediate injury. Indeed, the dissenters indicated that they would have difficulty ever finding standing for a litigant on this issue, because global warming 'is a phenomenon harmful to humanity at large' and redress sought is 'literally to change the atmosphere around the world'.[72] Such a view, had it prevailed, would have closed the courtroom doors to any judicial consideration of widespread environmental harm due to global climate change.

The very concept of global warming is thus inconsistent in their view with demonstrating a particularized injury, unless and until one of the petitioners can demonstrate actual or imminent harm other than by computer models. The dissenting opinion provides clear indications of the potential litigation hurdles for those seeking judicial action on global warming, by focusing on the widespread impacts, multiple contributions to the problem and the limitations of unilateral remedial action. As the dissenters see it, 'petitioners are never able to trace their alleged injuries back through this complex web [of greenhouse gas emitters] to the fractional amount of global emissions that might have been limited with EPA standards … [T]he connection is far too speculative to

[72] *Massachusetts* v. *EPA*, Roberts, J., dissenting at 7.

establish causation.'[73] Even more problematic in their view is redressability because of the increase in emissions predicted for developing countries.

On the merits, the EPA presented several reasons for the failure to regulate, in addition to its primary contention that it lacked statutory authority over GHGs. The Court rejected the EPA arguments and remanded the case to the EPA with directions to evaluate the regulation of GHGs according to the legal standards set forth in the Clean Air Act.

The Supreme Court's discussion of standing in *Massachusetts* v. *EPA* is of considerable interest to states contemplating the possibility of international judicial action to mitigate GHG emissions that increase climate change to their detriment. Although the petitioners were component states of a federal union suing the federal government rather than each other, the basis of their standing was the assertion of permanent sovereignty over their natural resources, with clear parallels to and an origin in international law. They also asserted the right and duty to protect their inhabitants. These same bases of jurisdiction could support an inter-state action before the ICJ, the Law of the Sea Tribunal, or a human rights tribunal, depending on the factual allegations. There is some support, in addition to the *Trail Smelter* arbitration, to suggest that an international tribunal could grant either interim measures of protection or injunctive relief to prevent actual or imminent harm.[74]

Assuming that jurisdiction and standing exist, what is the appropriate legal basis for determining inter-state responsibility and redress for injury caused or threatened by climate change? Thus far, transboundary pollution cases decided by international tribunals have largely relied on principles of equity, such as the abuse of rights doctrine discussed above, to assess and balance sovereign rights. Other equitable principles could be invoked, such as common but differentiated responsibilities, as these have emerged in international instruments. Furthermore, international human rights law, missing thus far in the inter-state claims presented for transboundary environmental harm, could add an important normative framework for judging the lawfulness of state acts and omissions. It is the addition of this framework that constitutes the proposed rights-based approach discussed below.

[73] *Massachusetts* v. *EPA* at 11.

[74] See, e.g., ICJ Rep. (1974), 312–71 and 494–523.

D Balancing sovereign rights: equity, human rights and climate change

Vast wealth disparities around the globe have resulted in a great variation in the nature of environmental problems, the contribution of each state to global environmental deterioration, including climate change, and each state's ability to prevent and remedy harm to the environment. At the beginning of the twenty-first century, the industrialized world, with 20 per cent of the global population, generated more than 80 per cent of the world's pollution and used about 80 per cent of global energy and mineral resources, but the environmental impacts that resulted from this production and consumption, particularly with respect to anthropogenic climate change, affected disproportionately the development of poorer countries. Developed countries also accounted for 83 per cent of the world's gross domestic product (GDP), with the gap between developed and developing countries in per capita GDP increasing during the last thirty years of the twentieth century, making developing countries relatively poorer than before.

Poverty itself has come to be seen both as a major source of environmental degradation[75] and as a human rights issue. Pollution from urban growth, lack of water quality and quantity, desertification and global climate change harm developing countries more than industrialized nations and the poor in every country. The natural resources of poor countries overwhelmingly bear the cost of unregulated development and the poor lack alternatives to using resources that would otherwise be conserved. Poverty also means that individuals lack an adequate standard of living, the food, shelter, medical care and education that are guaranteed by international human rights law.

The Stockholm Declaration recognized that the environment affects the well-being (and thus the human rights) of people throughout the world, with many environmental problems in poorer countries caused by a lack of development. Principle 5 called for sharing among all mankind

[75] In its preamble the Stockholm Declaration recognized that:

> [i]n the developing countries most of the environmental problems are caused by underdevelopment. Millions continue to live far below the minimum levels required for health and sanitation. Therefore, the developing countries must direct their efforts to development, bearing in mind their priorities and the need to safeguard and improve the environment. For the same purpose, the industrialized countries should make efforts to reduce the gap between themselves and the developing countries …'

the benefits from the use of non-renewable resources, while Principle 9 proposed the transfer of 'substantial quantities' of financial and technological assistance to supplement the domestic effort of developing countries to remedy environmental deficiencies. Several other principles gave particular attention to the needs of developing countries in meeting the costs of environmental safeguards.

Twenty years later, the Rio Declaration on Environment and Development echoed many of these principles and developed others in its effort to balance the two elements in its title. Principle 3 referred to equitably meeting the developmental and environmental needs of present and future generations. Principle 6 called for giving special priority to the situation and needs of developing countries, particularly the least developed and those most environmentally vulnerable. While these principles focused on elements of need as a basis for distributive justice, Principle 7 looked to responsibility for harm and capacity to redress: 'In view of the different contributions to global environmental degradation, States have common but differentiated responsibilities. The developed countries acknowledge the responsibility that they bear in the international pursuit of sustainable development in view of the pressures their societies place on the global environment and of the technologies and financial resources they command.' Thus, the declaration identifies at least three factors that could be taken into account in the equitable allocation of benefits and burdens: need, responsibility and capacity.

The first principle governing the 1992 Climate Change Convention is that the parties should protect the climate system for the benefit of present and future generations of humankind 'on the basis of equity and in accordance with their common but differentiated responsibilities and respective capacities' (Article 3). This principle not only calls for equity generally but identifies several factors to be taken into account in deciding what is equitable. The second principle adds the factor of need as a further element, reflecting a widespread emphasis on addressing the marginalized and most vulnerable. Respect for human rights, including principles of equality and non-discrimination, form part of the equitable construct.

Strictly legal, that is, formal equality demands rules of identical treatment to ensure full respect for the sovereignty of each state regardless of size or wealth. Where such equal treatment is unjust or impossible because the impacts of activities or circumstances are inherently different, equitable norms may adjust international benefits and burdens. For example, the norm of equitable utilization of transboundary waters by riparian states ensures the possibility of modifying the otherwise obligatory equal allocation of a shared resource between riparian states to ensure a 'fair' distribution. The concept of fairness may, and in the case of watercourses does, import human rights law as a critical factor in determining what is equitable allocation and utilization. Thus, the UN Convention on the Law of the Non-Navigational Uses of International Watercourses[76] provides in Article 10 that in weighing competing uses of freshwaters, 'special regard' is to be given to the requirements of vital human needs. In other words, equity cannot be achieved without respect for human rights.

With or without human rights, the role of equity and equitable principles in climate change is contested. Some industrial countries continue to press for strictly equal obligations in addressing climate change. In 1997, the US Senate voted 97 to 0 in favour of a resolution stating that the United States should not join any agreement on climate change that would require the industrialized countries to reduce their GHG emissions, unless the agreement imposed similar obligations on developing country parties. The resolution implicitly suggested that 'unequal' treatment would be discriminatory and thus unfair. In March 2001, President George W. Bush rejected the Kyoto Protocol, which, indeed, established targets and timetables for developed countries but not for developing ones.

Other countries, especially small island developing states, argue that equal treatment is itself inequitable, because they may be injured by future climate change to which they have largely not contributed and which is due to industrial and other processes from which they have not benefited. In their view, equity means the cost of clean-up or reduction of pollution must be borne by those who created the problem. Using principles of distributive justice, they seek to reconcile competing social and economic policies in order to obtain the fair sharing of a common resource: the earth's atmosphere. They also integrate human rights considerations, arguing that law and equity requires addressing the disproportionate human rights impacts of global climate change on vulnerable populations.

As the competing positions on climate change illustrate, equity is often used as a synonym for fairness or justice, both in procedure and substance.[77] The procedural aspect is concerned with reaching decisions by

[76] GA Res. 51/299, 21 May 1997, reprinted in 26 I.L.M. 700 (1997).

[77] Thomas Franck, for example, considers equity as subsumed in the concept of fairness, which in his view has both procedural and substantive dimensions. T. Franck, *Fairness in International Law and Institutions* (Oxford: Clarendon Press, 1995), at 7–9.

the 'right process', while the substantive dimension aims at distributive justice.[78] Imposing equal obligations on subjects of law who are unequal in relevant ways is perceived as unjust because it exacerbates inequalities and settles burdens on those least able to bear them. Legal systems, including the international legal system, therefore, sometimes base the distribution of goods and the burdens of society according to the principle of distributive justice, seeking substantive equality by treating like cases alike and unlike differently according to various criteria such as prior entitlement, just deserts or need.

'Sustainable development' incorporates this understanding of equity as distributive justice, in its effort to strike a fair balance between the goals of short-term economic development and long-term environmental and human rights protection. Many of the principles in the Rio Declaration on Environment and Development reflect this balance as they address inter- and intra-generational equity. Principle 6, for example, mandates particular priority for the special situation and needs of developing countries, particularly the least developed and those most environmentally vulnerable. Inter-generational equity, recognized in the Rio Declaration, is based on acknowledging two key facts: (1) that human life emerged from, and is dependent upon, the Earth's natural resource base, including its ecological processes, and is thus inseparable from environmental conditions; and (2) that human beings have a unique capacity to alter the environment upon which life depends. From these facts emerges the notion that humans that are alive today have a special obligation as custodians of the planet to maintain its integrity to ensure the survival of the human species. Those living have received a heritage from their forebears in which they have beneficial rights of use that are limited by the interests and needs of future generations. Current environmental goods, wealth and technology are owing to the progress of prior generations and because this debt cannot be discharged backward it is projected forward and discharged in the present on behalf of the future. The other side of this duty coin is, of course, the right of present and future generations to equitable resource use and preservation of the resource base.

In many states, the concept of inter-generational equity is manifest in the doctrine of the public trust, in which the state acts as trustee of natural resources on behalf of present and future generations. This notion is inherent in the *Massachusetts* v. *EPA* judgment, although

[78] Franck, at 7.

none of the states explicitly relied on public trust doctrine. Generally, the trustee's obligations require the conservation and maintenance of the trust resources against threats from inside the state as well as outside, thus placing constraints on the present generation of beneficiaries. Meeting the obligation does not mean that no harm is allowed, but it does call for the minimization or avoidance of long-term and irreversible damage to the environment. Thus, first, each generation is required to conserve the diversity of the natural resource base so that it does not unduly restrict the options available to future generations to satisfy their own values and needs. Second, the quality of ecological processes passed on should be comparable to that enjoyed by the present generation. Third, the natural heritage should be conserved so that future generations will have access to it. The state, as trustee, may protect these interests against threats from inside or outside the state, as an exercise of its sovereignty.[79]

Equitable burden-sharing, intra- and inter-generational, is justified for several reasons. First, all states share an interest in the conservation and sustainable utilization of the Earth's biological resources, many of which are the source of desired products as well as ecological processes (for example, tropical forests as carbon sinks) including the global climate. Second, the contribution of states to the current environmental crisis has not been equal and the industrialized nations are predominantly responsible for pollution. Third, developing states legitimately plead their inability to participate or comply in climate change mitigation due to poverty and the need to develop. Finally, developing countries cannot fulfil their obligations to respect and ensure the economic and social rights of their inhabitants without international cooperation and assistance, something required by Article 2 of the International Covenant on Economic, Social and Cultural Rights.

[79] Edith Brown Weiss posits that the rights-holders are generations, some of which are here and some of which are in the future. Generations hold these rights as groups in relation to other generations. Since the future individuals are indeterminate, the state as a guardian or a representative of the group may enforce these rights. E. Brown Weiss, *In Fairness to Future Generations: International Law, Common Patrimony and Intergenerational Equity* (Dobbs Ferry, NY: Transnational Publishers, 1989) at 95–7. Within states, there may be other representatives or guardians. The Philippine Supreme Court found that present generations have standing to represent future generations in large part because 'every generation has a responsibility to the next to preserve that rhythm and harmony for the full enjoyment of a balanced and healthful ecology'. *Minors Oposa* v. *Secretary of the Department of Environment and Natural Resources*, Philippine Supreme Court, reprinted in 33 I.L.M. 168 (1994).

These questions of allocation of shared resources, responsibility for conserving each state's natural resources and controlling pollution and the distribution of costs arising from pollution prevention and environmental degradation have brought the issue of equity to the fore. In particular, an understanding that unilateral, bilateral, or even regional solutions are likely to be ineffective in resolving global issues such as climate change have led to a recognition of the need to encourage full participation by states in environmental regimes through acknowledging different capacities and responsibilities. Equity can be seen in this context as a counterpoint to the classical insistence on the formal equality of states and as providing a compromise between permanent sovereignty over natural resources and a common concern such as the conservation of biological diversity. In this sense, negotiators must rely upon equity to resolve conflicts in applying the two parts of the Stockholm Principle 21: the right to use resources and the duty not to cause transboundary environmental harm.

In most legal systems, equity has traditionally played a major part in determining the distribution of rights and responsibilities in conditions of scarcity and inequality. However, the assertion that like cases must be treated alike and those that are different handled otherwise requires determining which similarities and differences are relevant in which situation. To take an example from within national legal systems, income differences are generally accepted as a proper basis for allocating tax burdens but not for voting in national elections. Thus, while the general value of equity or fairness is largely accepted in the context of scarcity and inequality, debate centres on the appropriate principle on which to determine equitable allocation – whether decisions should be based on need, capacity, prior entitlement, 'just deserts', the greatest good for the greatest number, or strict equality of treatment. The various factors may point toward allocation in one direction or in many different directions. In addition, a single factor, such as need, may be asserted by more than one actor or group of actors. The principle of common but differentiated responsibilities is generally included in environmental agreements, but still requires the determination of what is the equitable basis for allocation of rights and responsibilities.

E The role of human rights in equitable allocation

The environmental law principle of *common but differentiated responsibilities* does not establish any preference among the different factors relevant to equity. Instead, it reflects a multifaceted approach, with most formulations referring to different historical responsibilities as well as to different capacities and needs. Given this, states responsible for GHG emissions may plead for recognition of acquired rights or prior uses. In response, affected states could stress the polluter pays principle, the principle of good neighbourliness, or the human rights of its population as a basis for limiting activities, even when there are acquired rights. Affected states may invoke the rights of those persons within their jurisdiction, the duty to protect those rights and their sovereign right to be free from significant transboundary pollution as determinative factors in equitably allocating responsibility for addressing anthropogenic climate change.

As this section discusses, state sovereignty is limited by human rights obligations. Moreover, a rights-based approach to equitable distribution of benefits and burdens is the one that best serves to merge the three pillars of sustainable development: economic development; environmental protection; and human rights. International human rights law emphasizes each individual's right to a certain quality of environment because it is linked to the enjoyment of a host of internationally and domestically guaranteed rights that cannot be exercised otherwise. Former UN Secretary-General, Kofi Annan, in his 1998 Annual Report on the Work of the United Nations Organization spoke in favour of a rights-based approach to environmental protection because it 'describes situations not simply in terms of human needs, or of development requirements, but in terms of society's obligations to respond to the inalienable rights of individuals'. Environmental protection is undoubtedly a pre-condition to the enjoyment of some internationally guaranteed human rights, especially the rights to life, health, private and home life and cultural rights, but it also directly or indirectly impacts other rights as well.[80]

[80] One study has estimated that 40 per cent of the world's deaths can be attributed to environmental factors. D. Pimental *et al.*, 'Ecology of Increasing Diseases: Population Growth and Environmental Degradation', *Bioscience*, 48:10 (October 1998), 817–26. In addition, some 1.2 billion people in developing countries lack clean and safe drinking water, with the result that waterborne infections account for 80 per cent of all infectious diseases worldwide. In many areas industrial and household wastes are dumped directly into rivers and lakes. Air pollution adversely affects the health of 4 billion people. Some 2.5 billion kg of pesticides are used worldwide each year – a fifty-fold increase over the past 50 years – resulting in about 3 million cases of human pesticide poisonings being reported annually.

Other means of allocating benefits and burdens may be considered first. *Formal equality* (for example, per capita distribution) is one method of allocating resources and burdens. As noted earlier, rules are generally deemed just if they apply to all without discrimination, and non-discrimination is clearly a predominant obligation in international human rights law. Yet, equal treatment is required only for those equally situated. It may yield extreme outcomes when pre-existing economic or other inequalities exist in society. At the international level, when allocations are based on formal equality, moreover, the issue of whether the appropriate apportioning unit is the state or the individual may arise, as in determining permissible emission levels. Requiring all states to reduce GHGs in an identical fashion would make many developing countries, or groups in those countries, worse off, at least in the short term. From the perspective of equity toward the most vulnerable or least well off, environmental protection should not result in further deterioration of their well-being. In order to address this problem, non-equal or differential obligations can, and are, being imposed as an equitable means of fostering substantive equality in the long term.

Notions of *entitlement* uphold the existing distribution of goods if they were justly acquired according to the rules in force at the time of acquisition; it is a recognition of the right to property. Entitlement protection is contained in some environmental laws and agreements that 'grandfather' existing activities by exempting them from retrofitting to meet more exacting and newly enacted standards or allowing emissions to continue at pre-existing levels. The rewards that this system grants to those who have the goods may be too high to result in what is considered to be a fair distribution. An entitlement approach also may serve to deny essential goods to others. As a result, human rights instruments that contain a right to property also recognize that the public interest may justify some expropriation, provided it is non-discriminatory and fair compensation is paid.

Traditional international law affords some protection based on prior entitlement. All states, including those newly created, have equality of opportunity as sovereigns, but pre-existing natural endowment and activities make older states substantially stronger in wealth and power and developing states substantially stronger in natural (biological) resources. Since traditional international law entitles all states to an equal right to obtain or use common resources, from fish in the high seas to the geostationary orbit, technologically advanced states have the ability to, and may choose to, acquire the greatest part of the resources from the common area. Equality of rights, however, does not necessarily bring about equality of outcomes and the least favoured may find themselves in a continually declining position unless the rules are modified to afford preferential treatment to those most in need.

Different *capacities* (from each according to his or her ability) may be the decisive factor chosen to achieve distributive justice, as expressed in environmental agreements that require the Organisation for Economic Cooperation and Development (OECD) or other groupings of countries to finance poorer countries or transfer technology because they have the ability to do so. One problem that can arise is making the relevant determinations of ability to pay. States may argue that various factors make it fair for them to be grouped with the poorer countries. The Kyoto Protocol classifies Saudi Arabia and Singapore as 'developing', while Bulgaria is classified as developed, even though it is still an economy in transition state. Without objective criteria to determine the groupings, along with the flexibility to move states from one group to another, the problem will largely be a political one. Some treaties avoid this problem by incorporating notions of capacity generally, requiring each state party to take measures 'in accordance with its particular conditions and capabilities' or 'as far as possible and as appropriate'.[81]

Different *needs* (to each according to his or her need) as a basis for equitable allocation are recognized in the Rio Declaration and reappear, for example, in the UNFCCC. In implementing the Convention, the parties are to be guided by 'the specific needs and special circumstances of developing country Parties, especially those that are particularly vulnerable to the adverse effects of climate change, and of those Parties, especially developing country Parties, that would have to bear a disproportionate or abnormal burden under the convention'.[82] The question of what would be 'disproportionate' is left open. Article 4(8) adds that all parties are to consider what actions, including funding, insurance and transfer of technology, may be necessary to meet the specific needs of specially affected states. Determining need, like determining capacity, may require the development of objective criteria and the assessment of the situation over time of each state party.

Different *historical responsibility* or 'just deserts' – that is, past and present contribution to environmental harm, is deemed by developing countries to be one of the most relevant factors in allocating burdens.

[81] Articles 6–11 of the Convention on Biological Diversity.

[82] Article 3.2 of the UN Framework Convention on Climate Change.

The 1991 Beijing Declaration on Environment and Development stated the view of the developing world that 'the developed countries bear responsibility for the degradation of the global environment. Ever since the Industrial Revolution, the developed countries have over-exploited the world's natural resources through unsustainable patterns of production and consumption, causing damage to the global environment, to the detriment of the developing countries'. Fairness and a morally coherent response suggest that these states, which attained their current developed status through imposing non-internalized costs on the environment, take the major abatement actions, rather than demanding that everyone equally mitigate the externalities, including those not responsible for initially creating the problem. Equity, in this sense, is justified as a means of corrective justice, requiring remedial conduct to correct past wrongs.

The polluter pays principle, which requires that the entity causing environmental harm should bear responsibility for the costs ensuing from that harm, is an economic principle requiring the internalization of externalities. However, it is also compatible with corrective justice since it serves a reparative function by making those states that caused most environmental harm pay for the remediation or losses suffered by others. Similarly, compensatory or reparative justice for historical wrongs and takings may be a basis for equitable (preferential) treatment for developing countries, especially where colonizing states built their industrial development on the exploitation of the natural resources of their colonies.

Finally, human rights obligations must be considered to be a determining factor in equitable allocation. Human rights are maximum claims on society because rights are inherent attributes that must be respected in any well-ordered society. There are more than a dozen references to human rights in the UN Charter and member states have a clear obligation to take joint and separate action with the organization to promote respect for and observance of human rights and fundamental freedoms (Articles 55 and 56). The Universal Declaration of Human Rights set forth 'a common standard of achievement for all peoples and all nations'. Eleanor Roosevelt said it might well become 'the Magna Carta of all mankind'. The Declaration today represents an agreed definition of 'human rights' as that term is used in the United Nations Charter.

Respect for human rights is thus part of international law, contained in the nearly 100 global human rights agreements in force today and the regional systems in operation or emerging around the world.

Each state has a consequent right and duty to ensure that its population can enjoy the full panoply of internationally guaranteed human rights, even when the threats to them appear from another state. The International Covenant on Economic, Social and Cultural Rights (ICESCR) even contains a statement of obligation that explicitly encompasses transnational action: 'Each State Party … undertakes to take steps, *individually and through international assistance and cooperation, especially economic and technical, to the maximum of its available resources* …' to realize the rights contained in the Covenant (emphasis added). Article 11(2) reiterates the obligation to take measures individually and through international cooperation, in this instance to combat hunger, adding specific reference to the need for equitable distribution of world food supplies in relation to need, taking into account the problems of both food-importing and food-exporting countries. A third relevant statement is found in Article 15(4): 'The State parties recognize the benefits to be derived from the encouragement and development of international contacts and cooperation in the scientific and cultural fields.'

At a minimum, the stated obligations in the ICESCR encompass the duty 'to take steps' in two ways: to cooperate and to provide international assistance. Cooperation can be viewed as an obligation of conduct, while the provision of international assistance, to the maximum of a state's available resources, constitutes an obligation of result. The obligation of conduct means that it is not necessary to show that specific harm results from breach of a duty to cooperate; it is enough that a state refuses or fails to fulfil its obligation to cooperate imposed by treaty or otherwise.[83] From the perspective of international responsibility, it does not matter whether the obligation is one of conduct or one of result, because a breach of either duty can be considered a wrongful act.[84]

The critical question is whether or not the duties to cooperate and to provide international assistance are specific enough that breaches can be identified and give rise to state responsibility. The duty to cooperate, at least, has been held to give rise to enforceable rights. In the *Mox Plant* case (*Ireland* v. *United Kingdom*), Ireland invoked a duty to cooperate in

[83] ILC SR Articles, Commentary to Art. 2, para. (9). *Interpretation of Peace Treaties with Bulgaria, Hungary and Romania, Second Phase*, ICJ Rep. (1950), 228.

[84] In the *Gabçikovo-Nagymaros Project* (*Hungary* v. *Slovakia*), ICJ Rep. [1997], 7, at 77, para. 135, the ICJ referred to the parties having accepted 'obligations of conduct, obligations of performance, and obligations of result.' See C. Tomuschat, 'What is a "Breach" of the European Convention on Human Rights?', in R. Lawson and M. de Blois (eds.), *The Dynamics of the Protection of Human Rights in Europe: Essays in Honour of Henry G. Schermers* (Dordrecht: Nijhoff, 1994), 315 at 328.

the field of protecting the marine environment. The International Tribunal on the Law of the Sea (ITLOS), in its order on provisional measures issued on 3 December 2001, reprinted in 41 I.L.M. 405 (2002), opined that the duty to cooperate is a fundamental principle in general international law, as well as one contained in the relevant treaty provisions, and that rights may arise therefrom which the Tribunal may protect. The ITLOS provisional order mandated that the parties cooperate to exchange further information about the environmental consequences of the proposed project and devise measures to prevent harm that could result from proceeding with the project.

Even without an explicit transnational obligation, one can couple the duty not to cause or to allow significant transboundary environmental harm with the duty to respect and to ensure that international human rights develop a coherent rights-based approach to climate change. This approach builds on the discussion of standing in *Massachusetts* v. *EPA* to view state sovereignty not as a barrier to the implementation of human rights, but as a vehicle for ensuring their enforcement and enjoyment. Calling on responsible states to reduce GHG emissions and assist in adaptation, even by litigation if necessary, is a clear application of such a rights-based approach.

Conclusions

There is growing recognition of the interdependence of states and of problems that are insoluble through unilateral action, leading to acceptance of the moral principle of solidarity or partnership. Interdependence underscores the search for a just global society, which is a quest as old as human civilization. To many, a just society involves ensuring that the natural components of the environment continue to sustain life in all of its diversity, and that the natural benefits that humans enjoy are shared fairly among all those present and to come. The moral dimension of equity is such that it is often deemed synonymous with justice and is an end in itself. In addition, however, substantive equality over time is a goal intended to enhance the social and economic development of those worst off in the world and, thus, remedial measures are designed to achieve this aim. Temporary exceptions may, therefore, be created to allow actors to enjoy the rights established by the rules in force.

The recognition that global resources are shared or of common concern or heritage has given rise to a duty to assist those states unable to participate in the utilization of the resources. Equity in international environmental law thus means a rational sharing of the burdens and costs of environmental protection, discharged through the procedural and substantive adjustment of rights and duties. Equity in the sense of fairness also means warning states of imminent peril and cooperating to resolve problems that will impact the ecological processes or resources on which future well-being depends.

Equity is important and, with its emphasis on fairness, is more attractive to many than economic efficiency or open conflict as a means of deciding how to allocate and sustain limited common resources. Without a cooperative and equitable solution to the issue of allocation, competitive utilization of the resource may continue until it is depleted. Equitable or differentiated obligations may induce participation in action among the competing states as well as among states that may not have any direct interest in a specific environmental issue. Developing countries have noted that ozone depletion, which is of greater concern to developed countries, has been addressed more rapidly and seriously than desertification or other issues of greater interest to the South. Such observations may be a disincentive to cooperation, notably, the 2002 GEF decision to fund desertification projects, which seems at least in part a response to criticisms heard before and during the World Summit on Sustainable Development that the limited mandate of the fund was unfair.

Equity also may be justified on the basis of self-interest. Developed countries gain from secure access to primary resources situated largely in developing countries. More generally, environmental protection is in everyone's interest, and the adjustment of legal obligations to achieve better protection is self-interested. An allocation of burdens that takes into account the more vulnerable position of developing states may benefit all through inducing their cooperation to improve global environmental conditions. Moreover, Scott Barrett's work has indicated that agreements perceived to be fair are not only likely to induce greater participation but are more likely to be self-enforcing and thus successful over the long term.[85]

Finally, distributive justice is compatible with and can further the goals of promoting and protecting human rights. Any allocation of benefits and burdens that makes vulnerable populations worse off, even if the harm is felt outside the boundaries of the state, cannot be regarded as equitable or in conformity with international human rights law. States

[85] S. Barrett, *Environment and Statecraft: The Strategy of Environmental Treaty-Making* (Oxford University Press, 2003).

may invoke both their sovereign rights to exist and to freedom from significant environmental harm, as well as the human rights of their inhabitants, to demand reduction of GHG emissions by those states most responsible for anthropogenic climate change. Thus, in weighing the various factors to achieve an equitable balance, international human rights obligations must play a role. States have pledged to take unilateral and joint action together with the United Nations to achieve promotion and protection of human rights.

In sum, equitable approaches are not only based in morality and a sense of justice but may also foster more effective action on issues of common concern and more effective implementation of norms. Equity, as reflecting notions of fairness and legitimacy, may produce more or better compliance with environmental agreements. In practice, therefore, equitable differentiation has probably become the price to be paid to ensure universal participation in environmental agreements concerned with global problems. Yet, it should not be forgotten, as Thomas Franck has noted, that '[t]he law promotes distributive justice not merely to secure greater compliance, but primarily because most people think it is *right* to act justly'.[86]

[86] Franck, *Fairness in International Law.*

4

Climate change, human rights and corporate accountability

PETER NEWELL*

> It is about social justice and the human rights of the world's poor and marginalised. Failure to act on climate change would be tantamount to a systematic violation of the human rights of the poor. (Watkins 2007)

> climate change has clear and immediate implications of the full realization of human rights, including *inter alia*, the right to life, the right to take part in cultural life, the right to use and enjoy property, the right to an adequate standard of living, the right to food, and the right to the highest attainable standard of physical and mental health. (Malé Declaration, Human Dimension of Global Climate Change, Adopted 14 November 2007, Malé)

Businesses are key actors in the governance of the environment and climate change is no exception. From the late 1980s to the present day business actors have mobilised themselves effectively to shape national, regional and international policy on climate change. They have done this by organising themselves into industry organisations to express their collective interests and represent their concerns before, during and after rounds of international negotiations as well as actively engaging with all aspects of the issue, from funding scientific work and economic studies to working with the media and publishing position papers (Levy 2005; Newell 2000; Newell and Paterson 1998). Alongside this, from the 1990s onwards there has been growing interest in the role of business in efforts to protect human rights (Amnesty International 1998;

* Professor of Development Studies, School of Development Studies, University of East Anglia, James Martin Fellow, Oxford University Centre for the Environment and ESRC Climate Change Leadership Fellow. I am grateful to Stephen Humphreys and Scott Jerbi for reviewing an earlier version of this paper. Many of the arguments developed in this chapter were first developed for the International Council on Human Rights Policy meeting in Geneva (12–13 October 2007). I am grateful to participants at that meeting for comments and feedback.

Muchlinski 2001; ICHRP 2002; Ruggie 2007). This chapter connects these two developments through an analysis of the legal tools and political strategies that are being brought to bear upon corporations aimed at promoting positive human rights duties and addressing the human rights violations that flow from their contribution to climate change. The frame of accountability is used as it draws attention to the politics of rights-claiming which exist within and beyond the law as a vehicle for pursuing human rights protection. It is argued below that accountability incorporates both *answerability* (the right to demand and expect a justification of (in)action) and *enforceability* (the means to sanction unresponsiveness) (Schedler *et al.* 1999; Newell and Wheeler 2006). These dimensions help us to understand who has a right to make accountability claims and under what circumstances such claims are likely to be effective.

The chapter shows that, at the moment, corporations have responsibilities to act both on human rights and climate change, deriving from distinct sets of obligations which apply to states in the first instance. These are not yet clearly articulated as human rights duties with respect to climate change. Provisions in human rights treaties describe duties and obligations with regard to the protection of rights, many of which will be violated as a result of climate change. At the same time, provisions within the UN climate change agreements and within soft law instruments articulate responsibilities for the international community as a whole, both with regard to climate change and specifically with regard to the responsibility of the private sector for environmental protection measures.

The discussion analyses a series of parallel developments, where the conversation about the human rights responsibilities of business is not yet cognizant of the growing interest in the use of human rights tools to promote state and corporate action on climate change and the potential to hold actors to account for human rights violations through the climate change regime itself remains underexplored. The increasing importance of business actors as 'rule shapers' and governance actors in their own right through their creation of the forms of private regulation discussed below and their leading role in the day to day functioning of the clean development mechanism and emission trading schemes means, however, that they are likely to attract heightened attention from human rights and climate change activists alike. The question remains, however, whether responsibilities will apply directly to firms or are more likely to be framed as state duties to address corporate abuses of human rights resulting from climate change.

The chapter is structured as follows. The next section summarises the main arguments advanced in this chapter. The following section then analyses the human rights obligations of firms before exploring how these play out in the context of climate change. This frames the subsequent discussion of the legal strategies employed to date with regard to the responsibilities of corporations to act on climate change. Having explored some of the limitations of legal approaches, the final section of the chapter assesses the potential contribution of business-led initiatives in this area before concluding with some suggestions on possible ways forward.

The argument

This chapter advances several arguments in relation to the role of corporations as actors in debates about the relationship between human rights and climate change, and in particular concerning the tools and strategies that might best hold them to account for their responsibilities to act.

(1) We tend to assume in climate debates that the key decisions relating to climate change which affect peoples' lives are made in public arenas by public actors. They are not. They are made day to day by all of us, but especially significantly by actors that control the very processes of energy production and use that determine the degree of climate change that collectively we will subject ourselves to. The rights to food, water, health and livelihood that will be affected as a result will be violated principally and directly, not by states for whom we have highly developed frameworks of human rights law, but by corporations (public and privately owned) and, indirectly, individuals, for whom human rights obligations and tools are far more under-developed. We are faced with a situation in which states are both rights enforcers (in theory) and rights violators (in practice) and in which they are implicated in economic practices which we know are systematically undermining the human rights that states have obliged themselves to protect.

In thinking about the role of the law as a vehicle for protecting human rights violations associated with climate change we have to recognise the implication of states in the generation of climate change, which may profoundly affect their willingness and ability to confront those violations. Insofar as international measures aimed at protecting human rights undermine or challenge existing regimes of legal and informal resource control, we can expect to see resistance to their implementation.

The forestry sector is a good example in this regard (Seymour 2007). With large and increasing sums of money available for carbon offset schemes, there are strong incentives to 'protect' areas of forest for the absorption of carbon by those wealthy enough to pay for such offsets. The rush to make money from carbon sinks often brings human rights violations in its wake. For example, a Norwegian company operating in Uganda that leased its lands for a sequestration project is alleged to have resulted in 8,000 people in thirteen villages being evicted (Bachram 2004). The project in Bukaleba Forestry Reserve was meant to offset greenhouse gas (GHG) emissions of a coal-fired power plant to be built in Norway. International criticism at the time prevented the project from claiming carbon credits to 'offset' the power plant emissions, but the project continued and the trees were planted. When the duty to protect the human rights of vulnerable (often indigenous) groups,[1] communities with whom the state is in any case in conflict over land and property rights, conflicts with an opportunity to attract high levels of investment, those without a voice in the deal-brokering are likely to lose out.[2] As Seymour claims (2007, 4):

> As payments for conserving forests for carbon storage become increasingly likely, state and non-state actors alike will have strong incentives to passively ignore or actively deny the land and resource rights of indigenous, traditional and/or poor forest users in order to position themselves to claim compensation for forest stewardship in their stead.

This political reality does not negate the fact that states continue to be key actors in initiating action on climate change. As the political and legal entities with the power, resources and authority to engineer change, they are inevitably key to effective political change. It does, however, strike a note of caution about the extent to which effective solutions aimed at realising human rights are likely to come either from legal remedies alone or from states in isolation from the adoption of a range of other political strategies. In this regard it is useful to keep in mind Muchlinski's reflection that: 'it is not difficult to create technical legal solutions to the question of corporate responsibility for human rights violations. The real issue is whether the political will exists to put them in place' (2001, 47).

[1] For example under ILO Convention 169 on Indigenous and Tribal peoples.

[2] The World Bank estimates that 90 per cent of the 1.2 billion people living in extreme poverty around the world depend on forest resources for some part of their livelihood (Seymour 2007). See also Frances Seymour's contribution in this volume (Chapter 7).

(2) Histories of attempts to hold corporations accountable for their social, environmental and, indeed, human rights responsibilities through national and public international law do not provide many grounds for optimism about the effectiveness of public international law as an accountability tool for addressing climate change-related human rights violations. Though important in terms of articulating social expectations and defining the human rights responsibilities of the private sector, they so far fall short of being effective instruments for receiving and processing claims or providing redress to those whose rights have been violated. Moreover, though some instruments (as we will see below) contain reference to the environmental obligations of firms in general terms, they do not contain climate-related provisions which could be invoked by victims. Hence, it remains the case both that the human rights responsibilities of corporations are still in a process of being defined and that the climate-related aspects of these are contested and far from clear.

(3) A more general argument made throughout the chapter is that we should use the law as a political tool cautiously. The use of the law can crowd out and undermine the effectiveness of other strategies which might be employed to claim and enforce rights. It is also the case that the law provides a poor mechanism for realising social justice as it is also inaccessible to those who suffer the worst effects of climate change. As a mechanism of redress in the face of violations it is an important, but often limited, tool for the majority of the world's people. Attempts have been made to hold corporations to account for their climate impacts through tort and public nuisance litigation as I will discuss below. At the moment, however, these carry largely symbolic rather than substantive value.

(4) Accountability is weakly embedded in systems of private regulation which provide the preferred forum and means of addressing the climate issue for the private sector. Sanctions are weak for non-enforcement and there is little or no public oversight of the standards of protection afforded: that is, their appropriate level; who sets and enforces them; and on whose behalf. While they often provide a welcome attempt by the business community to engage with its responsibilities, they are not up to the task of addressing human rights dimensions of climate change because they are set by actors whose duties are poorly defined and who lack the capacity to recognise and mediate and let alone act upon rights-based claims.

(5) Viewing the relationship between climate change and human rights through the lens of accountability allows us to identify relations

of power and pressure-points for change, levers that generate responses from public and private actors that lead to action which reduces and, hopefully, ultimately reverses those actions which produce climate-related human rights violations. Accountability is a means to an end and we need to be clear about the end in order to understand which strategy is most likely to yield change (Newell and Wheeler 2006). If the goal is compensation for a violation, the pursuit of a legal claim makes sense. If the goal is a more proactive one in order to hold an actor to account *ex-ante* on the basis that continued contributions to global warming will result in unspecified and, as yet, unquantifiable future damages, then an array of other political strategies present themselves as alternatives. By focusing on relations of accountability, we can locate mechanisms of *answerability* and *enforceability* that will deliver short-term action on climate change. Human rights law provides one tool, one lever of change, but ultimately just one among many.

MNCs and human rights

> The traditional notion that only states and state agents can be held accountable for violations of human rights is being challenged as the economic and social power of MNEs appears to rise in the wake of the increasing integration of the global economy that they have helped to bring about. (Muchlinski 2001, 31)

There are a number of existing instruments and initiatives that seek to clarify the human rights duties of corporations,[3] though none that do so explicitly in relation to climate change. In a positive sense the aim is to mobilise the extensive economic weight and corresponding political influence of firms to lever progressive change when operating in countries and regions with poor human rights records, as well as demonstrating responsible leadership within the business community. More than an exercise in corporate social responsibility (CSR), however, the appeal of human rights is their universality so that 'companies cannot "buy offsets" to counterbalance harm to human rights for which they are responsible through philanthropic acts or by fulfilling rights in other areas' (SRSG 2007, 2). Increasingly, this attention is also focused on addressing a corporation's own role in exacerbating rights violations not only in the

[3] I do not discuss here the 'human rights' afforded to corporations, such as the right to property, free speech, to a free trial and to privacy by some legal instruments (Muchlinski 2001).

workplace, but beyond, for example, in fuelling conflict and violence through operations in volatile regions (Frynas 1998; Christian Aid 2000).

So what tools exist for addressing the human rights obligations of corporations? It is worth noting, first, that the Universal Declaration of Human Rights is addressed not only to states but to 'every individual and every organ of society', a formulation which is sufficiently broad as to include corporations. General Comment 31 by the Human Rights Committee also confirms that under the International Covenant on Civil and Political Rights 'the positive obligations of states … will only be fully discharged if individuals are protected by the state … against acts committed by private persons and entities' and that states should 'take appropriate measures to exercise due diligence to prevent, punish, investigate or redress the harm caused by such acts by private persons or entities'. The Committee on Economic, Social and Cultural Rights has also suggested that parties take steps to 'prevent their own citizens *and companies* from violating rights in other countries' (cited in Ruggie 2007, 15; emphasis added). By contrast, there is 'insufficient evidence at this time to establish direct corporate responsibilities under customary international law', the most binding and far-reaching source of international law (Ruggie 2007, 19).

Internationally, 'soft' law approaches have prevailed to date as the preferred way of articulating the 'global' social and environmental responsibilities of firms, despite various demands for international legally binding measures. An international code of conduct to regulate the activities of transnational companies (TNCs) was proposed as far back as the 1970s. The United Nations Centre for Transnational Corporations was set up in 1973 to draft a code, but after two decades of failed negotiations, the Centre was closed in 1993 and has been replaced by the Division on Transnational Corporations and Investment located within the United Nations Conference on Trade and Development (UNCTAD). There are, nevertheless, generic guidelines such as the OECD Guidelines for Multinational Enterprises of 1976 (revised in 2000) and the International Labour Organisation (ILO) Tripartite Declaration of Principles Concerning Multinational Enterprises and Social Policy of 1977. Both, however, are non-binding, creating no new legal duties as such and lack effective powers of enforcement leaving what Ruggie refers to as a 'sizeable protection gap' (2007, 21). Despite the fact that the OECD guidelines specify that 'Enterprises should … Respect the human rights of those affected by their activities consistent with the host government's international obligations and commitments', Ruggie notes that, 'their current human rights provisions not only lack specificity, but in key respects have fallen behind the voluntary standards of many companies and business organisations' (HRC 2008, 13).[4] With respect to environmental issues, the issue of the regulation of TNCs was dropped from the United Nations Conference on Environment and Development (UNCED) agenda at the insistence of the United States in particular responding to vocal business objections to the idea (Chatterjee and Finger 1994). Similarly, while Agenda 21 includes recommendations that affect TNCs, it does not take the form of a code of conduct. More recent efforts to elaborate the terms of a UN-led Corporate Accountability Convention, most recently at the World Summit on Sustainable Development (WSSD) in 2002, have not as yet come to fruition.

The preference instead has been to articulate the human rights responsibilities of firms in general terms through voluntary instruments such as the Global Compact.[5] The Draft Norms on the Responsibilities of Transnational Corporations and Other Business Enterprises[6] with Regard to Human Rights[7] adopted by the UN sub-Commission on the Promotion and Protection of Human Rights (but not by the UN Human Rights Commission) suggest the possibility of a non-voluntary formulation of private sector commitments.[8] The text notes that states are primary duty bearers of human rights but stipulates that firms, within their 'spheres of activity and influence', have corresponding legal duties. This means, it is declared in the article on General Obligations, that: 'Within their respective spheres of activity and influence, trans-national corporations and other business representatives have the obligation to promote, secure the fulfilment of,

[4] The guidelines are criticised for not making an impact on business behaviour because: (i) they are not widely known; (ii) they are entirely voluntary and without sanction; (iii) they are outdated even when compared with some companies' own codes of conduct; and (iv) they are only weakly implemented through National Contact Points which offer advice to businesses about the application of the guidelines. Such NCPs are also often hosted in government departments that are also charged with business promotion creating potential for conflicts of interest (McLaren 2000).

[5] This voluntary initiative became operational in 2000 and aims to promote UN principles in the areas of human rights, labour and the environment. It currently has 3,000 participating companies and forty national networks.

[6] This language was inserted to shift the focus from solely multinational corporations.

[7] The text comprises twenty-three articles setting out human rights standards for companies in areas including the environment with regard, for example, to the 'precautionary principle'.

[8] The Commission has now been replaced by the Human Rights Council.

respect, ensure respect of and protect' nationally and internationally recognised human rights. It also calls for compliance to be monitored by national and international agencies and for victims to be provided with effective remedies. The norms are described as 'the first non-voluntary initiative [in the area of business and human rights] accepted at the international level' (Ruggie 2007, 3). For its part, the business community in the form of the International Chamber of Commerce (ICC) and International Organisation of Employers (IOE) has firmly opposed the Draft Norms, and the Human Rights Commission instructed the sub-Commission not to engage in any monitoring of corporate activities. In July 2005 United Nations Secretary-General, Kofi Annan, announced the appointment of Professor John Ruggie as Special Representative on the issue of human rights and transnational corporations and other business enterprises.[9] The mandate includes identifying and clarifying standards of corporate responsibility and accountability with regard to human rights. The move was seen by some as an attempt to move beyond the divisions caused by the sub-Commission norms, noted above.[10] The outcome and impact of this initiative remains to be seen, but initial indications are that discussion about a 'corporate responsibility to respect' (SRSG 2007) will be emphasised alongside the duty of states to protect against corporate abuses of human rights as a result of climate change. Ruggie's recent formulations, for example, have comprised three elements: (i) the duty of the state to protect against human rights abuses by third parties including business; (ii) the corporate responsibility to protect human rights; and (iii) the need for more effective access to remedies (HRC 2008).

Such formulations do not, as such, help to navigate the practical grey zone of how to differentiate state from corporate responsibilities, how to disaggregate 'primary' from 'secondary' duties and where the boundaries of the corporate sphere of influence lie. This is often described in terms of a concentric circle of influence with core company operations placed at the centre and suppliers and communities included in outer circles.[11]

9 The creation of this mandate was requested by the United Nations Commission for Human Rights in its resolution 2005/69 and approved by the Economic and Social Council on 25 July 2005.

10 I am grateful to Scott Jerbi for this insight.

11 The SRSG clarifies further: 'Sphere of influence is not about what rights companies must respect, but rather about *when* and *where* companies must take steps to ensure that they respect human rights' (2007: 6).

Firms will be expected to exercise due diligence with regard to a range of activities still in the process of being defined,[12] but which will include 'do no harm obligations' that also imply proactive duties to 'do good'. General, as well as sector and firm specific tools, may help to ground the practical implications of such aspirational claims in the form of usable human rights assessment tools.[13]

As part of this discussion it is appropriate to note interest in the use of legal tools which can provide a measure of foreign direct liability: where firms are held legally accountable in courts where they are domiciled for acts of corporate irresponsibility committed overseas. The Alien Tort Claims Act (ATCA) in the United States and other litigation has been invoked to articulate the extraterritorial obligations of firms (Newell 2001, Ward 2002).[14] Attempts to make environmental claims through the ATCA have so far been rejected, however, and overcoming the barriers of both *forum non conveniens* where it has to be shown that the 'home' court of the company is the more appropriate venue to hear the case, and piercing the corporate veil which separates parent and subsidiaries even of the same firm for the purposes of liability, present enormous challenges for potential plaintiffs (Chinen 1987). Cases brought so far using the ATCA have rested on demonstrating corporate complicity in human rights abuses committed by states, such as in the case brought against UNOCAL in the United States concerning the firm's alleged use of forced labour in its operations in Myanmar. In the case of climate change, companies arguably cooperate, participate in and benefit financially from complicity in rights violations deriving from their contribution to climate change. But litigants invoking claims of foreign direct liability in relation to climate change would probably have to demonstrate state liability for the conduct of a firm within their jurisdiction even if operating in another territory. No such responsibility has been provided for to date for climate change *per se*, but the Special Representative's report makes clear that 'states have a duty to protect against human rights abuses by non-state actors, including by business,

[12] As the SRSG notes 'the *ex ante* specification of rights for which companies might bear some responsibility is an inherently fruitless exercise; in principle, all rights will be affected' (2007: 2).

[13] The 'Human Rights Impact Assessments' of the IFC as well as the work of the UN Global Compact, International Business Leaders Forum and the Danish Institute for Human Rights Compliance Assessment tool are worth mentioning in this regard.

[14] More than forty cases have been brought against companies under this statute since 1993 (HRC 2008).

affecting persons within their territory or jurisdiction' (SRSG 2007). Within their territory implies addressing violations that take place within their borders. Within their jurisdiction could imply responsibility for companies registered in their country when operating overseas.

In the case of climate change, the transnational element is also distinct. It is less the case that a company from one country commits an act of environmental negligence in another jurisdiction. From a climate point of view it does not matter where the emissions occur – the effect they produce is global even if it has uneven, localised consequences. Following this logic, it is not a transnational, extraterritorial mechanism that is needed *per se*. It is about nation-states controlling emissions within their sovereign borders irrespective of whether the company is national or foreign-owned. The question of the transnational duties of states becomes relevant, as it does in the field of human rights, when a state fails to act upon its responsibilities either to its own citizens or to the global public where their failure to act creates spill-over effects on other countries. If the process of drafting norms that Ruggie is overseeing proceeds in the direction indicated above, it may seek to formulate state duties to address corporate abuses within their jurisdictions and not just their territories. This may provide scope for triggering legal action against states failing to address the extraterritorial impacts of firms over which they exercise jurisdiction. Foreign direct liability may then be appropriate for emphasising corresponding 'home' country responsibility for the climate change (in)actions of their firms when they operate overseas. It may also compensate, as it has attempted to previously, for state incapacity or unwillingness (complicity) to address human rights violations taking place within their jurisdiction. This may be an important check on the ability of states to exercise a form of comparative advantage as a pollution haven: to attract and protect global firms seeking to relocate to jurisdictions where the GHG emissions reductions do not apply or to allow Annex 1 parties to outsource the most energy-intensive stages of their production processes to countries not bound to reduce their emissions under the Kyoto Protocol.

Alongside human rights legal tools, there are also conventions that might provide a supportive and enabling environment for holding corporations to account for their role in climate change. The Aarhus Convention (UNECE Convention on Access to Information, Public Participation in Decision-making and Access to Justice in Environmental Matters), adopted on 25 June 1998, for example, contains provisions on rights to information, public participation and access to justice which undoubtedly apply to the issue of climate change. It is suggested, moreover, that 'Although the Convention is not primarily focussed on the private sector, privatised bodies having public responsibilities in relation to the environment and which are under the control of the aforementioned types of public authorities are also covered by the definition' (UNECE 2007). Claim-making, would, in the first instance, however, be directed toward public authorities.

With regard to soft law articulations of the human rights responsibilities of companies, many of the usual incentives to adopt such measures do not apply in the case of climate change. For example, Mary Robinson argued when she was UN High Commissioner for Human Rights that, 'business needs human rights and human rights need business' on the basis that 'business cannot flourish where fundamental human rights are not respected', and that corporations that violate human rights will suffer loss of reputation (Robinson 1998). These arguments may stand up in the context of abuses of labour rights and lack of respect for basic civil and political rights. The embarrassment factor the company experiences in the wake of public exposure of its negligence can trigger company action because the focus is on the negligence of one company around a specific incident. With climate change, albeit to hugely varying degrees, we are dealing with the complicity of the bulk of enterprise engaged in industrial activity. Not one-off highly visible sensational breaches of human rights, but large-scale contributions over long time frames in which it is almost impossible to connect specific acts of culpability with tangible impacts. Dealing with such gradual, cumulative but systemic violations of human rights presents enormous political challenges. It may be true, as Muchlinski asserts, that 'in general corporations are unlikely to act in a manner that deliberately seeks to violate human rights' (2001, 44), and it is clear that there will be few long-term winners in a world of accelerated climate change. But such an assertion becomes problematic in a context in which the precise impacts upon human rights that flow from failing to act are increasingly known, if not easily quantified and apportioned.

Climate change and human rights

The intersection between human rights and the issue of climate change takes place at several levels that are described in more detail elsewhere in this book (*cf.* Sachs 2006). It is increasingly clear that through the volume of GHGs the private sector generates it will (indirectly and perhaps unwittingly) contribute to violations of rights to health, food, water and even the right to life. In a context in which the scientific consensus is sufficiently robust to anticipate extreme negative consequences for poor and marginalised communities, we know that unchecked climate change will lead to widespread deterioration of the means of survival and ultimately death for an increasing number of the world's poor. Insofar as corporations (as well as governments and individuals) continue to choose to emit large amounts of GHGs whose impact on others' human rights is known, they are culpable, albeit in a general political (rather than narrow legal) reading of responsibility. The condition of ignorance about the consequences of our actions, often invoked to delay action, no longer applies. The petition submitted to the Inter-American Commission on Human Rights by the Inuit Circumpolar Conference presents the issue concisely and in terms which could equally apply to the private sector:

> With full knowledge that this course of action is radically transforming the arctic environment upon which the Inuit depend for their cultural survival, the United States has persisted in permitting the unregulated emission of greenhouse gases from within its jurisdiction into the atmosphere.[15]

Leaving aside for the moment both data about volumes of emissions emitted by countries, calculations (which fail to differentiate where GHGs are produced from the consumers of the products and services for whom they are generated), and debates about some emissions (survival ones) being more legitimate than others (luxury ones) as discussed below, disaggregating responsibility for specific violations presents a complex political (and legal) challenge. Apportioning blame when GHGs are generated by all of us all of the time (albeit to different degrees), deriving from the basic activities which sustain human existence, that is, transportation, energy use, agriculture and industry, presents a gargantuan task. The task is not just hampered by the methodological difficulty of establishing which source of pollution brought about which rights violation and in what proportion, but ascertaining whether complicity lies with the provider of a source of energy (the energy company, for example), or the user and source of demand for that energy (the consumer). In most litigation it would be the source of the pollution that would be the target of a legal intervention. But we are

[15] Petition to the Inter-American Commission on Human Rights seeking relief from violations resulting from global warming caused by acts and omissions of the United States, 1.

not talking about the production of a particular toxin or chemical that damages human health in tangible and widely understood ways. We are talking about gases that are emitted in every aspect of everyday life that, when they interact in the atmosphere, produce an effect which has long-term and uneven consequences. So while it would be easy to suggest that polluters and those who benefit from pollution (all users of fossil fuels) are complicit, we need to establish who is most to blame and how responsibilities and damages can be meaningfully apportioned in order to assess the potential of human rights instruments, traditionally understood, to effect change.

Despite the many political and technical barriers that inhibit us from positing clear, undisputed and causal links between climate change and human rights, the issue is indisputably now firmly on the international agenda. Besides the Malé declaration cited above, there has been a strong human rights emphasis in the latest Human Development Report, with the UN Human Rights Council requesting before the UN General Assembly that the United Nations Office of the High Commissioner for Human Rights (OHCHR) conduct a detailed analytical study of the relationship between climate change and human rights (UNGA 2008).

The law and its limits[16]

There has been a great deal of interest in the potential of the law to protect those whose human rights are threatened by climate change. This has taken a number of forms. One interesting source of momentum has come from groups adopting a range of legally-based strategies to hold governments to account for their obligations to act on the issue. A few examples will serve to illustrate the potential and limitations of these legal cases as accountability strategies. Given the underdeveloped nature of the means for holding firms to account for their human rights obligations, it is unsurprising that activists have sought to hold governments to account for their support to private polluters. It is also unsurprising that the United States has been the key target of this legal activism, as a leading polluter with a weak track record on the issue, with perhaps the most litigious environment in the world and home to courts with an expansive view of their jurisdiction.[17]

[16] Parts of this section draw on Newell (2008b).

[17] Presentation at the ICHRP seminar 12–13 October 2007 (Chatham House rules).

In August 2005, Greenpeace and Friends of the Earth, together with a series of US cities[18] alleged that the Export-Import Bank and Overseas Private Investment Corporation illegally provided over $32 billion in finance and insurance for oil fields, pipelines and power plants for over ten years without assessing their contribution to global warming or their impact on the environment of the United States. In doing so, the claimants argue that these export credit agencies have not met their obligations under the US National Environmental Policy Act (NEPA). On 31 March 2007, the US District Court for the Northern District of California held that the NEPA *does* apply to major federal projects that contribute to climate change. A similar case has been brought by a coalition of German NGOs attempting to hold their government to account for the climate change externalities it permits through the support it provides to fossil fuel industries. In June 2004, GermanWatch and BUND brought a legal challenge against the German Federal Ministry of Economics and Labour in the Administrative Court of Berlin. They successfully secured an order from the Administrative Court which forces the German government to disclose the contribution to climate change (in terms of tonnes of CO_2 released) made by projects supported by the German Export Credit Agency, Eueler Hermes AG, since 1997. The German government contested the order in 2006 on the grounds that the export credit agency is not bound by European environmental law in this area.

Inuit groups in North America have taken a different approach in seeking to advance their claims regarding the impacts of US government inaction on climate change. On 7 December 2005, the Inuit Circumpolar Conference submitted a petition to the Inter-American Commission on Human Rights seeking relief from violations of the human rights of the Inuit people resulting from global warming caused by the GHG emissions of the United States. With the help of legal advisers, Sheila Watt-Cloutier, Inuk woman and Chair of the Inuit Circumpolar Conference, submitted the petition on behalf of herself, sixty-two other named individuals 'and all Inuit of the arctic regions of the USA and Canada who have been affected by the impacts of climate change'.[19] The petition calls on the Inter-American Commission on Human Rights to investigate the harm caused to the Inuit by global warming, and to declare the

[18] Oakland, Arcata, Santa Monica and Boulder.

[19] Petition to the Inter-American Commission on Human Rights seeking relief from violations resulting from global warming caused by acts and omissions of the United States, 1.

United States in violation of rights affirmed in the 1948 American Declaration of the Rights and Duties of Man and other instruments of international law, such as the International Convention on Civil and Political Rights and the International Convention on Economic, Social and Cultural Rights. Specifically the petition alleges:

> The impacts of climate change, caused by acts and omissions by the United States, violate the Inuit's fundamental human rights protected by the American Declaration of the Rights and Duties of Man and other international instruments. These include their rights to the benefits of culture, to property, to the preservation of health, life, physical integrity, security and a means of subsistence and to residence, movement and the inviolability of the home.

The rights that are threatened, therefore, refer to a range of political, economic (livelihood) and cultural rights.[20] The plaintiffs had to show that in bringing the case all domestic remedies had been exhausted. The Commission rejected the petition, perhaps because it 'wasn't ready to tell a government what to do ... advising a government of its human rights responsibilities ... it was uncomfortable demanding specific science-driven remedial steps'.[21] Importantly, the human rights issues raised by the case were not disputed by the Commission. A further positive outcome of the case has been that the Commission invited petitioners to request a public hearing on the matter, which took place on 1 March 2007.[22]

There are also many examples of litigation to hold private actors to account for their climate change responsibilities. In June 2004 New York Attorney General, Eliot Spitzer, eight states and New York City filed an unprecedented lawsuit against five of North America's largest power companies as contributors to public nuisance under common law, between them contributing over 10 per cent of the nation's CO_2 emissions.[23] Invoking liability claims that build on earlier judicial activism against the tobacco and asbestos industries, they demanded that these companies cut their CO_2 emissions in the light of global warming and the damage their emissions were causing in terms of impacts on human

[20] See online at: www.ciel.org/Climate/Climate_Inuit.html.

[21] ICHRP seminar, Geneva, 12–13 October 2007.

[22] Letter of invitation from Ariel Dulitzy, OAS to Sheila Watt-Cloutier, Martin Wagner and Daniel Magraw inviting attendance at the OAS hearing on global warming and human rights, 1 February 2007.

[23] The firms in question were AEP, Southern Company, Tennessee Valley Authority, Xcel Energy and Cinergy.

health, and economic impacts on agriculture and tourism (among other things). In September 2005, the District Court dismissed the case on the basis that regulating power companies was an issue for the political domain and not appropriately settled through judicial means. In time, cases such as this, if successful, may exercise a deterrent effect in persuading firms to internalise the negative externalities of their activities.

The State of California, represented by State Attorney Bill Lockyer, has also recently sought legal redress from car companies regarding their climate change responsibilities which collectively amount to 20 per cent of the United States' total emissions.[24] The action is based upon the common law principle of public nuisance, in this instance emissions of CO_2, and seeks monetary compensation for damages caused by these emissions.[25] The lawsuit is the first of its kind to seek to hold manufacturers liable for the damages caused by GHGs that their products emit. In September 2007, California lost the case against the car companies. The Court found that 'injecting itself into the global warming thicket at this juncture would require an initial policy determination of the type reserved for the political branches of government'.[26] This has not prevented corporations from using the law to challenge states' authority to regulate the fuel economy of cars, a subject they claim is appropriately addressed at federal level. In the *Green-Mountain Chrysler-Plymouth-Dodge et al.* v. *Crombie et al.* case, fourteen states were sued by a group of car makers for attempting to regulate CO_2 emissions from cars. A Vermont court ruled on 17 September 2007 in favour of the states on this occasion, though the companies look set to appeal the decision (AAM 2007).

Reflecting on the use of such cases, one activist lawyer put it the following way: 'our approach is to try and sue everyone we can. Most cases will fail but we may just do it anyway'.[27] In principle, the value of such cases, if successful, will lie in the catalytic effect they may have on the financial backers of industries and projects contributing to climate change, such as the insurance industry and banking sector, to reconsider their investments in these sectors (if the injuries are large enough) and to

[24] The six defendants are Chrysler Motors Corporation, General Motors Corporation, Ford Motor Company, Toyota Motor North America, Inc., Honda North America and Nissan North America.

[25] Under the law, a 'public nuisance' is an unreasonable interference with a public right, or an action that interferes with or causes harm to life, health or property.

[26] *People of the State of California* v. *General Motors et al.*, 17 September 2007.

[27] Legal environmental activist from US. Identity protected.

raise awareness of the range of harms being generated by climate change. They may also galvanise US support for the climate regime in the face of legal liabilities as a form of regulatory defence. This perspective reinforces the point about the broader political goals that legally-based strategies can serve.

Such cases, nevertheless, encounter a number of challenges: (i) *recruiting plaintiffs* – for example, likely candidates such as governments of small island, developing states may be reluctant to confront the world's largest economic power upon whom they are dependent for trade and aid; (ii) *harms remain speculative* – how, who, how much; and (iii) *assigning liability* – which actors you can get before a court, that is, direct emitters, such as power plants, producers of carbon (oil companies), or car makers as has happened in California. Responsibility is cumulative, that is, desegregating contributions, gases, current versus past emissions, especially given the long life cycle of these gases in the atmosphere, presents huge obstacles. It is virtually impossible in such a situation to apportion current responsibility. Establishing percentages for pay-outs would make judges very nervous. After all, as noted above, everyone contributes to the problem – even the plaintiffs.

Alongside these cases, there has been a wave of legal activism which does not explicitly invoke climate change as a rationale, but seeks forms of action which nevertheless constitute action on climate change and which also invoke rights-based claims as an accountability strategy to challenge public and private actors simultaneously; specifically their collusion in producing environmental harm. A relevant case would be that of the Iwerekan community of the Niger delta. The communities, supported by Earth Rights Action in Nigeria, filed a legal action against the Nigerian government, the Nigerian National Petroleum Corporation (NNPC) and the Shell, Exxon, Chevron, Total and Agip ventures in Nigeria to stop gas flaring.[28] The Federal Court of Nigeria ordered that the gas flaring must cease as it violates constitutional rights to life and dignity. When it did not stop, contempt of court proceedings were brought against Shell and NNPC. The case is currently adjourned but shows how legally-induced changes, prompted by non-climate concerns may, nevertheless, have a positive impact on action for climate change, drawing as they do on a long history of legally-based community activism to hold oil companies to account for their social and environmental responsibilities (Frynas 1999).

[28] See online at: www.climatelaw.org/media/gas.flaring.suit.

The law should, nevertheless, be viewed as just one among many strategies and tools that will help to achieve change and ultimately, contain and reverse those actions which continue to inflict human rights abuses on the poor. We have learnt from the experience of the environmental justice movement that processing all rights claims through legal processes removes an issue from the arenas where poorer groups have a right to participate and the capacity to make a difference and places them in a setting where resources and elite expertise shape outcomes. There have been many instances where the energy and dynamism that characterises a movement is sapped once it moves to a legal arena (Cole and Foster 2001). Indeed, the experience of the environmental justice movement suggests that the lack of protection afforded to poorer people by environmental law should not be considered a failure of that law, because it was not designed to protect those groups in the first place. Ruiters, writing about environmental justice in South Africa, argues that '[t]he emphasis (wrongly) falls on the distribution of environmental hazards; the struggle for improved regulations; stricter enforcement; and better access to information about industries, their products and workplace conditions. A deeper approach to environmental justice, however, requires a focus on the production and prevention of injustices' (2002, 112). Because they are risk-oriented, environmental laws tend to support decisions as long as emissions comply with minimal state regulatory thresholds. As Cole notes 'while we may decry the outcome, environmental laws are working as designed. Such a disproportionate burden is legal under US environmental laws ... Thus decisions to place unwanted facilities in low-income neighbourhoods are not made in spite of our system of laws, but because of our system of laws' (1992, 646). Added to this are the many barriers facing the poor when trying to ensure access to environmental justice, including low levels of legal literacy, financial resources to bring and sustain cases or to settle them in the event of losing a case, distrust of the legal system and high levels of scientific proof that are required in common law traditions to demonstrate beyond doubt the relationship between cause and effect (Newell 2001). If making connections between harmed individuals and communities, on the one hand, and industrial polluters, on the other hand, is hard in instances of toxic pollution, demonstrating causality in a way that would satisfy a court between desegregated and diffuse causes of climate change and effects which are rarely attributable directly and in and of themselves to climate change, presents even tougher challenges. From a strategic point of view, there is also the urgency of the issue which suggests that attempting to resolve an issue or secure short-term action from legal processes, which are often long, drawn out and subject to delays, implies political (and human) costs.

We also need to recall that, for many of the reasons spelled out, the law has tended to protect the interests and property of the wealthy while providing uneven protection for the poor (Cutler 2002; Newell 2005a). Reducing all rights claims to legally recognisable claims is to misunderstand the histories of struggle that precede the acceptance of a rights-based claim as one with legal traction and to fetishise the law inappropriately as a vehicle for bringing about change. Many governments already have strong but poorly enforced bodies of human rights law, while they are often simultaneously cast as rights enforcer and rights violator. Assuming that governments will defend the rights of groups affected by climate change against their infringement by powerful economic forces, which in some cases will be the state itself or in other cases will be corporations that provide state revenues through tax and employ large numbers of people, may be mistaken.

Private climate regulation and its limits

Given the limitations of legal approaches to reducing the corporate sector's contribution to rights violations associated with climate change, we need to consider the protection afforded by the private sector's own attempts to reduce its contribution to climate change through a range of voluntary initiatives and standards. Many of these have been adopted by firms as part of broader corporate social responsibility (CSR) programmes.

From a long history of opposition to action on climate change (Newell and Paterson 1998), climate change has now been repositioned by some elements within the business community as a business opportunity. The extent to which this is so depends on the region and firm in question (Levy and Newell 2000; Rowlands 2000; Sæverud and Skjærseth 2007). There is, nevertheless, a great deal of evidence of companies taking voluntary action on climate change to reduce their emissions, capitalising on the economic savings to be made and the public relations credit to be earned from being seen to take a lead on the issue. Chemicals giant Du Pont reduced its emissions by 65 per cent below their 1990 level, while IBM has saved \$115 million since 1998 through cutting its carbon emissions. A number of NGOs, such as the Climate Group in the United Kingdom and the Pew Center in the United States have played an important role in making the business case for action on climate change and publicising the benefits achieved by existing leaders in the field. Alongside individual efforts, companies have also been keen participants in schemes such as the Chicago Climate Exchange, the European Emissions Trading Scheme, the Carbon Disclosure Project and the establishment of a Voluntary Carbon Standard. The business benefits of such engagements are described in the following way in relation to the Carbon Disclosure Project:

> The Carbon Disclosure Project (CDP) provides a secretariat for the world's largest institutional investor collaboration on the business implications of climate change. CDP represents an efficient process whereby many institutional investors collectively sign a single global request for disclosure of information on Greenhouse Gas Emissions. More than 1,000 large corporations report on their emissions through this web site. On 1st February 2007 this request was sent to over 2400 companies.[29]

The CDP now covers US$57 trillion worth of assets from over 3,000 companies. The scope of private regulation is, therefore, impressive. Sometimes with private climate regulation, the emphasis is on 'beyond compliance'; going further than measures required by the climate regime. This is the case for CDM Gold Standard developed by a Swiss not-for-profit with the endorsement of a wide range of NGOs. It seeks to reward those projects which meet a broader range of sustainability criteria, such as renewable and energy efficiency projects. The CSR dimension is, nevertheless, apparent. The Gold Standard web site claims:

> The Gold Standard method lowers risk and boosts reputation. Extra costs are low, due to the fact that Gold Standard screens are harmonized with the regular CDM and JI project design documents. For renewable energy or energy efficiency projects, there is simply no better way to get the market distinction and reputational reward your project deserves.[30]

Also taking as its point of departure the distrust surrounding the weak regulation of carbon markets and the lack of mechanisms of accountability and enforcement, the voluntary carbon standard (VCS) claims to provide 'a robust, new global standard for voluntary offset projects. It ensures that carbon offsets that businesses and consumers buy can be

[29] See online at: www.cdproject.net.
[30] See online at: www.cdmgoldstandard.org/how_does_it_work.php?id=42.

trusted and have real environmental benefits'.[31] The VCS, initiated by the Climate Group, the International Emissions Trading Association and the World Economic Forum in 2005, includes among its objectives the standardisation and provision of transparency and credibility for the voluntary offset market, the creation of a trusted and tradeable voluntary offset credit and the provision of a clear chain of ownership over voluntary offsets that prevents them being used twice. This is achieved through multiple VCS registries and a central project database that is open to the public.

The rapid growth of the carbon economy and the emergent forms of private regulation which manage it and provide benchmarks of its success, whatever their environmental merits (or otherwise), create serious challenges with respect to regulation, legitimacy and accountability. Who sets and enforces the standards? Efforts to assess what voluntary efforts add up to have yielded indecisive results amid different timeframes employed by voluntary schemes, using different baselines and including different GHGs. A positive reading of this new landscape of private 'regulation' is that it enhances and goes beyond state-based action within the climate regime, creating new channels of pressure on key contributors to climate change. It may do this by plugging some of the weaknesses and gaps in the climate regime. As with CSR in general, social expectations about the responsibilities of firms sometimes far outstrip those that are expressed in legal instruments. Voluntary standards are also quicker to approve, respond to private and consumer needs and potentially provide wide coverage if adopted by large firms with extensive supply chain networks. When a company like the supermarket giant Tesco in the United Kingdom announces its intention to ultimately label all the products it sells in its supermarkets according to their carbon footprint, the global reach of the measure is significant.

It is important though to consider the relationship between forms of private regulation and the public regulation discussed above which has become a key target for groups seeking to advance action on climate change and human rights. It is often assumed that there is a disconnect between the respective responsibilities of public and private actors in climate governance where states have obligations and private actors have only voluntary commitments. It is true, of course, at the level of international law that it is states that formally signed the UNFCCC and Kyoto

[31] See online at: www.v-c-s.org/about.html. http://intranet.iucn.org/webfiles/ftp/public//ForumEvents//E1547/Final%20Document/CC-HR_leaflet.pdf.

Protocols. But they did so in negotiation with the private sector in their country about which emissions reductions were plausible and over what time frame. As David Levy (2005) notes, 'If an agreement cannot be crafted that gains the consent of major affected industries, there will likely be no agreement at all.' This is hardly surprising given the centrality of business to all aspects of the climate regime's work. As the ICC declared back in 1995:

> Industry's involvement is a critical factor in the policy deliberations relating to climate change. It is industry that will meet the growing demands of consumers for goods and services. It is industry that develops and disseminates most of the world's technology. It is industry and the private financial community that marshal most of the financial resources that fund the world's economic growth. It is industry that develops, finances and manages most of the investments that enhance and protect the environment. It is industry, therefore, that will be called upon to implement and finance a substantial part of governments' climate change policies (ICC 1995).

Governments, of course, also establish frameworks of national regulation to meet those agreements which require them in most cases to establish obligations for private sectors, often by sector. Firms do, therefore, have obligations even if these are passed on to them by states. Cap and trade regimes operate on this basis: states agree to the caps and the framework within which trading between utilities can take place. Indeed, markets always require property rights and rules of the game to operate even if advocates of market approaches prefer not to concede as much (Newell 2008a).

There is a key difference, nevertheless, in terms of issues of accountability, representation and the treatment of rights concerns between state regulation *of* and *for* the private sector and regulations established *by* and *for* private actors themselves. Voluntary regulation is discretionary and derives from incentives; some leading firms have to demonstrate leadership or seek to exploit a 'first mover' advantage in the market. The problem with relying on such responses is that only those firms under pressure (from within and outside the firm) and in the public spotlight will act. Given the choice, many will not, leading to problems of uneven obligations and free-riding by those that benefit from the actions taken by others without making sacrifices themselves. This is true within states and internationally where, in the absence of global coverage, action may be ineffective if key sectors and polluters are not on-board and unappealing for firms if private sectors elsewhere are not also making sacrifices.

Indeed, many firms have argued that unless Chinese firms sign up to CSR initiatives, they can no longer participate in them without putting themselves at a competitive disadvantage (Jenkins 2007). If we rely on making the 'business case' for action on climate change, sectors that make their profits from burning fossil fuels may not do anything by voluntary means alone. The same is true of the human rights obligations of firms. As Muchlinski notes:

> the more conscientious corporations that invest time and money in observing human rights, and in making themselves accountable for their record in this field, will be at a competitive disadvantage in relation to more unscrupulous corporations that do not undertake such responsibilities. (2001, 35–6)

Here we encounter the limits of private regulation. It is, as the name suggests, privately negotiated by for-profit actors and their counterparts in the NGO world. It is driven by market opportunities, failures or responses to legitimacy crises. It is unlikely to deliver effective *distributional justice*. Despite the best efforts of initiatives, such as the Voluntary Carbon Standard which attempt to steer market actors to desirable projects, investor-led initiatives are attracted to 'low hanging fruit' (easily accessible and abundant opportunities for gaining credit for carbon emissions with minimal capital input), visible projects which meet public relations needs and short-term returns to satisfy shareholders that the investment is worthwhile, as against longer-term investments that are more likely to bring sustainable development. Private regulation also generally performs poorly in relation to *procedural justice*. Indeed, its appeal for many participants derives from the possibility of establishing regulation, albeit non-binding, in a more proactive and speedy way that bypasses the high requirements for public participation and consultation that attend equivalent public processes. As a result decisions are made in a non-transparent manner, often closed to public participation, unaccountable to publics affected by decisions taken and operating largely without sanctions for non-compliance. As John Ruggie notes (2008, 9): 'how do most companies know they respect human rights? Do they have systems in place enabling them to support the claim with any degree of confidence? Most do not.' Moreover, non-judicial mechanisms for access to remedies at company level are 'seriously under-developed' (Ruggie 2008) while even international arbitration involving investors is conducted in strict confidentiality 'so that the public in the country facing a claim may not even know of its existence' (Ruggie 2008, 12). Most instruments of private climate regulation have also yet to demonstrate a regard for the human rights dimensions of climate change. A function of weak levels of procedural justice is the fact that affected groups have no means of redress with private regulation if they feel aggrieved or if it fails to deliver. This presents a serious problem as private climate governance increasingly outpaces and outgrows its public counterpart and as the governance of the carbon economy is increasingly conducted by the very actors that stand to gain from it most.

If accountability has two main elements, that is, answerability and enforceability; while private regulation demonstrates a perceived sense on the part of private actors that both claims can be made of them and that, under certain circumstances, they have a duty to respond, it performs poorly on enforceability. There are few accepted standards for reporting and many gaps in enforcement. Even initiatives such as the Global Reporting Initiative (GRI), which provides standardised approaches to improve the comparability of company social and environmental reporting (including human rights indicators), concedes that fewer than 200 firms report in accordance with its guidelines. For this reason, while private regulation may provide the tools of accountability through the provision of improved levels of information and transparency, enforceability, in the form of 'mechanisms to investigate, punish and redress abuses' will have to be provided through state regulation in the form, for example, of company law and duty of care legislation (HRC 2008, 25). An important role could be played in this regard by national human rights institutions, nearly half of which are able to handle grievances related to the human rights performance of companies according to the SRSG, John Ruggie (2007).

An important supplementary and complementary form of enforcement may derive from the efforts of civil society organisations to construct new forms of regulation of business in the absence of state willingness or capacity to do so, sometimes referred to as 'civil regulation' (Zadek 2001). Some of the initiatives described above, such as the Gold Standard, would fall under this heading as an attempt to go beyond existing public regulation and supplement gaps in its reach by trying to set new standards of performance for business. Civil regulation also includes a range of more confrontational approaches which include the use of shareholder activism, consumer boycotts and exposé campaigns (Newell 2001). Drawing on this, Michael Mason (2005) refers to 'civil redress'; alternative systems of sanctions to penalise irresponsible conduct and reward positive behaviour. Loss of market value or consumer confidence, tarnished public reputation and disaffection among shareholders are the tools of persuasion and coercion that serve to 'harden the environmental accountability demands levelled at corporations' (Mason 2005, 151). Not only do they provide 'an instrument of accountability for ecological performance', a critical dimension of their effectiveness derives from the construction of mechanisms of *civil redress* (Mason 2005, 150). Corporations are then, in effect, subject to a variety of accountability sanctions that go beyond the strict public regulation of their activities.[32] As Ruggie notes, 'the primary means through which grievances against companies play out are litigation and public campaigns'. Hence, 'For a company to bet on winning lawsuits or successfully countering hostile campaigns is at best optimistic risk management.' The incentive instead is to 'identify and address grievances early, before they escalate' (2008, 24).

Conclusion: strategic dilemmas

The above discussion raises a series of strategic choices for those keen to advance action on climate change because of the human rights violations it brings in its wake or for a range of other ecological, social and moral reasons. One option is to push for binding public regulation which spells out human rights obligations of corporations vis-à-vis climate change. This could occur, for example, as part of a UN Corporate Accountability Convention of the sort proposed by a number of activists at the time of the World Summit on Sustainable Development in 2002, mentioned above. Moves towards such a convention may be symbolically useful because ironically, private action is often driven by fear of impending public regulation. International public regulation also has the advantage of global reach and would, should the will exist, be able to settle interstate justice claims. A reading of the history of previous attempts at global business regulation, however, as well as the failure to advance such an approach at WSSD suggest that it is unlikely to be agreed any time soon. Such an agreement would also be unlikely to recognise and process individual claims against private actors. It is also possible that human rights language will increasingly feature in the UN climate

[32] These include the infliction of financial penalties as we saw above in relation to tort law, criminal prosecution, withdrawal or non-renewal of a licence to operate, termination of financial support and, on a more personal level, reduction in responsibilities or removal from a job.

negotiations, as occurred at the Bali meeting in 2007 reflected in the Malé declaration which opens this chapter. Whether it is possible to imagine a post-Kyoto arrangement which spells out the human rights obligations of companies and provides means for processing them is another question. Many would argue that human rights protection lies beyond the scope and competence of the climate regime as currently constituted.

A second option, which recognises that action on climate change needs to be advanced not just through international environmental law, but through trade and investment law and through those regimes that regulate key private actors, is to attempt to mainstream climate into business-friendly treaties. Rather than settling all these issues through public international environmental law alone, the climate obligations of firms could be written into bilateral trade and investment agreements, for example. Under such a scenario a right to market access would be conditional on concrete obligations regarding responsible investment with respect to climate change, such as the adoption of best available technology and subjecting investments to screening for their possible climate impacts. This would be an indirect approach to advancing the human rights obligations of firms on climate change through climate conditionalities, but it would be one way of legally securing action from companies through arenas and instruments which they value and in which they actively participate. Though such a proposal may seem far-fetched, it is worth recalling that firms contracted by large banks such as the World Bank already accept such screening as part of their business operations.

It is also important to extend the chains of responsibility and liability for climate impacts so that insurance companies, banks and shareholders are required to accept their duties to ensure that their financing is not undermining people's pursuit of their human rights through funding projects with a large climate footprint. Obligations could also be written into project financing along the lines of the Equator Principles[33] or through the use of International Financial Institution (IFI) safeguard policies. The screening of government export credit agencies for the climate-related human rights impacts they create, building on activist campaigns in this area (Newell 2008b), might provide another viable

[33] Banks adhering to the Equator Principles are responsible for more than 80 per cent of commercial project lending. The principles provide a financial industry benchmark for determining, assessing and managing social and environmental risk in project financing, see at: www.equatorprinciples.com/principles.shtml.

channel. This would improve upon the status quo. As Ruggie argues (2007, 16): 'Very few [governments] explicitly consider human rights criteria in their export credit and investment promotion policies, or in bilateral trade and investment treaties, points at which government policies and global business operations most closely intersect.'

There are a range of instruments by which the public sector governs private investment or the latter community establishes its own rules of conduct which could be used to advance climate responsibilities, mainstreaming obligations into trade and investment accords, voluntary and binding. The World Bank and other donors could also play a role here in articulating and enforcing investor obligations. These could be positive '*do good*' provisions about using clean technologies and production, rather than merely '*do no harm*' negative obligations, though the two go together. CSR conditions are making their way into bilateral trade agreements, such as that which exists between the United States and Chile and other agreements, such as North American Free Trade Agreement (NAFTA) contain environmental side agreements, so why not climate-related conditions? The challenge is both to ensure that these are not 'add-on' features of business as usual treaties whose overall effect is to significantly contribute to climate change, and to address the concerns which many developing countries party to such agreements may have about an additional layer of green conditionalities perceived as being designed to control the South's development.

Ultimately, an emphasis on the human rights responsibilities of firms must not absolve states from their primary responsibilities for putting in place systems to safeguard human rights (Muchlinski 2001). Corporations are poorly equipped, for the reasons outlined above, to provide procedural or distributive justice and have insufficient incentives to ensure that their actions do not exacerbate the general human condition vis-à-vis climate change. Emphasising the importance of the state as the appropriate venue for addressing this issue is not to overlook the desirability of rights obligations for firms, albeit mediated by the state, nor to underestimate the central role of the private sector in simultaneously producing and helping to tackle climate change.

This chapter has also emphasised the need to maintain a political and strategic view of the law. It is not a neutral vessel in terms of whose interests it represents, whose property it protects, who it benefits and to whom it affords protection. The law provides just one means of protecting human rights. As Ruggie concludes: '[m]any elements of an overall strategy lie beyond the legal sphere altogether. Consequently, the interplay between systems of legal compliance and the broader social dynamics that contribute to positive change needs to be carefully calibrated.' He cites Sen's warning not to reduce rights to 'proto legal commands' or even 'laws in waiting'. To do so would 'constrict … the social logics and processes other than the law that drive the evolving public recognition of rights'. For this reason 'any successful regime needs to motivate, activate and benefit from all the moral, social and economic rationales than can affect the behaviour of corporations' (2007, 29). Most people most of the time, particularly those most affected by environmental change and least able to protect themselves from its disruptive effects, use protest, direct action and coalition-building among social movements to achieve change. Climate change will be no different in that sense even if the forms of mobilisation required bring distinct challenges (Newell 2005b). NGOs and legal advocacy groups may achieve much by pressing broadly based social claims into the service of justiciable claims, but they should be wary of reifying legal arenas as the only or even most appropriate vehicle of change.

Rights-claiming can be done by those who benefit from not acting on climate change just as easily as by those who look set to suffer the worst consequences. Just as poorer communities contest the loss of their livelihood rights as a result of climate change so, too, oil producing OPEC states call for 'rights to compensation' for loss of revenue accruing from efforts to curtail CO_2 emissions. The political imperative, in this context at least, is to accept that some rights have more moral weight than others. Some rights are *luxury* rights and others are *survival* rights and in the area of climate and human rights the latter are expected to be sacrificed so that the wealthy can enjoy rights to travel and consume as they choose. In the early 1990s a battle ensued between the World Resources Institute and the Centre for Science and Environment (CSE) over the difference between 'luxury' and 'survival' emissions, a battle that continues today, albeit in a different form. CSE's Anil Agarwal once put it the following way:

> Is one tonne of a greenhouse gas produced by a New Yorker or a Londoner equal to a tonne of the same gas produced by a peasant in Guatemala, Chad or Bangladesh? The simple, moral answer is 'no'. The first tonne is the result of luxury. The second tonne of basic survival. Both of them go into the atmosphere. But one needs to be controlled and the other needs to be supported. (Agarwal 2000)

The 'right to development', in the absence of a sense of which forms of development take priority and which can be sustained in a carbon-constrained world, confers rights on high as well as low contributors to climate change. Rights do not neatly align themselves with those most deserving of them. They also conflict and so we need to place the political and institutional arenas which have to recognise and mediate such competing claims at the centre of analysis. It is imperative that they do so with a clear sense of social justice in mind. I will end with the words of Ambassador Lionel Hurst of Antigua-Barbuda (March 2003) who reminds us of what is at stake:

> The most populous and wealthiest of the world face a moral challenge greater than colonialism or slavery. They are failing in that challenge. Men have lost reason in the fossil fuel economy... Inhabitants of small islands have not agreed [to be] sacrificial lambs on the altar of the wealth of the rich.

References

Agarwal, Anil, 'Global Warming in an Unequal World', *Equity Watch* newsletter, 15 November (New Delhi: CSE, 2000). Available at: www.cseindia.org/campaign/ew/art20001115_2.htm.

Alliance of Automobile Manufacturers (AAM), Statement of Dave McCurdy on Vermont GHG decision, 12 September, 2007. Available at http://auto.mindville.net/media/pressarchive2007.php?PHPSESSID=cd59c8d746d1a7a9ea40328905a83f35, accessed 24 September 2007.

Amnesty International, *Human Rights Principles for Companies* (London: Amnesty International, 1998).

Bachram, Heidi, 'Climate Fraud and Carbon Colonialism: The New Trade in Greenhouse Gases', *Capitalism, Nature, Socialism*, 15:4 (2004), 10–12.

Chatterjee, P. and M. Finger, *The Earth Brokers: Power, Politics and World Development* (London: Routledge, 1994).

Chinen, M. A., 'Jurisdiction: Foreign Plaintiffs, *Forum Non Conveniens*, and Litigation Against Multinational Corporations', *Harvard International Law Journal*, 28 (1987), 202–9.

Christian Aid, *Scorched Earth: Oil and Water in Sudan* (London: Christian Aid, 2000).

Cole, Luke, 'Empowerment as the Key to Environmental Protection: The Need for Environmental Poverty Law', *Ecology Law Quarterly*, 19 (1992), 619–83.

Cole, Luke and Sheila Foster, *From the Ground Up: Environmental Racism and the Rise of the Environmental Justice Movement* (New York University Press, 2001).

Cutler, C., 'Historical Materialism, Globalization and Law,' in Rupert, M. and H. Smith (eds.), *Historical Materialism and Globalisation* (London: Routledge, 2002), 230–56.

Frynas, G., 'Political Instability and Business: Focus on Shell in Nigeria', *Third World Quarterly*, 19:3 (1998), 457–78.

'Legal Change in Africa: Evidence from Oil-related Litigation in Nigeria', *Journal of African Law*, 43 (1999), 121–50.

International Chamber of Commerce, Statement by the International Chamber of Commerce before COP1, 29 March 1995, Berlin.

International Council on Human Rights Policy, *Beyond Voluntarism: Human Rights and the Developing International Legal Obligations of Companies* (Versoix, Switzerland: International Council on Human Rights Policy, 2002).

Jenkins, R., 'China and CSR', paper presented at meeting of the International Research Network on Business, Development and Society, September 2007.

Levy, D., 'Business and the Evolution of the Climate Regime: The Dynamics of Corporate Strategies', in David Levy and Peter Newell (eds.), *The Business of Global Environmental Governance* (Cambridge, MA: MIT Press, 2005), 73–105.

Levy, D. and P. Newell, 'Oceans Apart? Comparing Business Responses to the Environment in Europe and North America,' with David Levy, *Environment*, November, 42:9 (2000), 8–20.

Mason, M., *The New Accountability: Environmental Responsibility Across Borders* (London: Earthscan, 2005).

McLaren, D., *The OECD's Revised Guidelines for Multinational Enterprises: A Step Towards Corporate Accountability*?, February (London: Friends of the Earth, 2000).

Muchlinski, Peter, 'Human Rights and Multinationals: Is There a Problem?', *International Affairs*, 77:1 (2001), 31–47.

Newell, Peter, *Climate for Change: Non-State Actors and the Global Politics of the Greenhouse* (Cambridge University Press, 2000).

'Managing Multinationals: The Governance of Investment for the Environment', *Journal of International Development*, 13 (2001), 907–19.

'Race, Class and the Global Politics of Environmental Inequality', *Global Environmental Politics*, 5:3 (2005a), 70–94.

'Climate for Change: Civil Society and the Politics of Global Warming,' in F. Holland *et al.* (eds.), *Global Civil Society Yearbook* (London: Sage, 2005b).

'The Marketisation of Global Environmental Governance: Manifestations and Implications,' in J. Park, K. Conca and M. Finger (eds.), *The Crisis of Global Environmental Governance: Towards a New Political Economy of Sustainability* (London: Routledge, 2008a), 77–96.

'Civil Society, Corporate Accountability and the Politics of Climate Change', *Global Environmental Politics*, 8:3 (2008b), 122–53.

Newell, P. and M. Paterson, 'A Climate for Business: Global Warming, the State and Capital', *Review of International Political Economy*, 5 (1998), 679.

Newell, Peter, and Joanna Wheeler (eds.), *Rights, Resources and the Politics of Accountability* (London: Zed Books, 2006).

Robinson, Mary, 'The Business Case for Human Rights', *Financial Times Management* Visions of Ethical Business (London: Financial Times Professional, 1998), 14.

Rowlands, I., 'Beauty and the Beast? BP's and Exxon's Positions on Global Climate Change', *Environment and Planning C: Government and Policy*, 18:3 (2000), 339–54.

Ruggie, John Gerrard, 'Business and Human Rights: The Evolving International Agenda,' Harvard Corporate Social Responsibility Initiative, Working Paper No. 38 (2007).

Ruggie, J., 'Protect, Respect and Remedy: a Framework for Business and Human Rights', Report of the Special Representative of the Secretary-General on the issue of human rights and transnational corporations and other business enterprises, UN Doc. A/HRC/8/5, 7 April 2008.

Ruiters, G., 'Race, Place and Environmental Rights: A Radical Critique of Environmental Justice Discourse,' in D. McDonald (ed.), *Environmental Justice in South Africa* (Athens, OH: Ohio University Press; Cape Town: University of Cape Town Press, 2002), 112–27.

Sachs, Wolfgang, 'Climate Change and Human Rights', *Interactions between Global Change and Human Health,* The Pontifical Academy of Sciences, Scripta Varia 106 (Vatican City, 2006) 349–69.

Schedler, A., L. Diamond and M. Plattner, *The Self-Restraining State: Power and Accountability in New Democracies* (Boulder: Lynne Rienner, 1999).

Sæverud, I. A. and J. B. Skjærseth, 'Oil Companies and Climate Change: Inconsistencies Between Strategy Formulation and Implementation?', *Global Environmental Politics*, 7:3 (2007), 42–63.

Seymour, Frances, 'Forests, Climate Change and Human Rights', Note prepared for the meeting of the International Council on Human Rights Policy, Geneva, 12–13 October 2007.

Special Representative of the UN Secretary-General (SRGS) (2007), 'Corporate Responsibility to Protect Human Rights', *Report of a Multi-Stakeholder Consultation*, Geneva, 4–5 December 2007.

UNECE, 'The Contents of the Convention', available at: www.unece.org/env/pp/contentofaarhus.htm, accessed 29 November 2007.

UN General Assembly, 'Promotion of all Human Rights, Civil, Political, Economic, Social and Cultural Rights Including the Right to Development,' 7th session, Human Rights Council, 26 March 2008 (A/HRC/7/L.21/Rev.1).

Ward, H., *Corporate Accountability in Search of a Treaty? Some Insights from Foreign Direct Liability*, Briefing Paper No. 4, Chatham House (London: RIIA, 2002).

Watkins, Kevin, Editor of the UNDP Human Development Report, at its launch in Brazil, 27 November. Quoted in Larry Elliot and Ashley Seager, 'Cut Carbon by up to Third to Save Poor, UN tells West', *The Guardian*, 28 November 2007. Available at: www.guardian.co.uk/environment/2007/nov/28/climatechange.

Zadek, Simon, *The Civil Corporation: The New Economy of Corporate Citizenship* (London: Earthscan, 2001).

5

Rethinking human rights: the impact of climate change on the dominant discourse

SAM ADELMAN*

In times of crisis it is necessary to think imaginatively, provocatively and radically as orthodoxies collapse around us, and to distinguish common sense from good sense. Who would have thought that the 'credit crunch' crisis would so rapidly and thoroughly destabilise the dogmas of neoliberalism, and provoke the re-emergence of nationalisation and regulation and a rediscovered role for the state? Every crisis is potentially also an opportunity. The unsustainability of capitalist globalisation has long been apparent to those on the receiving end, who will yet again bear the costs of bailing out the corporate culprits, as it has been to the environment. Indeed, finding solutions to both crises must involve a fundamental restructuring of global governance.

This chapter challenges certain shibboleths in the dominant human rights discourse and the efficacy of prevailing ideas about governance and markets in combating anthropogenic global warming. In particular, it examines the viability of market-based solutions, in which human rights rarely feature – much as if the two sets of principles existed in different domains. It raises disconcerting issues, for example, whether the right to a sustainable environment should, to echo Hannah Arendt, become a new principle of humanity that takes precedence over other rights.[1] Perhaps paradoxically, the chapter nonetheless assumes that human rights provide an important means of addressing climate change, either through the construction of a new overarching or meta-right to a sustainable environment or by deploying existing rights.[2] Universal human

* School of Law, University of Warwick, Coventry CV4 7AL, UK.

[1] See Peg Birmingham, *Hannah Arendt and Human Rights* (Bloomington, IN: Indiana University Press, 2006).

[2] There is, however, a third possibility: that the 'language of human rights ultimately may prove inadequate for encapsulating the problems posed by climate change' and that we may be asking them to achieve something for which they are not designed (Michael Depledge and Cinnamon Carlane, 'Sick of the Weather: Climate Change, Human Health and International Law', *Environmental Law Review*, 9:4 (2007), 238).

rights might be expected to be in the forefront of attempts to respond to the global scope and urgency of the risk – which threatens the very possibility of human rights fulfilment – but the literature on the relationship between them is sparse.[3] Among the reasons for this may be the difficulties in establishing causality and responsibility for carbon emissions, of identifying the form and content of any right and conflicts between existing rights.

I proceed by analysing the deficiencies inherent in the centrality of market-based attempts to deal with the problem under the current regulatory framework, bearing in mind that in no other field are law, policy and regulation so thoroughly contingent upon science and, more problematically, economics. Equally deficient is the international juridico-political framework in which the principle and rationality of sovereignty predominate.[4] Despite the widely held view that sovereignty is in decline it continues to define the context of human rights. It is within this juxtaposition of markets and sovereign states that the possibilities and dilemmas of using human rights to address climate change must be assessed.

If the scientific consensus is correct, global warming threatens not only human rights to health, food, property and culture but to life itself.[5] The consensus, embodied in the fourth report of the Intergovernmental Panel on Climate Change (IPCC),[6] is that a 2°C increase in global temperature will produce more floods, droughts and food shortages,[7] and that an increase of 4°C or more constitutes a threat to the human species. The

[3] See International Council on Human Rights Policy (ICHRP), *Climate Change and Human Rights: A Rough Guide* (2008), ch. 1.

[4] Sovereign rationality is predicated on conventional if tenuous principles of international law, such as formal equality, non-interference in the internal affairs of other sovereigns and voluntary assumption of obligations by treaty and, crucially, on state monopolies of law and legitimate violence identified by Max Weber. More than this, however, it is a rationality underpinned by the imperative of self-reproduction and, therefore, of resisting dilution.

[5] Climate change and global warming are not the same; the Earth has undergone numerous changes of climate, not all of which were due to global warming. However, since it is overwhelmingly likely that current climate change is due to rises in average global temperature, I use the terms interchangeably in this article unless the context demands otherwise.

[6] Intergovernmental Panel on Climate Change (IPCC), Fourth Assessment Report: *Climate Change: Synthesis Report* (2007). Available at: www.ipcc.ch/pdf/assessment-report/ar4/syr/ar4_syr.pdf.

[7] For an excellent analysis of the failure of the global food market and the disastrous consequences of the rush to biofuels, see Raj Patel, *Stuffed and Starved: From Farm to Fork, the Hidden Battle for the World Food System* (London: Portobello Books, 2008).

solution is obvious if difficult: in order to save the planet carbon emissions must be stabilised and then reduced to a sustainable level, preferably around 350 parts per million (ppm) of CO_2. According to some prognoses, we have a decade in which to do so, after which climate change is likely to be irreversible and, therefore, catastrophic.[8]

The threat posed by climate change is arguably qualitatively different to all other risks including nuclear proliferation, and is potentially incapable of being confronted within the confines of the dominant human rights discourse. Under the current regulatory regime based on the UN Framework Convention on Climate Change (UNFCCC) and the Kyoto Protocol, carbon emissions are measured in territorial terms rather than 'upstream' at points of emission and states are exclusively responsible under international law for policing carbon emissions by actors under their jurisdiction. Since carbon emissions do not respect borders there is a mismatch between this state-centric legal architecture and the global scope of the threat.

Environmental rights are widespread, but there is no generally accepted human right to a sustainable environment.[9] Human rights are distinguished from civil rights derived from membership of a territorially bounded community because they are inalienable, transcendental, universal moral claims or entitlements that individuals are argued to possess by virtue of their humanity. The problem is that they are honoured more in the breach than the observance and that they are not always fully justiciable. The assertion of a right implies that it is possible to identify the content of the entitlement or the nature of the harm it is designed to prevent and the duty bearer against whom it can be exercised. Historically, the sovereign state has been the primary bearer of the duty to promote and protect human rights and the bridge between human and civil rights. States are the primary subjects of international

[8] See especially the consensus reached by the IPCC as reflected in UNDP (2007), *Human Development Report 2007/2008 – Fighting Climate Change: Human Solidarity in a Divided World*, 22 (available at: http://hdr.undp.org/en/media/hdr_20072008_en_complete.pdf), which argues that '[w]e have less than a decade to ensure that the window of opportunity is kept open'. See also *The Guardian*, 27 November 2007, '10 years to change our ways, warns UN report'. Available at: www.guardian.co.uk/environment/2007/nov/27/climatechange.

[9] An overarching right against environmental degradation is most commonly constructed to as the right to a healthy environment, as the discussion below reveals. My preference is for a right to a sustainable environment, which I use in appropriate contexts. First, it logically includes the right to a healthy environment. Second, it highlights the issue of sustainable development, particularly in the global South.

law but non-state actors are equally, if not more, responsible for carbon emissions.

The principle of formal equality is belied by the historical reality of substantive inequalities between states, a recognition that forms the basis for a solution to climate change. Substantial resources must be redistributed to less developed countries, which have contributed least to the problem but are likely to suffer first and worst from its effects, to enable them to adapt and mitigate without threatening their right to development. At the same time, conventional measurements of development like per capita income are inadequate: the People's Republic of China (PRC) is a developing country soon likely to become both the largest emitter of greenhouse gases (GHGs) and the biggest economy in the world.

The context: market failures and sovereign rationality and human suffering

When markets fail, human suffering is increased and rights are threatened. The notion that the market is a private sphere subject to different rules from the public domain of politics was always specious, but the border between the state and civil society has become increasingly blurred during the past twenty-five years.[10] As areas previously in the domain of the state are privatised, they escape regulation and accountability. Treated as a private matter between the parties, the contract of employment has long been the basis for human rights violations, but under capitalism, markets violate rights obliquely as well as directly. It, therefore, borders on the perverse that a market in carbon trading forms the centrepiece of the regime regulating climate change but, *faute de mieux*, this is what the UNFCCC and the Kyoto Protocol provide.[11]

Among the numerous defects in the framework is the fact that the cost of carbon in the rapidly expanding market does not reflect the true social

[10] Saskia Sassen analyses the rise of 'private authority, including the privatizing of domains of domains once exclusive to the state' (223) in *Territory, Authority, Rights: From Medieval to Global Assemblages* (Princeton University Press, 2006), ch. 5.

[11] The Convention entered into force on 21 March 1994. The Protocol, which was adopted on 11 December 1997 and entered into force on 16 February 2005, sets binding targets for the reduction of GHGs for thirty-seven industrialised countries and the European Union amounting to an average of 5 per cent of 1990 levels between 2008 and 2012. The United States has withdrawn from the Protocol, which is due to be renegotiated by December 2009.

and environmental costs of carbon emissions.[12] The centrality of the market mechanism in the UNFCCC is in stark contrast to the 1989 Montreal Protocol, which successfully addressed the depletion of the ozone layer through direct regulation, demonstrating what is possible through voluntary adherence by states to a coherent regulatory regime.[13] The Kyoto process has suffered from delays, the withdrawal of the United States, the absence of targets for developing countries or long-term incentives for investment in cleaner technologies, the omission of aviation and shipping, a weak compliance regime, inadequate funding for adaptation and mitigation and ineffective, inefficient and abuse-prone flexibility mechanisms.[14] It is a market that appears to have been *designed* to fail.[15] Far from reducing GHG emissions, it has failed even to stabilise them. For example, the 'UN's main offset fund is being routinely abused by chemical, wind, gas and hydro companies who are claiming emission reduction credits for projects that should not qualify. The result is that no genuine pollution cuts are being made.'[16] Even if the market was working optimally, there is, as Tickell argues, 'a complementary role for direct regulation to constrain greenhouse gas emissions, and additional, targeted taxes, levies and subsidies'.[17] He calls for an end to territorial accounting and national allocations in favour of a global cap

[12] In other words, carbon is being traded below its true costs. In addition to carbon trading, the two other mechanisms enabling states parties to meet their reduction targets are the clean development mechanism (CDM), which allows a country with emission reduction or limitation commitments to implement projects in developing countries and earn saleable credits (CER), and joint implementation (JI), which enables countries to earn emission reduction units (ERUs) from such projects. Both mechanisms are flawed.

[13] The Montreal Protocol on Substances that Deplete the Ozone Layer, concluded under the 1987 Vienna Convention for the Protection of the Ozone Layer.

[14] Oliver Tickell, *Kyoto 2: How to Manage the Global Greenhouse* (London: Zed Books, 2008), ch. 2.

[15] The Dag Hammarskjöld Centre describes it as one of a number of American-inspired 'fixes' to climate change, a market designed to contain the 'political threats implied by climate change – while at the same time using it to create new opportunities for corporate profit' (Larry Lohmann (ed.), *Carbon Trading: A Critical Conversation on Climate Change, Privatisation and Power*, Development Dialogue, 48, September 2006 (Uddevalla, Sweden: Mediaprint), 45). Although it appears to be the most comprehensive regulatory regime, similar criticism can be made of the European Union Emissions Trading Scheme (ETS), in which the caps set on national allocations bear little relationship to the real social costs of carbon emissions.

[16] *The Guardian*, 26 May 2008, 'Billions wasted on UN climate change programme'. Available at: www.guardian.co.uk/en-vironment//2008/may/26/climatechange.greenpolitics. The idea that carbon offsetting is a viable means of dealing with climate change is extremely problematic.

[17] *The Guardian*, 26 May 2008, 11.

on GHG emissions applied upstream designed to stabilise emissions at 350 ppm, to make companies accountable under the 'polluter pays' principle, the allocation of GHG permits

> under a sealed-bid auction system, subject to a reserve price reflecting the social costs of carbon, and a ceiling price at which unlimited permits will be sold, sufficient to fund additional investments in low-carbon development that will more than offset the extra permits beyond the allocated cap being sold now, and so allow the extra permits to be clawed back in future years.

These funds should be used to 'finance solutions to climate change – accelerating the use of renewable energy, raising energy efficiency, protecting forests, promoting climate-friendly farming, and researching geoengineering technologies. And [to] commit hundreds of billions of dollars per year to finance adaptation to climate change, especially in poor countries.'[18]

> The market-based, economics-driven approach to tackling global warming ... is powerful and necessary. It does not, however, represent 100 per cent of the solution. Just as, according to Stern, climate change itself represents 'the greatest market failure the world has seen', so even the most sophisticated market mechanisms we might create will leave gaps and may contain implicit market failures of their own.[19]

There is obscenity in the idea that the solution to climate change lies in a market that encourages profit from pollution but Tickell's argument highlights both the absence of strong political leadership and the neoliberal dogma that economic actors respond most effectively to the price mechanism, and that the market is the indispensable means of delivering incentives to develop clean technologies,[20] the resources required to fund adaptation and mitigation and to change attitudes

[18] Oliver Tickell, *The Guardian*, 11 August 2008, 'On a planet 4°C hotter, all we can prepare for is extinction' (available at: www.guardian.co.uk/commentisfree/2008/aug/11/climatechange) and Tickell, *Kyoto 2*, 81–2.

[19] Tickell, *Kyoto 2*, 139.

[20] Market-based solutions fetishise the belief that we just have to wait for the cleaner technologies that will be delivered by the profit incentive. It is not certain that the market will deliver and when it does the law of unintended consequences is likely to kick in: the conversion of agricultural land for biofuels leading to devastating rises in food prices provides a cautionary tale. While Luddism is not the answer, we should not lose sight of the fact that while we wait for the market to deliver, the only viable approach is to stabilise and then reduce carbon emissions.

and behaviour with a minimum of social and political unrest.[21] Sovereign states came to be regarded as the handmaidens of markets, their powers in terminal decline.[22] The 'credit crunch' reveals the consequences of faith in the supposed infallibility of unregulated markets, an imbalance unlikely to be ameliorated when the United States rejoins the Kyoto process.

Paradoxes abound. States whose sovereignty is declining are rediscovered as central actors in the global political economy. Even as some commentators discern the demise of unconstrained globalism, climate change demands an undilutedly global response.[23] Methodological nationalism, which comprehends law, politics and economics primarily through the prism of states, is rendered redundant by the aterritoriality of markets, the globalisation of law, the attenuation of sovereignty and the transboundary nature of global warming.[24] Despite this, sovereign rationality continues to define the dimensions and possibilities of a rights-based approach to climate change. States will negotiate the role

[21] Inevitably, this is a peculiarly Western perspective that neglects the unrest in many parts of the world due to steep rises in food prices caused in part by the production of biofuels – including, as it happens, Italy. 'Crisis talks on global food prices', *The Guardian*, 27 May 2008. Available at: www.guardian.co.uk/environment/2008/may/27/food.internationalaidanddevelopment.

[22] The literature on the decentring and deterritorialisation of sovereignty is vast and flows largely from Michael Hardt and Antonio Negri's *Empire* (Cambridge, MA: Harvard University Press, 2001). Space precludes a fuller discussion, but sovereignty is theoretically distinguishable from statehood. Colonial history provides many examples of states without the full attributes of sovereignty. On the relationship between sovereignty and statehood, see Antony Anghie, *Imperialism, Sovereignty and the Making of International Law* (Cambridge University Press, 2004). In an important sense the aim of human rights is to circumscribe, limit and redefine sovereign power.

[23] See, for example, Martin Jacques, 'Northern Rock's rescue is part of a geopolitical sea change', *The Guardian*, 18 February 2008 (available at: www.guardian.co.uk/commentisfree /2008/feb/18/northernrock.alistairdarling) and John Ralston Saul, *The Collapse of Globalism* (London: Atlantic Books, 2005). On global risk and methodological nationalism, see Ulrich Beck, *Risk Society: Towards a New Modernity* (London: Sage, 1992); *Power in the Global Age: A New Global Political Economy* (Cambridge: Polity Press, 2005); *The Cosmopolitan Vision* (Cambridge: Polity Press, 2006). For a critique of liberal cosmopolitanism and humanitarian interventionism, see Costas Douzinas, *Human Rights and Empire: The Political Philosophy of Cosmopolitanism* (Abingdon: Routledge-Cavendish, 2007). As Immanuel Kant recognised, in resisting the idea of global government in his prescription for perpetual peace, any global regulatory regime is susceptible to hegemonic capture. His ideal form of international governance thus involved more law and less government.

[24] See Beck, *Power in the Global Age*, note 23, above.

of the market in Kyoto II, which must come into force by 2012. If it fails, the next opportunity is set for 2020, by which time it will be too late.[25]

The problem that sovereignty poses for human rights in the contemporary global order stems from an unresolved contradiction that was locked into place at the end of the Second World War. In 1945, the UN Charter reaffirmed the pre-eminence of states as subjects of international law that emerged in the 1648 Peace of Westphalia.[26] Three years later, in response to the most egregious manifestation of sovereign exceptionality,[27] the Universal Declaration of Human Rights (UDHR) ostensibly proffered an alternative paradigm in which the rights of individuals would be balanced against those of nation-states. The uneven history of the past six decades of human rights is marked by attempts to circumscribe sovereign immunity and impunity.

As Agamben argues, the prisons at Guantánamo Bay, Cuba and Abu Ghraib in Iraq are not exceptional but the logical consequences of sovereign rationality. The existence of sovereignty provokes human rights. As their antithesis, it generates a need for them.[28] It is their *sine qua non* and the trap from which human rights have yet to escape in order to vindicate their promise – to become, as it were, rights not merely in themselves but for themselves. Sovereignty provides the context of human rights and delimits their possibilities through its impulse to subordinate them. The natural rights that the classical social contract theorists discerned in the state of nature were reinvented as human rights necessary to curb the states called forth in the contract. As Freeman argues, the 'contemporary concept of human rights is intended to protect individuals from the abuse of power by governments'.[29] Human rights were conceived as the antidote to the propensity to arbitrariness, exceptionalism and violence of Hobbes's *Leviathan*. Sovereignty is the largest

25 Tim Flannery, 'Words of warming', *The Guardian*, 9 August 2008 (available at: www.guardian.co.uk/books/2008/aug/09/scienceandnature.climatechange).

26 Article 2 declares that the organisation is based upon sovereign equality and Article 2(7) prohibits UN intervention 'in matters which are essentially within the domestic jurisdiction of any state', thus affirming two of the cardinal principles on sovereignty in international law (the third being the voluntary assumption of treaty-based obligations).

27 The best analysis of sovereignty and the theory of the exception is provided by Giorgio Agamben in *Homo Sacer: Sovereign Power and Bare Life* (Stanford University Press, 1998) and *State of Exception* (University of Chicago Press, 2005).

28 I discuss sovereignty and the state of exception in greater depth in a forthcoming monograph provisionally entitled *The End of Sovereignty*.

29 Michael Freeman, *Human Rights: An Interdisciplinary Approach* (Cambridge: Polity Press, 2002), 15.

unresolved problem of political modernity and the biggest impediment to dealing with climate change. To the PRC and the United States, pollution appears to be a sovereign prerogative.

A rights-based approach to climate change must, therefore, confront the enduring pre-eminence of the principle of sovereignty in the international system.[30] Climate change constitutes a potentially unique opportunity to develop more adequate institutions of global governance, because the failure of any state to subordinate perceived national interest is ultimately self-defeating in that it will undermine that country's development along with that of everybody else. It will lead to increased competition for ever scarcer resources, to potentially catastrophic food problems and loss of coastal lands, to war, human rights violations and humanitarian crises. A regulatory regime that reproduces the contemporary sovereign-centric order is thus nihilistic. Humanity faces a threat that cannot be addressed as long as polluting the atmosphere is regarded as a sovereign prerogative. Since global risks transcend territoriality, resort to unilateralism and exceptionalism subordinates the rights of everyone to selfish national interest. The regulatory regime must: (i) be global; (ii) limit sovereign rights;[31] (iii) set coherent and enforceable targets; and (iv) promote and protect human rights.

The problems and possibilities of a rights-based approach

Resolving the contradiction between sovereignty and human rights is a necessary but insufficient condition for dealing successfully with global warming. Today, few struggles are not fought under the banner of human rights but, initially at least, they are often a somewhat blunt instrument. The key lies in translating human rights as aspirations or moral claims into enforceable demands. Their effectiveness is sharpened by justiciability, which in turn depends on clarifying the content and identifying

[30] It is generally difficult for individuals or groups to petition international bodies because of the requirement that local remedies must be exhausted before complaints can be brought before meta-national institutions.

[31] The European human rights regime provides a contradictory model. It demonstrates what is possible when states agree to subordinate their sovereignty, albeit in problematic ways – one example being the lack of socio-economic or environmental rights in the European Convention on Human Rights. Unlike defiance of the ICJ by Israel and the United States, no state party to the Convention has contravened a decision of the European Court of Human Rights (ICJ, *Nicaragua* v. *United States of America*, Judgment of 27 June 1986 and ICJ, *Legal Consequences of the Construction of a Wall in the Occupied Palestinian Territory*, Summary of the Advisory Opinion of 9 July 2004).

the bearers of the rights in question and their concomitant duties.[32] They are invariably works in progress whose goal, unlike all other forms of sovereignty law, is self-redundancy. They are a contradictory mixture of transcendence – universal, inherent and inalienable – and, as Baxi and Santos insist, immanence in struggle and resistance.[33]

If the science is correct, we have until approximately 2016 to stabilise carbon emissions at sustainable levels, at which point we are no longer likely to be dealing with options but inevitabilities.[34] The urgency of dealing with climate change suggests that a tight state-based regulatory regime is the immediate requirement. Ideally, Kyoto II should prioritise human rights, but this is unlikely.

For Merrills, if 'rights are a good way of ensuring that something is taken seriously, designating an entitlement a *human* right is even better, on account of the status of this class of rights in legal and moral discourse'.[35] The challenge, as Alston sees it, 'is to achieve an appropriate balance between, on the one hand, the need to maintain the integrity and credibility of the human rights tradition, and on the other hand, the need to adopt a dynamic approach that fully reflects changing needs and perspectives and responds to the new threats to human dignity and well-being'.[36] Lee argues that 'an environmental violation becomes significant enough to become a human rights violation when, as a result of a specific course of state action, a degraded environment occurs with either serious health consequences for a specific group of people or a disruption of a people's way of life'.[37] The problem, as he acknowledges, is that although states may bear ultimate responsibility for environmental damage, pollution is often caused by non-state actors who regularly evade responsibility for

32 See the debate between Sen and Baxi on human rights as law or moral claims. Amartya Sen, 'Elements of a Theory of Human Rights', *Philosophy & Public Affairs*, 32:4 (2004), 315 and Upendra Baxi, *The Future of Human Rights*, 2nd edn. (Delhi: Oxford University Press, 2005).

33 Baxi, *The Future of Human Rights*; Boaventura de Sousa Santos, *Toward a New Legal Common Sense* (London: Butterworths, 2002).

34 IPCC, note 6, above.

35 John G. Merrills, 'Environmental Rights', in Daniel Bodansky, Jutta Brunnée and Ellen Hey (eds.), *The Oxford Handbook of International Environmental Law* (Oxford University Press, 2007), 666; emphasis in original.

36 Philip Alston, 'Conjuring Up New Human Rights: A Proposal for Quality Control', *American Journal of International Law*, 78 (1984), 609.

37 John Lee, 'The Underlying Legal Theory to Support a Well-Defined Human Right to a Healthy Environment as a Principle of Customary International Law', *Columbia Journal of International Law*, 25 (2000), 285.

violations of human rights law.[38] The virtue of a human right to a healthy environment is that it would apply universally, irrespective of the arbitrariness of economic situation or geographic location.[39] Lee calls for an independent, internationally recognised human right to a healthy environment, setting a universal legal standard to address the limitations of relying on the environmental laws of individual nations, on the one hand, and the deployment of the 'environmental components' of existing human rights, on the other hand.[40]

Environmental rights are usually classified as third generation solidarity rights, different to peremptory first generation civil and political rights, on the one hand, and progressively realisable second generation socio-economic rights, on the other hand. They are a hybrid which Birnie and Boyle argue 'do not fit neatly into any single category or "generation" of human rights; rather, they straddle all three ... categories'.[41] International law provides thin gruel for those seeking an unambiguous enforceable right to a clean or healthy environment.[42] Human rights instruments tend to address the environment in broad and aspirational terms. The 1972 Stockholm Declaration states that 'Man [*sic*] has the fundamental right to freedom, equality and adequate conditions of life in an environment of a quality that permits a life of dignity and well-being, and he bears a solemn responsibility to protect and improve the environment for

38 For example, the damage caused by Shell Oil in Ogoniland and Union Carbide in Bhopal.

39 Lee, 'Legal Theory', 288; Merrills, 'Environmental Rights'.

40 Lee, 'Legal Theory', 292.

41 Patricia Birnie and Alan Boyle, *International Law and the Environment*, 2nd edn. (Oxford University Press, 2002), 253.

42 The right to a satisfactory environment is recognised in more than fifty jurisdictions, but this does not amount to a generally recognised human right. See, for example, Jan Hancock, *Environmental Human Rights: Power, Ethics and Law* (Aldershot: Ashgate, 2003), ch. 4; James R. May, 'Constituting Fundamental Environmental Rights Worldwide', *Pace Environmental Law Review*, 23 (2005–2006), 113; Janelle P. Eurick, 'The Constitutional Right to a Healthy Environment: Enforcing Environmental Protection through State and Federal Constitutions', *International Legal Perspectives*, 11 (2001), 185; Ernst Brandl and Hartwin Bungert, 'Constitutional Entrenchment of Environmental Protection: A Comparative Analysis of Experiences Abroad', *Harvard Environmental Law Review*, 16 (1992), 1; Lee, 'Legal Theory', 314. The right to health is contained in the Universal Declaration of Human Rights, Article 25(1): 'everyone has the right to a standard of living adequate for the health and well-being of oneself and one's family, including food, clothing, housing, and medical care'. See also the American Declaration on the Rights of Man (Article XI), the International Covenant on Economic, Social and Cultural Rights (Article 12) and the African Charter on Human and Peoples' Rights (Article 16).

present and future generations.'[43] Similar terminology was noticeably absent from the 1992 Rio Declaration, which merely acknowledges that '[h]uman beings are at the centre of concerns for sustainable development. They are entitled to a healthy and productive life in harmony with nature.'[44] The closest we come to a fully-fledged right to a healthy environment are the provisions in Article 24 of the 1981 African Charter ('All peoples shall have the right to a generally satisfactory environment favourable to their development') and Article 11 of the 1998 Protocol of San Salvador ('Everyone shall have the right to live in a healthy environment and to have access to public services'). The African Charter formulates the right in collective terms, which seems logical but implicitly raises problems arising from conceiving it as an individual right, on the one hand, and the difficulties involved in seeking to enforce group rights under the dominant discourse, on the other hand.[45]

The Aarhus Convention, another instrument that links human rights and the environment, also acknowledges the importance of intergenerational equity.[46] It is a regional treaty adopted by the UN Economic Commission for Europe with global implications. It declares that sustainable development can be achieved only with the participation of all stakeholders and views state accountability and transparency as necessary for protection of the environment. Article 1 states that in 'order to contribute to the protection of the *right of every person* of present and future generations to live in an environment adequate to his or her health and well-being, each Party shall *guarantee* the rights of access to information, public participation in decision-making, and access to justice in environmental matters in accordance with the provisions of this Convention'.[47] Whereas Giorgetta regards Aarhus as 'the clearest statement in international law to date of a fundamental right to a

[43] Principle 1, Declaration on the Human Environment, *Report of the UN Conference on Environment and Development* (New York, 1973), UN Doc. A/CONF.48/14/Rev.1.

[44] Principle 1, Declaration on Environment and Development, *Report of the UN Conference on Environment and Development* (New York, 1992), UN Doc. A/CONF.151/26/Rev.1.

[45] On the difficulties involved in interpreting this right see *The Social and Economic Rights Action Centre for Economic and Social Rights* v. *Nigeria* (2001), reprinted in (2003) 10 I.H.H.R. 282 (*Ogoni*) and on collective rights generally, Merrills, 'Environmental Rights', 670.

[46] Convention on Access to Information, Public Participation in Decision-Making and Access to Justice in Environmental Matters, adopted 25 June 1998 and entered into force in 2001.

[47] My emphases.

healthy environment',[48] Nadal argues that, although the Convention 'has an untapped potential for empowering those suffering environmental injustice, it is overall a weak pillar of empowerment in the absence of an environmental justice rationale owing to certain inherent faultlines of disempowerment'.[49] She questions the efficacy of the Convention in facilitating the two pivotal aims of grassroots movements, 'to challenge the institutional causes of environmental injustice and empower those suffering environmental injustice to be "agents for environmental justice"'.

The number of cases in which courts have accepted a link between human rights and the environment that might form the basis for the assertion of the emergence of customary international human rights law is growing, albeit slowly. In *Subhash Kumar* v. *State of Bihar*, the Indian Supreme Court held that the right to a safe environment was integral to the right to life under Article 21 of the Indian Constitution.[50] In *Lopez Ostra* v. *Spain*, the European Court of Human Rights (ECtHR) accepted that environmental degradation may affect the right to enjoyment of private and family life under Article 8 of the European Convention.[51] The Inter-American Court of Human Rights decided that Brazil had violated the rights to life, liberty and personal security of the Yamomani Indians by failing to take measures to prevent environmental degradation.[52]

Human rights bodies have gradually begun to take a more expansive view and to interpret basic human rights like those to life and health as necessarily encompassing a right to a healthy environment, but progress has been patchy and hardly reflects the urgency of the threat.[53] The UN Human Rights Council has acknowledged that climate change 'poses an immediate and far-reaching threat to people and communities around the world and has implications for the full enjoyment of human rights'

[48] Sueli Giorgetta, 'The Right to a Healthy Environment, Human Rights and Sustainable Development', *International Environmental Agreements: Politics, Law and Economics*, 2:2 (2002), 171.

[49] Caroline Nadal, 'Pursuing Substantive Environmental Justice: The Aarhus Convention as a "Pillar" of Empowerment', *Environmental Law Review*, 10 (2008), 29.

[50] AIR 1991 SC 240.

[51] 303-C E.Ct.H.R. (Ser. A) (1994).

[52] Inter-Am. C.H.R. 7615 OEA/Ser. L.V/II/66 doc. 10 rev. 1 (1985). Compare this with the decision below in which the Court refused to find the United States in breach of Inuit rights by failing to take measures to reduce carbon emissions, below.

[53] Margaret E. Middaugh, 'Linking Global Warming to Inuit Human Rights', *San Diego International Law Journal*, 8 (2006), 181.

and that is a global risk requiring a global solution.[54] Nevertheless, at risk of repetition, there is a disturbing paucity of human rights instruments linking human rights and climate change.[55]

Fitzmaurice discerns three main schools of thought on the relationship between human rights and the right to a clean environment.[56] The first argues that there can be no human rights without environmental rights, the second that the right to a clean environment – whether existent or prospective – is highly questionable, while the third argues that such a right exists but is derived from other rights, such as those to life and health.

Human rights may need to be reconstructed in the context of climate change in one of two ways. The first is the construction of a 'meta-right' to a sustainable environment which would make carbon emissions above a certain level a human rights violation. This approach has the virtue of simplicity and directly targets the problem, but its chances of success appear to be less than those of the second, which involves recourse to existing human rights.

A 'meta-right'

I use the term 'meta-right' to suggest the desirability not merely of the right to a sustainable environment, but of a foundational right that would, where necessary, take precedence over other rights. Since the exercise of virtually all other rights is contingent upon a sustainable environment this seems logical, but is clearly contentious because it implies that a right to a sustainable environment could trump other rights under certain circumstances.[57] The idea that some human rights are core, fundamental or more basic than others is itself problematic because it implies a hierarchy at odds with the assertion that all rights are equal and indivisible,[58] and opens up the possibility of hegemonic

[54] The Council commissioned a detailed study by the Office of the High Commissioner on Human Rights on the relationship between human rights and climate change (Human Rights Council, A/HRC/7/L.21, 20 March 2008).

[55] The EU, which has sought to take the lead on climate change, decided not to include the right to a healthy environment in the 1999 Treaty of Amsterdam despite numerous calls from environmental groups to do so.

[56] Malgosia Fitzmaurice, 'The Right of the Child to a Clean Environment', *Southern Illinois University Law Journal*, 23 (1999), 612–13.

[57] No right is absolute and it is not uncommon for rights to conflict.

[58] 'All human rights are universal, indivisible and interdependent and interrelated.' Article 5, Vienna Declaration, World Conference on Human Rights, Vienna, 14–25 June 1993, U.N. Doc. A/CONF.157/24 (Part I).

determination of the relative weight of rights. In the dominant discourse this has resulted in the alleged superiority of civil and political rights over social, economic and cultural rights. Climate change requires a different approach.

A universal environmental right cannot emerge as long as the West privileges individual rights over group rights and solidarity or third generation rights, which must be made fully justiciable.[59] Non-state actors, especially transnational corporations, must be brought fully within the ambit of human rights law as duty bearers. Indeed, it may be necessary to go further in two ways: first, by developing a new category of species right which transcends traditional categorisations, highlights the truly universal nature of the threat, and which we hold not simply as individual human beings but rather by virtue of our membership of the species *homo sapiens*; and, second, by reconceptualising legal subjectivity in ways that bear no direct relationship to the category of human at all. Intergenerational equity dictates that we must consider the rights of the unborn, the not yet human and the possibly never to become human.[60] Depledge and Carlane argue that:

> In the vein of the customary international law principle of 'good neighbourliness' and the English common law notion of 'breach of peace', the creation of a collective right based on the notion of common concern could bestow the global community as a whole with a duty of protection for the global commons – here, the atmosphere – that any member – or, at least, any sovereign state – could enforce *ex post*, or *ex ante* if there were reasonable grounds for supposing that a breach has been or is about to be committed or renewed. This type of right is unprecedented and raises valid questions of standing, harm, causation and redressability … [T]he trans-boundary, inter-generational and cross-sectoral nature of climate change creates a strong case for developing a new category of right(s) that recognises that individual human rights are intrinsically tied to the health of the global commons.[61]

[59] See Lee, 'Legal Theory', 296, who notes that the disasters in Bhopal and Ogoniland suggest that it may be more productive to assert a right to a healthy environment on a group rather than an individual basis.

[60] See *Minors Oposa* v. *Department of Environment and Natural Resources*, I.L.M., 33 (1994), 173, in which the plaintiffs relied on the right in the Philippine Constitution 'to a balanced and healthy ecology in accordance with the rhythm and harmony of nature' to argue that deforestation was causing environmental degradation. Uniquely, the claim was made on behalf of the plaintiff's generation and 'generations yet unborn', at 180.

[61] Depledge and Carlane, 'Sick of the Weather', 238–9

The scale and urgency of the threat means that consideration must be given to making climate change the foundational organising principle of the global juridico-political order and, therefore, giving it an unprecedented primacy in human rights and humanitarian law tantamount to a new global *Grundnorm*,[62] the yardstick by which all human activity is measured. The right to development is at once the best example of the forging of a meta-right and the problems involved in doing so.[63] Like the right to self-determination, it illustrates the possibilities of a subaltern jurisprudence that proceeds not from the transcendent categories of the dominant discourse but by identifying the violations and harms that such rights are designed to address.[64] Justice Weeramantry has argued in the International Court of Justice that protection of the environment is 'a vital part of contemporary human rights doctrine, for it is a *sine qua non* for numerous human rights such as the right to health and the right to life itself ... damage to the environment can impair and undermine all the human rights spoken of in the Universal Declaration and other human rights instruments'.[65] He went further in his dissenting opinion in the *Nuclear Weapons* case, arguing that certain principles of environmental law are now customary international law independent of treaty provisions.[66] The implications are profound.

[62] The concept of a basic or grounding norm is borrowed from Hans Kelsen, *Pure Theory of Law* (Berkeley, CA: University of California Press, 1967). Hannah Arendt's demand for a right to rights underpinned by a new principle of humanity (*The Origins of Totalitarianism* (London: George Allen & Unwin, 1967), 295–6) is echoed in Santos' call for a *jus humanitatis* which 'expresses the aspiration to a form of government of natural or cultural resources which, given their extreme importance for the sustainability and quality of life on earth, must be considered as globally owned and managed in the interests of humanity as a whole, both present and future' (Freeman, *Human Rights*, 302).

[63] From the voluminous literature, see, for example, Philip Alston, 'Revitalizing United Nations Work on Human Rights and Development', *Melbourne University Law Review*, 18 (1991), 216, and Jack Donnelly, 'In Search of the Unicorn: The Jurisprudence and Politics of the Right to Development', *California Western International Law Journal*, 15 (1985), 473. The meta-right that is most necessary is against the ravages of capitalism, which systematically reproduces impoverishment and underdevelopment – although carbon emissions are not, of course, peculiar to this mode of production.

[64] Although a right to a sustainable environment must reflect the colonial history of global warming and its differential impact on the global South, climate change will make victims of all of us even if, as ever, the wealthy will have greater access to resources for adaptation and mitigation.

[65] Unjoined separate opinion in the *Case Concerning the Gabcikovo-Nagymaros Project (Slovakia–Hungary)*, Judgment of 25 September 1997, B. ICJ/6929210707575.

[66] ICJ, *Legality of the Threat or Use of Nuclear Weapons*, Advisory Opinion of 8 July 1966, ICJ Reports.

The International Monetary Fund (IMF) and the World Bank would be precluded from imposing conditionalities which reduce the capacity of developing states to adapt and mitigate, and the World Trade Organization (WTO) subordinated to a world environment organisation with – in yet another paradox – similarly extensive powers.[67] Alston and Petersmann engaged in a lively debate about the desirability of requiring the WTO's Appellate Body to take human rights into consideration.[68] It is hard to disagree with Alston's observation that the WTO:

> is an institution which is dominated by producers, and in which economic, social, cultural, political and various other interests of a great many people are not, in practice, represented. Its institutional structure, its processes and the outcomes it sanctions are far from what would be required of a body to which significant human rights authority could be entrusted.[69]

Although addressing climate change is a question of a different order to that of trade, the parallel is instructive.

Last, but not least, it is necessary to clearly define the content, holders and duty-bearers of any such meta-right, including: (i) deciding whether it should be framed in positive terms as the right to a sustainable environment, negatively as a right against global warming, or a combination of the two; and (ii) what acts would constitute a violation. It should not be limited to climate change but encompass all polluting activities as well as the preservation of global commons and the Earth's biodiversity.[70]

67 See, for example, Bharat Desai, 'UNEP: A Global Environmental Authority?', *Environmental Law and Policy*, 36 (2006), 3–4. A strong enforcement institution is essential to the success of the Kyoto process and, like the WTO, its rules and dispute settlement mechanism will have to take precedence over domestic, regional and even other international provisions. Ironically, this is one of the reasons that the WTO is so problematic.

68 See Philip Alston, 'Resisting the Merger and Acquisition of Human Rights by Trade Law: A Reply to Petersmann', *European Journal of International Law*, 13:4 (2002), 815 and Ernst-Ulrich Petersmann, 'Time for a United Nations "Global Compact" for Integrating Human Rights into the Law of Worldwide Organizations: Lessons from European Integration' and 'Taking Human Rights, Poverty and Empowerment of Individuals More Seriously: Rejoinder to Alston', both in *European Journal of International Law*, 13 (2002); see also Sol Picciotto, 'Private Rights vs Public Standards in the WTO', *Review of International Political Economy*, 10:3 (2003), 377. Another ominous parallel between the nascent climate regime and the WTO is the extent to which market fundamentalism is likely to supersede human rights and a viable regulatory regime in Kyoto II.

69 Alston, 'Resisting the Merger', 836.

70 The ECtHR has identified four components regulating the justiciability of a right: (i) the right must benefit recognised individuals; (ii) it must impose duties on a recognisable group of actors for the benefit of the rights holders; (iii) a causal link must exist between the right and the duties; and (iv) the duties must be capable of being recognised and enforced by the court (Lee, 'Legal Theory', 299).

Existing rights

The alternative to a meta-right is the innovative use of accepted rights, such as those to life or property.[71] The difficulties involved in doing so were demonstrated in 2005 when the Inuit Circumpolar Conference (a non-governmental organisation representing approximately 150,000 aboriginals in Alaska, Canada, Greenland and Russia) petitioned the Inter-American Commission on Human Rights for 'relief from human rights violations resulting from the impacts of global warming and climate change caused by acts and omissions of the United States', and the refusal of the Bush administration to alter policies which violated the American Declaration: *inter alia* the rights to life, residence and movement, health and well-being, the inviolability of the home and Inuit cultural and indigenous rights.[72] In the absence of a meta-right they were forced to disaggregate their claim into the assertion of these various rights. The Commission rejected the petition on the basis that it contained insufficient information to make a determination but permitted testimony on the relationship between climate change and human rights.[73] The absence of a globally enforceable regulatory regime was highlighted by the fact that the Court does not have the power to compel the United States even if it had reached a favourable decision. The Inuit sought to create a precedent by establishing a legal nexus between global warming and the violation of human rights along the lines of class suits against tobacco companies.

Another form of legal insurgency is possible within the existing human rights framework, namely, reclassifying carbon emissions resulting in global warming as a crime against humanity alongside torture and

[71] See Birnie and Boyle, 252.

[72] Petition to the Inter-American Commission on Human Rights Seeking Relief from Violations Resulting from Global Warming Caused by Acts and Omissions of the United States, submitted by Sheila Watt-Cloutier with the Support of the Inuit Circumpolar Conference on Behalf of All Inuit of the Arctic Regions of the United States and Canada, 7 December, 2005 (www.inuitcircumpolar.com/files/uploads/icc-files/FINALPetitionICC.pdf), 1. The rights in question are contained in Articles I, VIII, IX and XI of the Declaration. The Court and the Inter-American Commission are responsible for implementing the American Convention on Human Rights, which the Organization of American States (OAS) adopted in 1969 and which came into force in 1978. Individuals can file petitions with the Commission against states which, like the United States, have not ratified the Convention.

[73] Jessica Gordon, 'Inter-American Commission on Human Rights to Hold Hearing After Rejecting Inuit Petition on Climate Change', *Sustainable Development Law and Policy*, 7 (2007), 55.

genocide.[74] This would once again involve extending international legal subjectivity to non-state corporate actors, establishing legally acceptable forms for measuring violations and adequate forms of punishment.[75]

What is clear is that limping along under the current framework is not an option. Darfur is the first conflict in which climate change is an apparent contributory factor but it will not be the last. Competition for food and water will create masses of environmental refugees who do not qualify for asylum under the existing regime, which only recognises political refugees. Unless this matter is addressed millions will suffer the consequences of a problem for which they are not primarily responsible, especially in the global South, whose inhabitants have historically been the smallest emitters of carbon but who will suffer first and most from global warming.

Ultimately, dealing with climate change is a matter of politics. Human rights are always vulnerable to appropriation and depoliticisation by hegemonic forces and, although rights-based struggles are commonly counter-hegemonic, they tend toward the aporetic when they become ends in themselves rather than means towards more substantive justice ... or saving the planet.

Conclusion

The Brundtland Commission defined sustainable development as '[d]evelopment that meets the needs of the present without compromising the ability of future generations to meet their own needs' and, therefore, as a matter of intergenerational equity.[76] It argued that human laws

[74] Calls have been made for poverty (more accurately, impoverishment) to be regarded as a crime against humanity – by Thomas Pogge, for example (see Andreas Follesdal and Thomas Pogge (eds.), *Real World Justice: Grounds, Principles, Human Rights, and Social Institutions* (New York: Springer-Verlag, 2005). Support for climate change to be classified in this way received implicit support in the declaration by Jean Ziegler, the UN Special Rapporteur on the Right to Food, that the use of agricultural land for the production of biofuels is a crime against humanity (available at: http://news.bbc.co.uk/1/hi/world/americas/7065061.stm).

[75] Corporate liability poses difficult but not insurmountable problems. If science is able to set targets for states there should be relatively little difficulty in setting them for transnational corporations. Similarly, it is not beyond the wit of humanity to extend the jurisdiction of the International Criminal Court (assuming it does not collapse in incoherence) or grant a world environmental organisation similar powers over non-state actors.

[76] World Commission on Environment and Development (WCED), *Our Common Future* (Oxford University Press, 1987). 43.

'must be reformulated to keep human activities in harmony with the unchanging and universal laws of nature'.[77] However, a just solution to global warming also requires *intra*generational equity through an equitable redistribution of resources from the countries primarily responsible for the problem to those with inadequate access to the resources needed for adaptation and mitigation. Climate change is the sticking point at which repeated post-colonial demands for a fairer international order – such as the call in the 1970s for a New International Economic Order and the right to development – must finally and irrevocably be met. By threatening catastrophe, nature may finally provide the impetus for global justice. It is for this reason that climate change ironically provides an opportunity for the construction of a more rational and egalitarian global order based on legality, normativity and cooperation rather than unilateralism and sovereign exceptionalism.

All models of development are threatened by global warming, from the carbon-fuelled Western myth of endless economic growth and rising standards of living heedless of the environmental consequences to the PRC's authoritarian capitalism or Russia's so-called sovereign democracy. For capitalism, development is sustainable if it produces ongoing profits, but combating climate change is incompatible with turbo-capitalism and hyperglobalisation. It signals an end to business as usual, and to the ideology of developmentalism based on fossil-fuelled industrialisation.[78] It demands radical changes in attitudes, behaviours, cultures and, above all, economic activity. It will be hypocritical for the highly industrialised countries, whose environmentally degrading economic development caused the problem, to demand that developing economies forego the opportunity for carbon-based development yet this is precisely what is required.[79] Accusations of neocolonialism will

[77] WCED, *Our Common Future*, 330.

[78] As Westernisation, modernisation, structural adjustment and good governance, developmentalism became a dirty word to its victims. See Arturo Escobar, *Encountering Development: The Making and Unmaking of the Third World* (Princeton University Press, 1995). The problem is not industrialisation *per se*, but the pursuit of rampant growth based on the use of fossil fuels. On the need to rethink the spaces of development, see Sam Adelman, 'Between the Scylla of Sovereignty and the Charybdis of Human Rights: The Pitfalls of Development in Pursuit of Justice', *Human Rights and International Legal Discourse (HR&ILD)*, 1 (2008), 2.

[79] In terms of climate change, imitating the West is the worst possible path – apart, that is, from statist models of development such as those of contemporary PRC and the former Soviet bloc.

be justified and can be addressed only by tackling the inequities in the global political economy.

Climate change has both potentially progressive and deeply unpalatable implications. The hard won subaltern human rights to self-determination and sovereign control over natural resources may have to be reconceptualised. As the common heritage of mankind, it is no longer tenable to regard carbon sinks as sovereign property, but rather as global commons placed in the trusteeship of the state in which they are sited for the common benefit of humanity. Despite the fact that it does not assert the right to a healthy or sustainable environment, the right to development provides the most coherent link between climate change and human rights.[80] It certainly never included the right to endless unsustainable economic growth, but it is imprecise. Similarly, the right to self-determination cannot be construed as a right against the rest of humanity. Things have come to a pretty pass when, confronted with an unprecedented risk, we are forced to consider contradictions such as resorting to the institutional structures of the WTO or the legacy of the League of Nations for possible solutions to climate change.

James Lovelock, who was among the first to identify the threat of climate change, believes that we have already passed the point of no return and that climate change is already irreversible. Asked what he thinks we should do, his response was: 'Enjoy life while you can. Because if you're lucky, it's going to be 20 years before it hits the fan.'[81] We must hope and act as if he is wrong but according to the precautionary principle and the worst case scenario, a global temperature rise of 4°C or more is more likely than not unless we take urgent action against the sovereigns who fiddle while the earth literally begins to burn. In Lynas' words, 'the question now is whether humanity can summon up the courage and foresight to save itself, or whether business as usual – on climate policy as much as economics – will condemn us all to climatic oblivion'.[82]

80 Declaration on the Right to Development, G.A. Res. 41/128, UNGAOR (1986).

81 *The Guardian*, 1 March 2008 (available at: www.guardian.co.uk/theguardian/2008/mar/01/scienceofclimatechange.climatechange).

82 Mark Lynas, 'Climate Change is inevitable. We can Only Avert Oblivion', *The Guardian*, 12 June 2008 (available at: www.guardian.co.uk/commentisfree/2008/jun/12/climatechange.scienceofclimatechange).

PART II

Priorities, risks and inequities in global responses

6

The Kyoto Protocol and vulnerability: human rights and equity dimensions

PHILIPPE CULLET*

Given the broad-based nature of climate change, the focus of law and policy has never been exclusively environmental. Economic and trade issues have, for instance, played a key role from the outset. While environmental and economic considerations have been central to the climate change legal regime, the same cannot be said for its human rights aspects.

The existing regime provides a number of entry points for the consideration of human rights. The notion of vulnerability is an effective starting point since it has been an important component of the regime since its inception.[1] This important concept must, however, be given more specific content if it is to be effective in shaping the climate change regime in the future. Vulnerability applies both at the level of states and people. Concerning states, differential treatment already provides the conceptual basis for addressing vulnerability. However, the framework needs to be thought afresh to make it more effective at capturing the varying vulnerabilities of different developing countries as well as to take into account changed circumstances since the adoption of the Climate Change Convention (UNFCCC). Concerning people, the climate change regime needs to move beyond its traditional international environmental law model to encompass consideration of the specific vulnerabilities of individuals and communities. This is best achieved through the language of human rights that has already been widely discussed in environmental law contexts.

* Reader in Environmental Law, School of Oriental and African Studies, University of London and Programme Director, International Environmental Law Research Centre. Contact: pcullet@soas.ac.uk.

1 Article 3(2) of the Framework Convention on Climate Change, New York, 9 May 1992 (hereafter Climate Change Convention). See Jon Barnett, Chapter 9, below.

This chapter first examines the broad context within which human rights can be examined in the climate change regime. It then focuses on equity, one of the core concepts of the existing regime that provides direct and indirect links with human rights. It examines two dimensions of equity, first, in the context of emission reduction commitments and, secondly, in the context of the Kyoto mechanisms. The chapter then considers ways in which vulnerability could be given a much more central role in the future and examines a series of issues that concern the vulnerability of both states and individuals.

Human rights, vulnerability and climate change

Despite numerous links, human rights have not been a significant dimension of climate change policy debates to date. The link between greenhouse gas (GHG) emissions and economic growth has directed debate toward economics, trade and finance. The addition of a human rights dimension has the potential to completely change the way in which law and policy are conceived in this area. Indeed, the human rights consequences of climate change are potentially so severe that, if taken seriously, they must prevail over economic and related considerations. Placing human rights at the centre of law and policy on climate change is a precondition for ensuring the legitimacy of climate change law and ensuring that measures taken on environmental grounds do not have negative human rights consequences.

Human rights concerns arise in the context of both mitigation and adaptation. Mitigation issues arise for developing countries with regard to taking on emission stabilisation or reduction commitments. Indeed, commitments are justifiable only if their consequences are completely offset for the majority of the poor. This is a direct consequence of the principle in human rights law that while countries can take progressive measures to realise socio-economic rights, they cannot backtrack.[2] Therefore, climate change commitments should not lead to any reduction in measures currently taken to progressively realise human rights. Thus, if steps were undertaken to reduce GHG emissions in the generation of electricity, they must be accompanied by measures to increase access to electricity for those who do not have access at present. This may require a reduction in consumption from wealthier individuals

[2] Article 2(1) of the International Covenant on Economic, Social and Cultural Rights, New York, 16 December 1966.

and economic actors or the installation of alternative, CO_2-free sources of electricity in villages.

Conversely, the realisation of the human rights to life, health, food, water and environment for the majority of the poor should be put at the centre of climate change policies. In other words, any shift away from a carbon-based economy must be conceived with the priority of realisation of human rights in mind.

In the context of adaptation, human rights consequences are easier to identify since there is an immediate connection between ongoing climate change-related damages and the realisation of human rights. Again, since the poor are the most vulnerable to climate change, they are also the most affected by ongoing damages. Thus, food shortages and floods induced by climate change invariably affect the poor first and need to be given priority.

The climate change regime and equity

The notion of vulnerability, a central element of the climate change regime, emphasises the fact that countries and people are not similarly placed when it comes to making choices that influence their contribution to climate change or when it comes to the impacts of climate change. Vulnerability informs the development of differential treatment, the more specific legal measure on which the climate change regime is based. It is also directly related to human rights since people's vulnerability to climate change is a primary cause of the threats posed by climate change to the realisation of human rights.

Differential treatment and emission reduction commitments under the Kyoto Protocol

The international legal regime is premised on the neutrality of a system based on the formal legal equality of all states. As a consequence, rules are usually deemed just if they apply to all without discrimination.[3] Existing economic or other inequalities are in principle not taken into account. The notion of differential treatment refers to instances where, because of pervasive differences or inequalities among states, formal legal equality and reciprocity are sidelined to accommodate extraneous

[3] See, for example, H. Peyton Young, *Equity in Theory and Practice* (Princeton University Press, 1994).

factors.[4] These include divergences in levels of economic development, different contributions to the creation of a problem or unequal capacities to tackle existing problems.

The legal entrenchment of 'differential treatment' is intended to progressively modify the existing status quo between states without seeking a completely new international framework. The principle of progressive realisation of socio-economic rights, whereby states are required to fulfil these rights only within the parameters of their resources, is an example of differential treatment. There is thus a close link between the progressive changes that differential treatment seeks to achieve at the international level and the progressive realisation of socio-economic human rights.

Differential treatment has a central role in the climate change legal regime.[5] The historical responsibility for causing climate change is clearly borne by a limited number of countries broadly corresponding to those now classified in UN terms as developed countries. In terms of current emissions per capita, responsibility falls on the same group of countries. Further, it is also these countries that have the greatest economic and technological capacity to take measures to mitigate and adapt to climate change.

This relatively clear baseline for addressing climate change through international legal measures provided the basis for states negotiating the UNFCCC to agree on the principle of common but differentiated responsibility. The Convention is thus premised on the principle that:

> Parties should protect the climate system for the benefit of present and future generations of humankind, based on equity and in accordance with their common but differentiated responsibilities and respective capabilities. Accordingly, the developed country Parties should take the lead in combating climate change and the adverse effects thereof.[6]

This principle was further developed in the context of the negotiations for the Kyoto Protocol, which led to the adoption of separate commitments for developed and developing countries.[7] The fact that only one

[4] See generally, P. Cullet, *Differential Treatment in International Environmental Law* (Aldershot: Ashgate, 2003).

[5] See, for example, L. Rajamani, 'The Nature, Promise and Limits of Differential Treatment in the Climate Change Regime', *Yearbook of International Environmental Law*, 16 (2007).

[6] Article 3(1) of the Climate Change Convention, footnote 1, above. See also, Article 10(1) of the Kyoto Protocol to the United Nations Framework Convention on Climate Change, Kyoto, 11 December 1997 (hereafter Kyoto Protocol).

[7] Article 3, Kyoto Protocol, footnote 6, above.

group of countries takes on emission reduction commitments based on the principle of common but differentiated responsibilities (the 'CBDR principle') is noteworthy because few international treaties have gone so far in the realisation of the implementation of differentiation. While only developed countries take on emission reduction commitments, this does not mean that developing countries are doing nothing to address climate change under the existing legal regime.[8]

Differential treatment constitutes an acknowledgement that the existing status quo is not conducive to achieving the basic fairness and solidarity goals that international law sets for itself. It thus suggests progressive changes through a system of specific rules that contribute to change in the existing pattern of inequality and bring about substantive equity in inter-state relations. Differential treatment is also relevant in a human rights context because an equitable international legal order cannot be conceived exclusively in terms of the relations between states. Indeed, equity between states and between individuals are but different sides of the same coin. This observation is reinforced by the fact that the grip that states have traditionally had over international legal relations is slowly being eroded with the increasing importance of a variety of other actors. Neither equity nor human rights require overnight changes but both require progressive change in a definite direction. In both cases, the focus is on the improvement of the situation of the poorest or most disadvantaged.

Equity in the context of flexibility mechanisms

The relatively progressive nature of the Kyoto Protocol from the point of view of emission reduction commitments in terms of equity was not achieved without some compromises. One of the major concessions that was made in the process of negotiating the UNFCCC, and more particularly the Kyoto Protocol, was the introduction of 'flexibility' under the guise of what are now known as Kyoto mechanisms.

Flexibility includes two distinct components. First, it provides an escape clause for developed countries that allows them not to implement the commitments they have taken on at home. This is novel in international law because countries are supposed to implement commitments

[8] See, for example, P. Cullet, 'Equity and Flexibility Mechanisms in the Climate Change Regime – Conceptual and Practical Issues', *Review of European Community and International Environmental Law*, 8 (1999), 168.

they take by themselves. The rationale for allowing this flexibility is that what matters most is the global environment. Since emission reduction or emission avoidance has the same impact anywhere on the planet, flexibility provides a way of achieving emission reduction commitments through the cheapest emission reduction opportunities available anywhere in the world.[9] As part of the difficult process for the adoption of the Kyoto Protocol, flexibility was even seen as a potential alternative to the multilateral negotiations able, in particular, to bring the United States on board.[10] The clean development mechanism (CDM) met with approval from developing countries because it was seen as an instrument that would ensure additional foreign direct investment in host countries. As defined under Article 12 of the Protocol, the CDM seeks to facilitate joint emission reduction projects between Annex I countries and developing countries. It specifically emphasises the fact that projects must assist developing countries in realising sustainable development.[11] From the point of view of Annex I countries, its main interest is that certified emission reductions (CERs) accruing from CDM projects are credited to them so that they can use them as an additional means to comply with their commitments.[12]

Secondly, flexibility is novel because it gives much increased prominence to the private sector in the implementation of an international treaty. There is no necessary congruence between the 'outsourcing' of compliance and the private sector since the former could happen without the latter. Yet, in the context of the Kyoto Protocol, the two are intrinsically linked. This novel dimension calls for new safeguards to ensure that the focus on finding the cheapest emission reduction opportunities and the involvement of private sector actors in doing so do not compromise environmental and social objectives.

Equity and the clean development mechanism

The Kyoto mechanisms, and in particular the CDM, raise a number of questions concerning equity. First, the focus on finding the cheapest emission reduction opportunities raises questions concerning the

[9] See, for example, A. G. Hanafi, 'Joint Implementation: Legal and Institutional Issues for an Effective International Program to Combat Climate Change', *Harvard Environmental Law Review*, 22 (1998), 441.

[10] See, example, T. C. Heller, 'Environmental Realpolitik – Joint Implementation and Climate Change', *Indiana Journal of Global Legal Studies*, 3 (1996), 295.

[11] Article 12 of the Kyoto Protocol, footnote 6, above.

[12] Article 3(12) of the Kyoto Protocol, footnote 6, above.

justification of the CDM. The CDM was meant to be a subsidiary mechanism in achieving the commitments that developed countries had taken up. The underlying logic was that developed countries would be reducing their emissions and that a part of that reduction would come from CDM projects. However, between 1990 and 2005 emissions have significantly risen in many countries with commitments. Some of the worst increases are in Spain (61 per cent) and Portugal (57 per cent), but countries in other regions of the world, such as New Zealand (41 per cent) and Australia (37 per cent), are not far behind.[13] In fact, the list of countries that have actually reduced their emissions numbers only six and includes only two of the G8 countries, Germany (– 15 per cent) and the United Kingdom (– 6 per cent).[14] The very logic of the CDM is thus undermined because it can (in principle) be used by countries with commitments as an authorised loophole to show formal compliance with their international obligations. Countries with commitments can safely rely on the fact that Article 12 of the Protocol, unlike Articles 6 and 17 and the decision setting up Activities Implemented Jointly under the Climate Change Convention in 1995, does not even mention that CDM projects must be supplemental to domestic action.[15] However, it is not legitimate to use the CDM merely to achieve formal compliance, even if it does not go against the letter of the regime. Indeed, if developing countries signed up to the CDM in a spirit of global solidarity and partnership to contribute 'to the ultimate objective of the Convention',[16] this was part of a balance based on the common but differentiated responsibilities (CBDR) principle which specifically implies that developed countries take the lead in mitigating climate change rather than rely on cheap emission

[13] In the case of Spain and Portugal, while under the EU 'bubble', they are allowed, respectively, a 15 per cent and 27 per cent increase, both are still much above these redistributed commitments. For the intra-EU allocation, see, for example, Assigned Amount Report of the European Union, Report from the Commission, COM(2006) 799 final (2006).

[14] See, for example, 'A Joke on the World', *Down to Earth* 16/14 (2007) 30, 32. In the case of Germany and the United Kingdom, these reductions are also much less than what they have to achieve under the EU bubble, respectively, a 21 per cent and 12.5 per cent reduction. See Report from the Commission, footnote 13, above.

[15] See Articles 6 and 17 of the Kyoto Protocol, footnote 6, above and Decision 5/CP.1, Activities Implemented Jointly Under the Pilot Phase, in *Report of the Conference of the Parties on its First Session*, Framework Convention on Climate Change, Conference of the Parties, First Session, Berlin, 28 March–7 April 1995, UN Doc. FCCC/CP/1995/7/Add.1.

[16] Article 12(2) of the Kyoto Protocol, footnote 6 above.

reduction opportunities in developing countries. The necessity for massive investments in renewable energies, such as solar energy, constitutes one of the ways in which developed countries can demonstrate their leadership. While the CDM could be one of the vehicles used for such changes, this has not happened because incentives for the same have not been provided.

Secondly, the CDM has been conceived from the point of view of short-term mitigation gains. While Article 12 of the Protocol provides a basis for reducing the overall cost of compliance with emission reduction commitments, it does nothing to steer the world economy towards a low or zero carbon economy. This is due to the fact that the CDM, in effect, provides an escape route for developed countries unwilling to implement drastic energy policy changes. As a result, significant investments in new or existing alternative technologies are not being undertaken.[17]

Additionally, the CDM does not include a framework that would ensure that projects are prioritised in accordance with their impacts on the poor and vulnerable and the environment in general. This is of concern because there are many climate change friendly activities that are neither environmentally nor socially progressive and can thus have negative impacts on the realisation of human rights. One example is that of large dams.[18] By the mid-1990s, it had become widely recognised that large dams had significant social and environmental costs that required at the very least reconsidering their place in the context of the drive towards making development more sustainable.[19] In the course of the present decade, the difficult learning curve of the previous two decades seems to have all but evaporated. Big dams have found a new justification because they are a climate change friendly source of electricity.[20] Yet, this does not answer any of the questions raised earlier concerning the justifications for big dams from a social or

[17] See, for example, G. Eklöf, *Broken Illusions – CDM in Practice* (Stockholm: Swedish Society for Nature Conservation, 2006), 19.

[18] On dams and the CDM, see, for example, Lori Pottinger, *Bad Deal for the Planet: Why Carbon Offsets Aren't Working ... And How to Create a Fair Global Climate Accord – Dams, Rivers and People Report 2008* (Berkeley, CA: International Rivers, 2008).

[19] See, for example, World Bank, Operations Evaluation Department, Learning from Narmada (Précis No. 88, 1995) noting that '[t]he broad lesson is that the social dimensions of civil works projects need much more attention from both the Bank and its borrower governments'.

[20] See, for example, World Bank, Water Resources Sector Strategy – Strategic Directions for World Bank Engagement (Washington, DC: World Bank, 2004), 21 arguing that

environmental point of view. In other words, while big dams may be better than coal-fired power plants from a GHG emissions perspective, this is insufficient to justify them.[21]

Thirdly, the CDM has perverse long-term side-effects for developing countries. Indeed, the search for the cheapest possible emission reduction opportunities means that developing countries are exhausting these options for the benefit of developed countries' compliance with their own commitments. Such options will no longer exist once developing countries take on commitments, something that is unavoidable in the long term from a global environmental point of view. In the case of land-use projects, other issues may arise in the future. Where the positive climate change impact of a project is premised on the potential of timber to store carbon, two scenarios may arise. If the host country does not ensure that carbon absorbed under CDM projects is kept stored, the question may arise as to whether these emissions are to be attributed to the host country. If they were, this would constitute a double loss for the country affected. If the host country ensures that timber is maintained in the form of forest the issue that arises is the lack of recognition of the trade-off that this long-term land use for climate change purposes implies from the point of view of development opportunities for local people.[22] There are also direct implications in terms of impacts on livelihoods and the realisation of human rights.

Fourthly, while the CDM can theoretically be an instrument of the public as well as the private sector, in practice it has largely been conceived as an instrument used by the private sector. This novel way to implement an international law agreement calls for specific safeguards to ensure that all the environmental and social conditions are complied with. The lack of an international body capable of such enforcement – the CDM Executive Board does not have such powers – implies that each country has to fulfil this at the national level. Additionally, this also means that there is no international supervision of the extent to which sustainable development is promoted through the CDM and vulnerability addressed. This is problematic for two reasons. First, the poor and vulnerable who would benefit from a levy on CDM projects

'[h]ydropower can, in principle, make a major contribution to reducing the greenhouse gas intensity of energy production'.

21 See, for example, P. Cullet, 'The Sardar Sarovar Dam Project: An Overview', in P. Cullet (ed.), *Sardar Sarovar Dam Project: Selected Documents* (Aldershot: Ashgate, 2007), 1.

22 See, for example, P. Cullet and P. Kameri-Mbote, 'Joint Implementation and Forestry Projects – Conceptual and Operational Fallacies', *International Affairs*, 74 (1998), 393.

for sustainable development activities have little capacity to influence a process that is led by governments and private sector interests. Secondly, the international framework guiding the CDM fails to provide effective guidance on technology choice and project focus.[23] The extent of the CDM's contribution to sustainable development and to long-term energy policy changes is thus left to individual host countries' decisions. The People's Republic of China (PRC) has, for instance, decided to tax different types of projects differently. Thus, for HFC-23 (trifluoromethane) and PFC (perfluorocarbon) projects – extremely potent GHGs that are relatively easy to eliminate – the government takes 65 per cent of the revenue generated by the transfer of CERs, while for energy efficiency improvement and renewable energy projects it only takes 2 per cent.[24] Similar measures must be adopted at the international level because governments may have their own reasons to favour their private sector industry over sustainable development and fail either to differentiate between types of projects or to tax projects for investment in measures favouring the most vulnerable.

Where there are no rules for distributing the revenue generated by CDM projects between different actors, cheap mitigating opportunities are used by private sector actors for their own individual benefits, as in the case of any other commercial transaction. This is problematic because without investments toward a low carbon economy it is citizens who will suffer the negative consequences of any emission stabilisation or reduction commitments that will be taken in the near- or medium-term future. In other words, private sector actors make money on account of climate change but since the projects for which CERs are obtained are not guided by a broader policy to reorient the economy toward a low carbon economy, the gains for broader society in either environmental, social or financial terms are negligible. Similarly, the CDM has the unfortunate consequence of pushing host countries to

[23] In fact, even on the use of nuclear energy projects under the CDM, nuclear facilities projects are not barred but Annex I countries are to refrain from using the certified emission reductions generated. See Preamble, Decision 17/CP.7, Modalities and Procedures for a Clean Development Mechanism as Defined in Article 12 of the Kyoto Protocol, UN Doc. FCCC/CP/2001/13/Add.2 (2001); and Decision 3/CMP.1, Modalities and Procedures for a Clean Development Mechanism, as Defined in Article 12 of the Kyoto Protocol, UN Doc. FCCC/KP/CMP/2005/8/Add.1 (2005).

[24] Article 24, China – Office of National Coordination Committee on Climate Change, Measures for Operation and Management of Clean Development Mechanism Projects in China, 2005.

delay climate change measures because any measure required by law makes it then nearly impossible for a project to obtain CDM approval.[25]

Toward a central role for vulnerability

Negotiations for new measures to address climate change after 2012 are ongoing. Yet, the framework within which this is taking place is inadequate. As a result, a number of elements need to be either rethought or given new content. This section focuses on some of the many issues that need rethinking in the continuous search for an effective climate change regime. It highlights the need for a new understanding of differentiation. It also emphasises the primacy of human rights and vulnerability as a necessary foundation of further measures on climate change. Further, it argues that air and the atmosphere should be recognised as a common heritage to ensure that the benefits of climate mitigation are not appropriated by private actors, but rather ploughed back into renewable energy or other measures that are sustainable and primarily benefit the most vulnerable. Finally, it argues that a new basis for allocating entitlements must be found to ensure that the poor and vulnerable are not indirectly dispossessed of something that is in essence humankind's primary survival resource.

Toward a new understanding of differential treatment for future emission reduction commitments

The basis for differentiation remains as strong as it was at the time of the negotiation of the UNFCCC. Indeed, on the whole it is the same small number of countries that contribute most to climate change in per capita terms. At the same time, there is still a majority of countries whose contribution to climate change is negligible, starting with all least developed countries. These countries are also the most vulnerable to the impacts of climate change.

Yet, rapid economic development in some parts of the world over the past decade has altered the balance of overall contributions that countries make. In particular, the share of big developing countries like India and the PRC in global GHG emissions has increased since 1990. This is due to the fact their emissions have been growing by at

[25] Eklöf, footnote 17, above at 19–20.

least 4 per cent per year, faster than any other region of the world.[26] Since the climate change legal regime is primarily about achieving a global environmental benefit, any substantial increase in emissions is to be taken into account, wherever the additional emissions are generated.

The case of India

The position of India is particularly noteworthy with regard to the need to rethink differential treatment for subsequent commitment periods. On the one hand, India remains without any possible doubt a developing country. India's position in the ranking of the Human Development Index at number 128, just ahead of several least developed countries like Laos and Cambodia, reflects the reality that the majority of Indians experience. On the other hand, India has experienced fast economic growth in recent years. Additionally, it has increasingly sought to flex its political muscle on the world stage by seeking recognition as a major power.

In terms of climate change, as in many other dimensions, India is today two countries. The India that shines has standards of living that often match those of developed countries with a concomitant negative environmental impact in terms of climate change. The India of the majority of the population has made little progress since 1990. Thus, 77 per cent of the population has an income of less than $2 a day.[27] In fact, while there has been some reduction in the percentage of people in 'extreme poverty', the overall number of poor and vulnerable people has increased from 733 to 836 million between 1993–4 and 2004–5.[28]

From an equity standpoint, India must be analysed from these two different perspectives. On the one hand, from the perspective of climate change, an international problem requiring the collaboration of all states to address it, India has a duty to contribute to efforts to mitigate climate change. In fact, India is already contributing to climate change mitigation through its involvement in the CDM like all other developing

[26] Central Pollution Control Board, Newsletter (October 2002). Available at: www.cpcb.nic.in/News%20Letters/Archives/Climate%20Change/ch9-CC.html.

[27] National Commission for Enterprises in the Unorganised Sector, Report on Conditions of Work and Promotion of Livelihoods in the Unorganised Sector (2007), 6.

[28] National Commission for Enterprises, at 7.

countries. Yet, progressively, more needs to be done. Additionally, from the perspective of a big country that shows no signs of overall vulnerability, it is increasingly difficult to justify that India should hide behind the veil of its 'developing country' status since it has little in common with countries like Malawi or the Maldives in terms of vulnerability.

On the other hand, the overwhelming majority of India's population is as vulnerable as the average inhabitants of other developing countries, including in many cases people in least developed countries. India's rank of 94 on the Global Hunger Index (out of 118 countries listed) reflects this other reality.[29] Equity, as realised through differential treatment in international law, cannot justify the imposition of emission reduction or stabilisation commitments in a way that would increase the vulnerability of the already vulnerable majority of the population. This would go against the idea of progressive realisation of fundamental rights.

Revisiting differential treatment

Differential treatment for the future needs to be rethought, since the legal regime must reflect the changes that have taken place since the early 1990s in the position of some developing countries, must reflect the increasingly central role that climate change plays among environmental issues and must reflect the fact that climate change is much more than an environmental and economic issue but also a core human rights issue.

First, it is increasingly difficult to attribute emissions on the basis of the fiction of legal equality of states alone. On the one hand, the direct or indirect contribution of each individual country varies, according to wealth and other factors. On the other hand, questions arise concerning the responsibility of a country for all emissions arising from its territory. The case of special economic zones (SEZ) is a telling example. Where companies invest under particularly beneficial conditions and where they export all the products they manufacture, equity requires that emissions be at least partly allocated to the actors that take advantage of the lax legal regimes that increase profits on products that are marketed in wealthier parts of the world. Beyond SEZs, a number of other

[29] International Food Policy Research Institute, The Challenge of Hunger 2007 – Global Hunger Index (2007).

situations may call for similar treatment, for instance, where deforestation is undertaken to use the cleared land to produce cash crops that are mostly exported. In this case, it is necessary to find new ways to allocate responsibility for climate change. These should take into account not only countries' contributions but also that of actors that directly benefit in economic terms from GHG emitting activities. This would constitute a useful application of the polluter pays principle. The issue can not, therefore, be reduced to a simple dichotomy between taking and not taking commitments. It is also not a simple case of whether developing countries as a block (the G77 group) should or should not take on commitments under the Kyoto Protocol.

Secondly, differential treatment is not in itself an instrument that seeks to favour developing countries. It so happens that under most existing treaties, differentiation has been approved based on countries' classification as developed or developing. Yet, since there is no generally agreed definition of which country is a developing country and since the decision is often left to self-identification, this is not in itself an effective guide. Further, the simple division into two groups is only for convenience's sake but is increasingly itself inequitable since it does not take into account the complete lack of congruence between the respective situations of Malawi and South Korea or Vanuatu and India. The real purpose of differential treatment, which is to foster substantive equality and a partnership among all countries in solving problems of a global nature, cannot be equated with the division of the world between developed and developing countries. There are thus a number of situations where developing countries should either be individually targeted for preferences or at least clubbed in smaller groups so that, for example, small island states that are going to disappear as a side-effect of climate change would not be put in the same category as OPEC countries that have become much wealthier because of the growth of the global carbon economy.[30]

Thirdly, differential treatment goes beyond the granting of preferences based on differences in levels of economic development. In fact, differential treatment in environmental treaties seeks primarily to further the overall environmental goals of the agreement by fostering

[30] Least developed countries are, for instance, frequently targeted for preferential measures. See, for example, Articles 4(9) and 12(5), Climate Change Convention, footnote 1, above and Article 66 of the Agreement on Trade-Related Aspects of Intellectual Property Rights, Marrakech, 15 April 1994.

the participation of countries that may have little incentive to participate. Thus, in the case of climate change, developing countries as a whole would have had little incentive in 1992 to join a global legal regime to address a problem to which they had hardly contributed to.[31]

The implication is that differential treatment in the context of any subsequent commitment period under the Kyoto Protocol needs to be much more closely tailored to the overall environmental goals of the regime while providing a much needed equity angle. This means that differentiation must be an instrument that takes into account both the contribution of each country to the problem, its capacity to mitigate and adapt and the vulnerability of its population. In the case of a country like India, this also requires going beyond a simplistic decision on commitments versus no commitments. What differential treatment calls for is that big countries like India and the PRC, whose emissions grow faster than any other regions of the world, take up their responsibilities as members of the international community and more specifically as aspiring military and political global powers. At the same time, the focus of differential treatment on equity clearly bars the imposition of any commitment that would harm the majority of the vulnerable population of these countries. Mechanisms thus need to be devised at the international and national levels to ensure that the burden of any commitments falls exclusively on polluting industries, on the people whose lifestyle makes a significant contribution to climate change and on government to ensure that climate change friendly policies are implemented. In other words, commitments should go alongside new forms of international technology transfers and new forms of resource redistribution at the national level.

It is clear that countries like India cannot simply curb their economic growth in a bid to satisfy the North. These countries must, nevertheless, urgently reorient their growth and find alternative economic development paths. One of the possible solutions is to rely on technology transfers where the North provides the more environmentally friendly technologies it has already developed to ensure that economic growth in developing countries is not hampered by taking climate change friendly measures. This could include, for instance, wind and solar energy technologies. Another solution lies in focusing on renewable energy, something that can easily be fostered by reallocating resources

[31] See, for example, D. Bodansky, 'The United Nations Framework Convention on Climate Change: A Commentary', *Yale Journal of International Law*, 18 (1993), 451, 463.

away from carbon intensive energy sources. In other words, addressing climate change does not have to be a costly proposition in terms of economic growth. It may, in fact, provide an excellent opportunity to rethink failed economic development strategies. Thus, climate change does not provide a basis for promoting just any energy source that is not harmful from a climate change point of view. Current efforts to suggest that nuclear energy is an apt alternative to carbon-based energy do not take into account the fact that nuclear energy has no justification from an environmental point of view. Indeed, while the actual production of energy may be harmless in terms of GHG emissions, nuclear energy is unacceptable from the point of view of its other impacts, particularly because there is no environmentally acceptable solution to nuclear waste at present and because a number of side-effects of nuclear power generation on human health are either unknown or not in the public domain.[32]

With regard to resource redistribution, two main points can be made. First, one option may be for some developing countries like India and the PRC to take on commitments with a view to ensure that climate change is effectively averted. This would give a strong signal that the world cannot tolerate more emissions and that further economic development strategies need to be rethought throughout the world. The commitments taken by such countries benefit the global environment through climate change mitigation and reduced costs of climate change adaptation, and so the costs of these commitments should be borne in part or entirely by developed countries under the CBDR principle. Secondly, any form of compensation that is provided by developed to developing countries with commitments should be carefully targeted. Resources made available should be invested primarily in mitigation and adaptation measures for the poor since they are the most vulnerable and least able to adapt, as well as in measures that put the poor at the centre of any new economic development strategies. This is a matter of equity and human rights since both focus on the situation of the most disadvantaged. Together, this will ensure that differentiation contributes to global and local environmental benefits as well as to poverty alleviation and the realisation of human rights. This new framework is imperative to redirect climate change law toward being more environmentally friendly and more equitable.

[32] See, for example, A. Katz, 'Chernobyl: The Great Cover-up', *Le Monde Diplomatique* (April 2008). Available at: http://mondediplo.com/2008/04/14who.

Recognising air as a common heritage

Air has until recently been of little interest to lawyers, economists or policy makers. Indeed, while air is a basic element that allows us to survive, it was for all practical purposes beyond appropriation. This situation changed relatively quickly over the course of the twentieth century with the introduction of aviation, which led states to assert control over their airspace.[33] At the same time, the question of air pollution led to the realisation that while air may be beyond legal control, humankind was able to impact on air in various negative ways. Yet, a treaty like the Convention on Long-range Transboundary Air Pollution does not address the question of air pollution from the point of view of the right of states to pollute.[34] As a result, it proposes a series of measures to reduce air pollution without trying to ascribe entitlements or addressing the status of air or the atmosphere. Beyond airspace, which cannot be directly compared with air or the atmosphere, the only other dimension that states have addressed is that of outer space where the consensus is that it is a common heritage of humankind.[35]

In the context of the climate change regime, the international community has agreed that the climate and its adverse effects are a common concern of humankind.[36] This implies an acknowledgement that the climate can be addressed only through the common action of all states, but it does not indicate whether states or individuals are in a position to lay specific claims on air or on air pollution. The Kyoto Protocol does not address this issue directly either. However, the Protocol indirectly provides the most polluting nations on Earth specific polluting entitlements. In other words, while no legal claims to air or the atmosphere are staked by any state, an indirect appropriation takes place. This is problematic because science has clearly shown that the global sink that is the atmosphere can only absorb a limited amount of carbon. Above a certain limit, consequences which are extremely harmful will most likely take place. In other words, the polluting rights indirectly given to developed countries under the Kyoto Protocol constitute entitlements that affect all nations on Earth.[37]

[33] Article 1 of the Convention on International Civil Aviation, Chicago, 7 December 1944.

[34] Convention on Long-range Transboundary Air Pollution, Geneva, 13 November 1979.

[35] Article 11 of the Agreement Governing the Activities of States on the Moon and Other Celestial Bodies, New York, 18 December 1979.

[36] Preamble, Climate Change Convention, footnote 1 above.

[37] See, for example, L. Lohmann, 'Carbon Trading – A Critical Conversation on Climate Change, Privatisation and Power', *Development Dialogue*, 48 (2006), 1, 74ff.

The approach taken in the Kyoto Protocol is problematic. The starting point for regulating emissions is grandfathering, which indirectly rewards industries that have done least to cut back pollution before the adoption of the new regime. Grandfathering also rewards countries that industrialised early because their high level of pollution becomes the baseline against which reductions are debated. Countries that have lagged in industrial development suffer the double disadvantage under a grandfathering scheme of having lower levels of economic development and lower pollution levels that in turn entitle them to lower future polluting levels. Both equity and environmental concerns call for a different type of response to climate change. In terms of equity or environmental conservation, the shortcomings of grandfathering call for giving the climate change regime new bases. One of the starting points for a differently conceived regime is to rethink the legal status of air and the atmosphere.

The Kyoto Protocol is in principle a treaty focusing on an environmental problem. Yet, in reality because of the nature of the problem being addressed, the real focus has been on economic development and the impacts that addressing climate change will have on economic growth. The debate has thus been framed mostly as an economic development issue within the broader context of environmental quality. This is unfortunate because it sidelines increasingly important impacts of air pollution on human health and thus the realisation of the human right to health. More generally, the current regime fails to take into account the human impacts of air pollution and thereby fails to directly acknowledge that vulnerability is not just an issue in terms of the impacts of climate change but also in terms of the causes of climate change. For instance, the urban poor in developing countries are much more likely to be affected by air-related health issues than the middle classes.

Since air pollution cannot be regarded as being limited to a dichotomy between environmental quality and economic growth, the legal status of air must be conceived in a broader perspective. Given that there is only one atmosphere, it follows that it needs to be managed as such. Individual control over air is physically impossible and would go against the need for a global solution. Air, the atmosphere and the global climate should thus be seen as a common heritage of humankind that needs to be commonly conserved and managed. The most obvious starting point for developing this concept is the notion of common heritage developed in the context of the law of the sea.[38] Common heritage status implies first of all that no sovereign claims can be made on the area or resource covered.[39] It also prohibits unilateral appropriation and requires international cooperation in the exploitation of resources, for instance, by giving an international body the necessary authority.[40]

The introduction of common heritage status for air and the atmosphere would make a significant contribution to policy debates on the future climate change regime. Indeed, it would provide a new solid basis for rethinking the allocation of emission reduction commitments and for regulating the use of flexibility mechanisms according to priorities focused on differential treatment and vulnerability rather than in terms of economic efficiency and the indirect allocation of individual property rights over a global heritage.

Common heritage status would, for instance, lead to setting a new framework for the CDM. At present the CDM provides essentially economic benefits to project partners. The CDM policy framework itself does not indicate how these benefits should be used. As a result, they can be used simply to foster the partner's business. Since benefits accruing through CDM projects are linked to climate change mitigation, under a scheme where air is a common heritage, there is no reason for project partners to receive unconditional benefits. Indeed, there are a number of social and environmental priorities that must be addressed in the context of climate change. The resources raised in the name of climate change mitigation should thus be used for activities that specifically contribute to addressing the global heritage since no one should be able to acquire direct or indirect rights to pollute something which is vital for survival for all living things. The use of CDM proceeds to address issues related to the global good is even more important in a context where governments often claim that they have insufficient resources to implement effective environmental and social policies.

[38] See United Nations General Assembly, Declaration of Principle Governing the Sea-Bed and the Ocean Floor, and the Subsoil Thereof, Beyond the Limits of National Jurisdiction, GA Res. 2749 (XXV), 17 December 1970, Resolutions Adopted by the General Assembly During its 25th Session, 15 September–17 December 1970, GAOR 25th Sess., Supp.28 (A/8028).

[39] See, for example, E. Holmila, 'Common Heritage of Mankind in the Law of the Sea', *Acta Societatis Martensis*, 1 (2005), 187, 195.

[40] See Part XI of the United Nations Convention on the Law of the Sea, Montego Bay, 10 December 1982 and Agreement Relating to the Implementation of Part XI of the United Nations Convention on the Law of the Sea of 10 December 1982, New York, 28 July 1994.

Turning the air, the atmosphere and the global climate into a common heritage will no doubt be fiercely resisted by a number of actors who have and still benefit immensely from the absence of clear concepts determining who is entitled to 'use' air and 'pollute' the atmosphere. Yet, this is in fact but a small extension of the notion of public trust, a concept widely used.[41] Interestingly, the Indian Supreme Court has already declared more than a decade ago that air is a public trust in India.[42] The notion of public trust implies that the state has to act as a trustee on behalf of all individuals, must take a long-term view of its protection and must ensure socially equitable and environmentally sustainable access to and use of the resource.[43] It also implies that the state is not in a position to trade away or sell pollution rights or carbon credits in its role of trustee.[44] These safeguards include fostering the realisation of human rights and ensuring that no violations of existing protection levels take place, as well as the respect for environmental law in general and not just of climate change law.

Toward new forms of entitlements on air

The basis for today's climate change law is, on the whole, the grandfathering of existing emission patterns. In political terms, this can be easily explained since any other formula would affect existing polluters more than the economic actors or the countries that contribute less to climate change. Yet, this is an ineffective way to address climate change. Indeed, while a baseline determined by existing energy use puts the burden on developed countries and on polluting industries, it does not provide any compensation mechanism to non-industrialised countries and to people who have not benefited from the standards of living achieved while causing climate change.

[41] In the case of water, see, for example, for California, *National Audubon Society* v. *Department of Water and Power of the City of Los Angeles*, Supreme Court of California, 17 February 1983, 658 P.2d 709; and for South Africa, Section 3, National Water Act (1998).

[42] *M. C. Mehta* v. *Kamal Nath*, 1997 1 SCC 388. For a similar example in the United States, see Article 1(27), of the Pennsylvania Constitution.

[43] *M. C. Mehta* v. *Kamal Nath*. See also Shelton, Chapter 3, above.

[44] Compare D. Takacs, 'The Public Trust Doctrine, Environmental Human Rights, and the Future of Private Property', *New York University Environmental Law Journal*, 16 (2008), 711, 733.

As long as existing levels of economic development and existing pollution patterns constitute the basis for regulation, climate change law will largely reflect the priorities of the economically and politically more powerful states. An equitable and effective climate change regime needs to be based on a different paradigm that takes into account a broader variety of factors. The starting point is the common benefit that a healthy global environment represents for the whole of humankind and for life on Earth in general. Basic principles of environmental law, such as precaution and equity, are thus at the centre of efforts to define entitlements. Today, environmental protection is conceived by all states as encompassing human rights, social and economic aspects. This implies that it is not only the realisation of the right to a clean environment recognised in nearly 120 countries that is at stake but also the realisation of all human rights.

This broad framework leads to the development of a regime which does not give economic growth and economic development the kind of importance they have under the Climate Change Convention and Kyoto Protocol. It is human development, and not economic development, which should be the starting point for a climate change regime. Human development gives primacy to human rights and environmental considerations but does not *per se* deny the necessity of economic development. In fact, the link between economic development and the realisation of human rights, in particular socio-economic rights, is well established. This is important because it recasts economic development as a tool for the realisation of the human rights of the poor and marginalised. In this context, the success or failure of policies and laws is rated according to their impact on the poor.

In terms of climate change the first step would be to move away from a system that allocates polluting rights based on past or present emissions. Indeed, any such scheme rewards long-term polluters – developed countries – and provides incentives to the few countries among developing countries, such as some Southeast Asian countries, India and the PRC to increase their pollution levels as fast as they can so that their own emissions levels will be grandfathered the day they take on commitments under the Kyoto Protocol. This is unjustifiable in environmental terms and inequitable for the majority of developing countries and all least developed countries that will be made to suffer the consequences of their lower levels of economic development twice over.

The most widely proposed alternative to grandfathering allocations is one based on per capita entitlements.[45] The basis for an equitable climate change policy should indeed take into account that every single human being has a right to a certain quantity of emissions. These include subsistence emissions such as emissions related to the growing of food or the use of firewood to cook meals or purify water.[46] This also includes livelihood emissions, which relate to everyone's right to benefit from the fruits of economic and technological development, for instance, by having access to electricity. Thus, there should be a basic human entitlement to a certain level of emissions. This level needs to take into account the requirements of the global environment and may thus imply reduced emissions by the minority of the world's population that directly or indirectly emits much more than that which the global atmosphere can support.

This entitlement is to be conceived from two related but distinct perspectives. At the international level, it provides a new way to allocate emission rights, which is fairer to countries that have not benefited from the fruits of economic growth. At the national level, it provides a similar mechanism whereby the poor and marginalised that do not have access to the amenities that their wealthier counterparts benefit from obtain a right to benefit from existing resources. In other words, the developed world and the minority of wealthy citizens within each country each have a debt to the poorer segments of the community.

While the measure for entitlements should be on a per capita basis, this cannot be the only criterion. Two reasons, at least, call for a more selective approach. First, a per capita entitlement may have the negative impact of fostering population policies, which may not otherwise be in the interests of the concerned countries. Secondly, an equitable legal framework should also take into account that some countries have a low population density because their environment is already degraded to such an extent that population has failed to grow over time. Since these countries usually happen to be among the poorest as well, recognition of their situation must also be taken into account.

The entitlement proposed here must differ from a Kyoto Protocol entitlement in an additional respect. The debt that rich countries and

[45] See, for example, International Institute for Sustainable Development, Per Capita Emission Rights (1998). Available at: www.iisd.org/didigest/sep98/sep98.2.htm.

[46] Compare Henry Shue, 'Subsistence Emissions and Luxury Emissions', *Law and Policy*, 15 (1993), 39.

rich people within each country have accumulated toward the poor cannot be redeemed by simply stabilising emissions or reducing them.[47] The entitlement scheme must be based on the premise that the only way in which emissions can be accessed from the poor who do not use their quota is by accepting a duty to invest an equivalent amount of money toward developing non-carbon development paths. If that is not undertaken, the entitlement system will simply end up being another market mechanism through which the poor will sell their entitlements but without any policy framework imposing the necessary changes for effectively mitigating climate change in the long term. Thus, any future CDM should fund only projects that provide zero carbon emissions so that the CDM itself becomes a vehicle for technology transformation and not just a cheap compliance mechanism that, at best, does nothing for the poor and, at worst, contributes to harming them further where already discredited development options are reintroduced in the guise of climate change friendly policies.

The new entitlement framework is thus conceived as a mechanism through which the poor and vulnerable can demand new technologies or emissions convergence. In other words, this entitlement framework imposes on the rich parts of the world (rich countries and rich segments of the population) to either reduce their own emissions or invest in ways and means so that the poor do not follow the rest of the world in increasing their own emissions as economic development eventually reaches them. In India, where the richest classes produce four and a half times more CO_2 than the poorest class and almost three times more than the all-India average, this convergence is also required.[48] A number of different initiatives could be taken. For instance, in a situation where, in India, only 31 per cent of rural households use electricity, there is untold potential for emissions increase if the poor are provided with the same kind of amenities from which the rich benefit.[49] The entitlement framework based on human rights indicates that the poor also have in principle a right to the lifestyle that the rich enjoy. As a result, the only way to ensure that poverty eradication does not harm the global environment more, while at the

[47] Compare Pia Halme, 'Carbon Debt and the (In)Significance of History', *Trames*, 11:4 (2007), 346.

[48] Greenpeace, *Hiding Behind the Poor* (A Report by Greenpeace on Climate Justice, Greenpeace India, 2007).

[49] 'What Equals Effective', *Down to Earth*, 16:14 (2007), 62.

same time providing alternative economic development paths for the rich and poor alike, is for the rich to invest in new ways to deliver development benefits. For instance, electricity generation in India could easily be focused on local solutions, in particular solar energy. Similarly, technological research should focus on new forms of public transport rather than on private vehicles with a lower negative climate change impact. Simply improving or changing the fuel on which private vehicles run may have a positive contribution on the global environment. However, as witnessed in the case of Delhi and its shift to compressed natural gas (CNG) on a large scale, this neither solves the environmental pollution caused by vehicles *per se* nor addresses the huge social and other problems caused by increasing reliance on private modes of transportation.[50]

Conclusion

Climate change must be addressed in earnest urgently. This requires measures that go beyond the existing Kyoto Protocol. It is critical to ensure that climate change is conceived in a broader manner that goes beyond the environmental and economic dimensions that have been central to the existing regime. Giving a central place to human vulnerability and incorporating the human rights language in climate change law is crucial. This must be achieved alongside a broader rethinking of the place of differential treatment in the climate change regime to ensure that it better reflects countries' and people's vulnerabilities in the future.

[50] See, for example, Naresh Kumar and Andrew D. Foster, 'Have CNG Regulations in Delhi Done Their Job?', *Economic and Political Weekly*, 42:51 (2007), 48.

7

Forests, climate change and human rights: managing risks and trade-offs

FRANCES SEYMOUR*

In the mid-1980s, tropical deforestation splashed onto the international agenda as the world became aware of threats to the survival of the human and biological diversity sustained by tropical forests. Activists protested the road-building and transmigration projects that were catalysing deforestation in Brazil and Indonesia; bilateral and multilateral donors mobilized funds for investment in forest protection; conservation organizations established alliances with indigenous and traditional peoples; and governments launched negotiations toward an international agreement on forests. Interest in tropical forests peaked in the aftermath of the United Nations Conference on Environment and Development (UNCED) in Rio in 1992, and steadily declined over the next decade as national and international efforts to reverse deforestation proved disappointing.

Tropical deforestation has now reappeared on the international agenda due to its newly-appreciated link to climate change. In 2006, a review commissioned by the Government of the United Kingdom (Stern 2007) called attention to the fact that some 20 per cent of current annual global greenhouse gas (GHG) emissions is due to land use change – most of which is deforestation in developing countries – a share greater than the emissions produced by the transport sector globally. The review asserted that controlling deforestation could provide one of the least expensive strategies for reducing emissions, and that such efforts must

* Frances Seymour is Director General of the Centre for International Forestry Research (CIFOR) in Indonesia. This chapter builds on a note prepared for a roundtable on climate change and human rights convened by the International Council on Human Rights Policy in Geneva, 12–13 October, 2007. The author is indebted to Stephen Humphreys, John Mutter, Carol Colfer, Marcus Colchester, Jessica Campese and Cecilia Luttrell for providing insights and/or comments on drafts, and to Lucy Heffern for research assistance.

be a key element of any future climate protection regime. As a result, reducing emissions from deforestation and degradation (REDD) is now central to discussions of global and national mitigation strategies.

While getting less attention than the importance of forests to the mitigation agenda, the potential impacts of climate change on forests, and the role of forests in adaptation to climate change are increasingly appreciated by the scientific community and relevant policy arenas. For example, higher temperatures and changes in rainfall patterns will affect the resilience of forests to fires, pests and disease. At the same time, maintaining natural forest vegetation can be seen as a key component of adaptation to climate change in other sectors. For example, forest-based sources of food and income can continue to provide a 'safety net' for agricultural households as crop-based food security is undermined by increasingly unpredictable weather patterns.

Any change in the condition or management of tropical forests is relevant to human rights, posing both risks of increased human rights violations and opportunities for improvement. Many of the world's poorest and most politically marginalized people are dependent on forests for their livelihoods. Compared with other economic and social sectors affected by climate change, forested areas and forest-related institutions tend to be especially characterized by unclear property rights, remoteness from public scrutiny and a history of repressive state actions. As such, forest governance has profound implications for the rights and welfare of indigenous, traditional and other forest-dependent peoples, and vice versa.

The purpose of this chapter is to provide an overview of the human rights issues likely to arise at the intersection of forests and climate change. These issues include the likely direct impacts of climate change on forest-based livelihoods, and consequent undermining of economic, social and cultural rights. They also include the risks to civil and political rights that could be posed by the implementation of various policy responses to climate change: forest-related adaptation measures; REDD; and other schemes to mitigate emissions from land use change and agrofuel development in forested areas. Procedural rights are also at risk if forest-related climate policies at national and international levels are developed without meaningful participation by key stakeholders. The chapter concludes with some reflections on the challenges to equity and justice posed by alternative forest-related climate policies, and the policy implications of taking human rights into account in their design and implementation.

The relevance of climate change to forests and forests to climate change

Forests and climate change are interrelated in a number of ways. While many of these relationships are complex and poorly understood (Bonan 2008), at least four linkages are generally accepted. First, forests and forest-based livelihoods will be directly affected by climate change. As described in greater detail below, forest ecosystems are vulnerable to a warming climate as well as to increased variability, and to an increased incidence and severity of extreme weather events. As a result, both government policies and community practices governing forest management will have to adapt to climate change in order to maintain the provision of direct goods and services that local communities and broader society derive from forests. For example, as higher temperatures and prolonged droughts render natural forests more vulnerable to forest fires, forest managers will need to invest more in fire control efforts to ensure that intentionally set fires do not escape and become wildfires, and that accidentally set fires are quickly detected and suppressed.

Second, maintenance of forest-based ecosystem services that support other economic sectors can strengthen societies' resilience to climate change. For example, forests play an important role in moderating the quantity and quality of water that flows out of watersheds. As rainfall patterns change, the hydrological services provided by forests will be increasingly important to maintaining municipal drinking water systems, agricultural water supplies and the production of hydroelectric power. This set of potential contributions to climate change adaptation has been ranked as especially important in Central America (TroFCCA 2008). In parts of Southeast Asia, where droughts are anticipated to be more severe and episodes of heavy rainfall more likely, maintaining the role of intact natural forest vegetation in controlling forest fires and landslides is a priority (2008). Accordingly, the adaptation strategies of other economic sectors, such as agriculture and hydropower (which are affected by forest hydrology) and air and land transportation (which are affected by haze from forest fires and landslides, respectively), need to be linked to sustainable forest management.

Third, forests are already being affected by the rapid development of agrofuels,[1] which is being driven in part by ostensibly 'climate friendly'

[1] The term 'agrofuels' rather than 'biofuels' is used to distinguish between crops grown for production of liquid fuel, such as sugar cane for ethanol and oil palm for biodiesel, and other forms of bioenergy, such as firewood, charcoal and dung.

subsidies on the part of the European Union, the United States and other governments (IATP 2007). In some cases, the impact is direct, as when intact tropical forests are converted to agrofuel plantations. In other cases the impact is indirect, as when agrofuel development displaces other land uses into forest areas. The extent to which future climate policy includes agrofuels as an emissions mitigation strategy will have a significant impact on forests and forest peoples.

Fourth, deforestation and forest degradation are a significant source of the GHG emissions that drive climate change, and are among those that can be mitigated at relatively low financial cost. The Intergovernmental Panel on Climate Change (IPCC) estimates that about half of the potentially avoidable emissions at a cost of less than US$100 per ton of CO_2 equivalent could be achieved by reducing emissions from deforestation (IPCC WG III 2007, 14). The potential additional social costs are discussed below.

However, the mitigation potential of forests could itself be affected by climate change. A warmer, drier climate could trigger a positive feedback loop that results in the dieback of forests, and thus increased emissions and further warming (Bonan 2008). In other words, warmer, drier weather could lead to a vicious circle in which increased incidence of burning renders forests less able to recover and sequester carbon in forest vegetation, which in turn would accelerate climate change. Some models predict that a significant portion of the carbon-rich Amazon rainforest will be replaced by carbon-poor savannah ecosystems if global warming is allowed to proceed beyond a certain threshold, thus releasing significant amounts of carbon into the atmosphere (WHRC 2008; Mayle *et al.* 2007, 299).

Two forested nations – Indonesia and Brazil – currently account for some two-thirds of total annual emissions from land use change. As a result of those emissions, estimates now place those countries as the third and fourth largest overall GHG emitters, after the United States and the People's Republic of China (PRC) (PEACE 2007). The drainage, clearing and burning of peatland forests – much of which are concentrated in Indonesia – is particularly emissions-intensive, due to the carbon-rich organic matter below the surface vegetation that can extend to a depth of several metres (Hooijer *et al.* 2006). Brazil's emissions are driven by high rates of deforestation in the Amazon.

While the linkage between deforestation and climate emissions has been on the global agenda for more than a decade, for a number of methodological and political reasons, mechanisms to address 'avoided deforestation' were not included in the Kyoto Protocol of the United Nations Framework Convention on Climate Change (UNFCCC 2007a).

Afforestation and reforestation activities were included in the Kyoto Protocol, but very few tree-planting projects have been approved under the Protocol's clean development mechanism (CDM). The conventional wisdom is that the procedures for certifying such projects were too complex and transactions costs too high to justify investment (FAO undated, 17). As a result, the current discourse on forest-related mechanisms stresses the need to 'streamline' the so-called 'safeguard policies', which set minimum substantive and procedural standards for attention to social and environmental impacts.[2] Such streamlining could imply less stringent attention to human rights implications.

In late 2005, the politics of linking forests to climate protection changed with an official submission by the Coalition for Rainforest Nations (led by Costa Rica and Papua New Guinea) to the UNFCCC. The submission called on parties to the UNFCCC and the Kyoto Protocol to open a dialogue toward addressing emissions resulting from tropical deforestation, in recognition of their significant contribution to overall emissions, and the fact that the Kyoto Protocol did not provide developing countries with a vehicle to reduce emissions through reduced deforestation (UNFCCC 2005).

A plan for negotiating positive incentives for REDD was one of the key features of the so-called 'Bali Road Map' negotiated at the Thirteenth Conference of the Parties (COP 13) to the UNFCCC in December 2007 (UNFCCC 2007b). Under a REDD regime, industrialized countries would make financial transfers to developing countries – through market and/or fund-based mechanisms – to compensate them for the opportunity and other costs of avoiding emissions from deforestation. COP 13 set in motion a process to resolve outstanding methodological issues related to the measurement and monitoring of forest-based carbon emissions, and encouraged parties to support REDD 'demonstration activities'. As a result, nations with significant areas of natural forest are now key players in international climate negotiations.

Forest governance and human rights

Debates about REDD have arisen in a context in which forest governance at both national and international levels is contested and dynamic. Under such conditions, the impact of forest policy changes and programmatic

[2] This is part of a larger discourse questioning the utility of safeguard policies for reaching development objectives. For an analysis in the context of the World Bank, see Seymour 2006.

interventions to promote climate change mitigation and adaptation objectives could either accelerate or retard efforts to mainstream a rights-based approach into forest-related law and management practices.

Forest governance

Forest governance in many tropical countries reflects a legacy of colonial era rights and management regimes, in which the state claims ownership of most forest areas, and forests are exploited primarily for commercial timber. State claims are often contested by indigenous peoples and/or traditional communities that reside in and around forest areas. In respect to forests:

> [t]he dominant pattern of government intervention has been one of increasing central control over forest resources, the denial of access to forest resources by groups that have traditionally or historically depended on them, and control over trade in (and thus the ability to benefit from) forest species and products. (Peluso and Vandergeest 2001, as cited in Menzies 2007, 6)

As a result, who owns the forest and what constitutes legal uses of forest resources are often unclear.

Over the last three decades, the international forestry profession has progressively embraced the rhetoric, and sometimes the practice, of 'social forestry', and what has now become known as 'community-based forest management' or CBFM. CBFM departs from traditional forest management practices by including forest communities in both the decision-making and sharing of the resulting benefits (Menzies 2007). Although CBFM has succeeded in providing some communities with new roles in and incomes from forest management, the progress of CBFM has been slowed by the reluctance of governments to cede real authority to communities (Menzies 2007).

While private and community-based forest management in the tropics is increasing modestly (White and Martin 2002), most forest areas in developing countries continue to be claimed by the state, whether or not the state exercises effective management of those areas (Sunderlin *et al.* 2008). Even in countries where indigenous and community rights over forests have been recognized on paper, local people have often failed to realize expected benefits due to inadequate enforcement of new forest tenure rights and other complementary rights – including citizenship, free prior and informed consent and the right to redress (Sunderlin *et al.* 2008, 12–14).

Despite the importance of forests to the income, health and identity of the communities that live in and around them (described further below), forest governance has tended to be dominated by the interests of political and economic elites. Laws and policies governing access to forest land and resources tend to be systematically biased against rural communities, and to grant special privileges to commercial interests (Larson and Ribot 2007). In Honduras, for example, the implementation of forestry regulations has favoured logging companies, while erecting bureaucratic hurdles to the legal exploitation of forests by communities and smallholders (Larson and Ribot 2007).

Proximate causes of deforestation and degradation include both need as well as greed. Much forest destruction is driven by commercial-scale economic activity that enjoys implicit or explicit state subsidies. For example, poor logging practices by domestic and transnational corporations can open up forest areas to colonization and hunting, while wood waste left behind can make forests vulnerable to forest fires (Laurance *et al.* 2001). Structural overcapacity in the wood-processing industry in Indonesia has created a demand for fibre that cannot be met by legally and sustainably produced wood from plantations, thus creating a demand for illegally produced wood from natural forests (Barr 2001; World Bank 2006). Commercial-scale agribusiness, ranging from cattle ranching in the Amazon region to oil palm plantations in Southeast Asia, has also caused large-scale conversion of natural forests to other uses (Kanninen *et al.* 2007). It is often the case that such enterprises do not face the true costs of forest loss to the local, national, or global economy.

But some forest conversion and degradation is effected by the rural poor, and such activities often make the poor better off (Chomitz 2007). Conversion of forest frontiers to agriculture crops or tree crop plantations provides a livelihood for peasants across the tropics, and depending on resource tenure, market access and commodity prices, a good living. As a result, in the absence of appropriate compensation, blunt policy instruments to protect forests can block a pathway out of poverty. Thus, simple formulations such as 'poverty causes deforestation' or 'deforestation hurts the poor' are both misleading generalizations and are insufficient to inform policy in particular circumstances.

A recent review of forest governance (Agrawal *et al.* 2008) highlighted three current trends:

(1) decentralization of forest management, especially of low-value forests, which has brought an additional 200 million hectares under some form of community management;

(2) the dominance of private companies extracting timber from state-owned tropical forests under logging concession agreements; and
(3) the growing significance of forest certification as a market mechanism, especially for temperate forests.

The review stressed that the effectiveness of forest governance is only weakly associated with the type of formal ownership. The impacts of forest interventions on local people depend on clear user rights and responsibilities, enforcement of property rights, participation by forest users in decision-making and downward and horizontal accountability of decision-makers (Agrawal *et al.* 2008, 1462; Wells *et al.* 2006).

Forest governance at the international level is also contested and dynamic. Over the last two decades, the international community has repeatedly failed to negotiate a binding agreement on forests. According to one recent analysis of those efforts, the United Nations Forum on Forests and other intergovernmental attempts to address deforestation are doomed to failure as long as they are subservient to current neoliberal trade and investment regimes (Humphreys 2006). Most recently, the emergence of avoided deforestation as a key climate protection instrument has shifted the centre of gravity of international forestry discussions away from forestry, agriculture and biodiversity-related forums and into the UNFCCC.

Forests and human rights

Forest-based goods and services are central to the economic, social and cultural (ES&C) rights of hundreds of millions of people around the world. The World Bank estimates that 90 per cent of the 1.2 billion people living in extreme poverty depend on forest resources for some part of their livelihood (World Bank 2004; UNDP *et al.* 2005). In Indonesia, for example, more than 10 million poor people live in state forest zones with good forest cover, while millions more depend on forests for their income (Wollenberg *et al.* 2004). In the Democratic Republic of Congo, 40 million people rely on forests for food, medicines, energy and income (Debroux *et al.* 2007).

Forests have proven to serve as important 'safety nets' for communities in times of economic stress. During the financial crisis in the late 1990s in Indonesia, many households turned to the forest for supplementary income sources (Sunderlin 2002). For example, thousands of people went into remote peat forests in Kalimantan to gather turtles and tree bark for sale in urban markets (Chokkalingam *et al.* 2005). Households unable to afford high prices for modern energy sources revert to collecting fuelwood from the forest. Forests also play a safety net role for vulnerable individuals: research in sub-Saharan Africa suggests that bushmeat from the forest provides an important source of protein to children orphaned by AIDS (Shackleton *et al.* 2006).

Forests are also important to the maintenance of cultural identity (Colfer *et al.* 1997). In East Kalimantan, Indonesia, research conducted with local communities identified more than 2,100 forest species with 3,642 different uses, including food, traditional medicine, hunting equipment, construction materials and culturally-significant ornamentation. Of these species, 119 had no known substitute for the particular use (Sheil *et al.* 2001).

Despite the importance of forests to the realization of ES&C rights, forest communities are often denied access to forest resources. In many countries, there is a rich history of repressive measures taken by both state and non-state actors to control forest access and use. There is a large literature on the human rights implications of this history, which includes allegations of violations of civil and political (C&P) rights and procedural rights as well as ES&C rights (Peluso 1993; Colchester 2006; Alcorn and Royo 2007).

For example, commercial timber companies have relied upon military and paramilitary assistance to deal with local opposition to their logging activities (Colchester 2006, 49). Conservation organizations have cooperated with law enforcement authorities to police access to protected areas, and in some instances communities have been forcibly evicted from those areas (Seymour 2008). Case studies on forest law enforcement from around the world indicate that high profile 'crackdowns' on illegal logging tend to be targeted against the rural poor rather than against the business people and officials who are often behind forest crime (Colchester 2006).

Such examples suggest that in many countries, current forest governance regimes are inadequate for upholding international human rights standards. Combining various strands of international law – including such instruments as the UN Declaration on the Rights of Indigenous Peoples and the Convention on Elimination of All Forms of Discrimination – it has been asserted that such standards:

> recognize the right of forest peoples to '*own, control, use and peacefully enjoy their lands, territories and other resources, and be secure in their means of subsistence*'. (Colchester 2007, emphasis in original)

There is often a disconnect between the recognition of such rights and standards in ratified treaties and national constitutions and their realization through law and practice in the forestry sector (Colchester 2007).

A key implication of this disconnect – and the broader characterization of the state of forest governance offered above – is that new initiatives designed to harness forests in the service of climate change adaptation and mitigation risk exacerbating existing weaknesses and inequities in current forest governance regimes.

The human rights implications of climate change related to forests

Direct impacts

As mentioned above, forests will be affected by increasing global temperatures and climate variability. Warmer surface temperatures and longer periods of drought will increase the risk of forest fires (IPCC WG II 2007, 18). In September 2007, Hurricane Felix devastated large swathes of forest in Central America, demonstrating the vulnerability of forests to extreme weather events, which are likely to increase in frequency and severity. To date, very little attention has been given to the policies and practices needed to maintain the adaptive capacity and productivity of natural or planted forests in the face of climate change (Guariguata *et al.* 2007).

Disruption of forest ecosystems will in turn lead to disruption in the provision of forest-based ecosystem goods and services. Such goods include timber, fuelwood, forage, fruits, medicines and materials for handicrafts, which are often of particular importance to poor communities in developing countries (Millennium Ecosystem Assessment 2005; Sunderlin *et al.* 2005). For example, research in Uganda indicates that up to 26 per cent of rural household income comes from forest resources (Jagger 2007). More generally, forest-based ecosystem services include cultural, spiritual and aesthetic services, as well as hydrological, pollination and pest control services important to other sectors of national economies (Millennium Ecosystem Assessment 2005).

The degradation of forest ecosystems – and associated resilience to the impacts of climate change mentioned above – will thus reduce forest-based incomes. Women, whose household responsibilities and income sources often include the gathering of forest products, are likely to be particularly disadvantaged by their loss (Colfer, forthcoming). The impact of climate change on forests will also render already vulnerable communities even more vulnerable to 'natural' disasters such as forest fires, landslides and floods that will result from human-induced climate change.

As with other direct impacts of climate change on the lives and livelihoods of vulnerable people, those mediated through their impacts on forests pose a challenge to the traditional human rights framework. 'Duty bearers' are widely dispersed in time and space, and it is virtually impossible to trace the responsibility for climate change from individual sources of emissions through to impacts on particular individuals. Nevertheless, the impacts are real, and in principle can be mitigated. Accordingly, the loss of forest-based income sources and ecosystem services due to climate change could be seen as violations of economic, social and cultural rights. Further, the exacerbation of those losses (through adaptation options foregone) due to poor forest management could be similarly understood.

Impacts of forest policy responses to climate change

The impacts of climate change on human rights through the disruption of forest ecosystems is a subset of a wider and increasingly well-understood set of effects resulting from the *direct* impacts of climate change. Less prominent in the discourse to date linking climate change and human rights is the potential for policy initiatives taken in *response* to climate change to have unintended negative consequences for human rights.[3] Specifically, if such responses were to be implemented in a repressive manner under conditions of weak governance, violations of civil and political rights could be at stake.

The forestry sector provides several illustrations of such risk. As mentioned above, climate-related interventions risk exacerbating existing weaknesses in forest governance regimes. Human rights are potentially at risk from policy responses to reduce GHG emissions from

[3] For example, a March 2008 request from the United Nations Human Rights Council to the Office of the UN High Commissioner for Human Rights for a 'detailed analytical study of the relationship between climate change and human rights' highlighted the effects of climate change, not the effects of efforts to address climate change (UN Human Rights Council, 2008).

forests, to increase production of 'climate friendly' agrofuels and to harness forests for adaptation to climate change.[4]

Mitigation of forest-based emissions

The prospect of a global REDD regime is already having an impact on forest-related decision-making around the world. In anticipation of REDD finance, national governments have begun making high-level commitments to the approach, and are making efforts to improve their capacity to monitor deforestation and forest degradation.[5] In addition, sub-national governments have announced initiatives ranging from moratoria on logging in Aceh and Papua in Indonesia (Reuters 2007) to pilot payments for ecosystems services schemes in Brazil (Government of Amazonas and CC-AI 2007, 16–17).

Public and private sector project proponents have announced the initiation of a number of REDD-related initiatives. In Madagascar, for example, the Swiss Agency for Development and Cooperation (SDC) in collaboration with GTZ (the German government agency for technical cooperation) and Intercooperation are implementing a demonstration activity called 'Committing Forests as Carbon Reservoir' (Intercooperation 2006). In 2007, the investment firm Merrill Lynch announced a $9 million deal brokered by Carbon Conservation, an Australian company, to protect forests in Aceh, Indonesia. According to the press release announcing the deal, the firm is betting that the avoided carbon emissions will generate $432 million in carbon financing over the next 30 years (Merrill Lynch 2008; Wright 2008).

Donor governments have also made significant commitments to advancing the REDD agenda: in March 2007, the Government of Australia committed Aus$200 million to be focused in Southeast Asia and the Pacific (Howard *et al.* 2007); at COP 13 in Bali, the Government of Norway announced a commitment of up to US$500–600 million annually (Halvorsen 2007); Germany has committed US$59 million to the World Bank's Forest Carbon Partnership Facility (World Bank 2007).

[4] Efforts to sequester carbon through tree planting (afforestation/reforestation in the context of the CDM, or ARCDM) could also pose human rights risks, but are beyond the scope of this chapter.

[5] See, for example, presentations by the governments of Papua New Guinea and Costa Rica presented at the UNFCCC Workshop on Methodological Issues relating to Reducing Emissions from Deforestation and Forest Degradation in Developing Countries, in Tokyo, Japan, 25–27 June 2008. Available at http://unfccc.int/methods_and_science/lulucf/items/4289.php.

As payments for conserving forests for carbon storage become increasingly likely, state and non-state actors alike will have strong incentives to passively ignore or actively deny the land and resource rights of indigenous, traditional and/or poor forest users in order to position themselves to claim compensation for forest stewardship in their stead.

While some representatives of indigenous peoples groups have cautiously welcomed REDD, others have denounced its potentially devastating impact on their communities. The International Forum of Indigenous Peoples on Climate Change has stated that:

> REDD will not benefit Indigenous Peoples, but in fact, it will result in more violations of Indigenous Peoples' Rights. It will increase violation of our Human Rights, our rights to our lands, territories and resources, steal our land, cause forced evictions, prevent access and threaten indigenous agricultural practices, destroy biodiversity and culture diversity and cause social conflicts. (Forest Peoples Program 2007)

A second human rights concern raised by REDD is the prospect of increased law enforcement efforts to deal with illegal logging, which is currently a significant, if poorly understood, driver of forest degradation in many countries (Tacconi 2007). High profile 'crackdowns' to deal with forest crime tend to discriminate in favour of those with the means to pay off law enforcement and judicial officials (Larson and Ribot 2007, 8). As a result, REDD-inspired law enforcement efforts could lead to an increase in arbitrary arrest and detention.

The prospect of REDD has raised concerns that rural communities that currently exercise stewardship over forests will be pushed aside by local elites, private investors, or others seeking to position themselves to receive new revenue flows in exchange for protecting the forest (Griffiths 2007). However, it is also plausible that the prospect of REDD could accelerate long overdue reforms in forest governance. Because REDD payments are likely to be linked to performance in actually reducing rates of deforestation and forest degradation, governments and private sector proponents could be forced to negotiate with rural communities who are in a position to control whether or not forests are functionally protected from fire, theft, conversion and other threats.

The prospect of REDD financial incentives will certainly put a premium on resolving questions about who owns (and therefore has a right to sell) forest carbon; the question is whether or not forest communities will be made better or worse off as a result. Creating the conditions for a rights-based approach for resolving these questions is thus a priority on the human rights agenda.

Promotion of agrofuels

Another issue linking forests, climate change and human rights is the rapid expansion of agrofuel plantation development. The use of agrofuels, including bioethanol (principally from sugarcane) and biodiesel (principally from oil palm), is being promoted in developed and developing countries alike as a 'climate-friendly' alternative to fossil fuels. As demand for agrofuels increases, it is driving land use change to substitute sugar cane, oil palm, or other agrofuel crops for current land uses, including forestry as well as food production (Fargione *et al.* 2008, 1).

Due to the unclear and contested property rights over forest areas mentioned above, the human rights implications of such agrofuel development are profound for indigenous and traditional forest users, who risk alienation from their land. The rapid expansion of oil palm plantations in Indonesia illustrates the risk. Social factors are rarely assessed prior to project initiation, and conflicts over land are widespread and have resulted in intimidation and violence (Sheil *et al.* forthcoming). In 2007, affected communities in West Kalimantan alleged that oil palm development had led to the takeover of indigenous peoples' customary lands without due process, resulting in conflict, in turn triggering repressive actions by companies and security forces (Colchester *et al.* 2007).

Ironically, to the extent that agrofuels development comes at the expense of natural forests – as is occurring – the net impact on GHG emissions is a significant net *increase* (Searchinger *et al.* 2008). Research suggests that it would take more than 840 years to repay the 'carbon debt' from converting Indonesia's carbon-rich peatland forests to oil palm plantations (Fargione *et al.* 2008, 2). Agrofuel development can thus pose a 'lose–lose' proposition from the perspectives of both climate protection and the rights of forest communities.

Adaptation

The forest management policies of many countries likely to be affected by climate change do not yet take into account the role of forests in supporting societies' adaptation strategies, nor their importance to sustaining livelihoods with the onset of climate change (Kalame *et al.* 2008). In the future, governments and the general public are likely to become increasingly aware of the importance of protecting natural forest cover as a way of maintaining ecosystem resilience to climate change. While such increased awareness would be a positive development overall, it could also prompt the implementation of policies that could result in human rights violations such as those described for forest-related mitigation interventions above.

For example, deforestation has often been blamed for massive landslides and flooding of the sort likely to become more frequent with climate change.[6] In the past, governments have been quick to announce logging bans and other controls on forest use in response to such catastrophes (FAO and CIFOR 2005). For example, catastrophic floods in the PRC, Thailand and the Philippines prompted logging bans that put millions of people out of work. There is a risk that in the name of adaptation to climate change, governments will limit settlement and farming in sensitive watersheds, which could in turn displace the poor without offering adequate compensation.

Further, the incidence of forest fires in the Amazonian and Southeast Asian rainforests is likely to increase along with longer droughts and higher temperatures (TroFCCA 2008; WHRC 2008). In the past, catastrophic forest fires have led governments to impose blanket bans on burning forests, without discriminating between commercial and traditional uses of fire, as happened in Indonesia (Barber and Schweithelm 2000). Some governments already have policies in place to limit shifting cultivation (Fujita and Phengsopha 2008). For example, the government of Lao PDR has an explicit policy goal of 'stabilizing' shifting cultivation, on which 39 per cent of the population was dependent in 2000 (Thomas 2003). As fires become more frequent, others will be tempted to ban traditional agricultural burning practices altogether. If enforced, such policies would disrupt the livelihoods of poor forest communities.

The challenge of respecting procedural rights

As described above, to the extent that policies intended to protect or exploit forests in the name of climate change mitigation or adaptation result in reduced livelihoods for the rural poor, economic, social and cultural rights are at stake. To the extent that such policies are enforced using repressive measures, civil and political rights are at stake as well. This section highlights the special challenge of respecting *procedural* rights in forests and climate-related decision-making.

[6] Despite the conventional wisdom linking forests and floods, the science is controversial. Bradshaw *et al.* (2007) have demonstrated significant correlation, but Bruijnzeel *et al.* (2007) dispute causality.

It is increasingly recognized that so-called 'access rights' – access to information, participation and justice as codified in the Aarhus Convention[7] – constitute human rights that are rooted in civil and political as well as economic, social and cultural rights (Foti *et al.* 2008, 22). Procedural equity – in terms of participation in decisions regarding what actions are to be taken and who will implement them – will in turn affect what options are available (Thomas and Twyman 2005, 116–17).

Changes in forest policy and implementation in response to the new imperatives of climate change mitigation and adaptation will challenge existing forest governance mechanisms at both national and global levels. Governance mechanisms more generally, and mechanisms for forest governance in particular, are for the most part poorly equipped to respect procedural rights in a consistent manner.

At the national level

National systems for providing public access to information, participation and justice related to decisions that affect the environment are still a work in progress. In case studies from around the world, the Access Initiative found a consistent lack of public participation in project-level decision-making. More than half of the cases of policy-level decision-making studied were characterized by insufficient lead time or inadequate information for meaningful participation by affected stakeholders (Foti *et al.* 2008, 62–5).

Extractive sectors (including forestry) tend to lag behind others in good governance practices such as transparency of information, public participation in decision-making and access to justice, in part due to the vested interests that oppose increasing openness (Foti *et al.* 2008, 43). Communities whose rights are most at risk from decision-making related to forests are often poor, located in remote areas and members of ethnic and linguistic minorities subject to discrimination. For example, communities with unclear land rights and those less integrated into markets are less likely to benefit from REDD (Macchi 2008, 45).

Establishing the Free Prior and Informed Consent (FPIC) of affected communities is increasingly recognized as a procedural standard to be achieved by governments and private corporations prior to the

[7] The United Nations Convention on Access to Information, Public Participation in Decision-Making and Access to Justice in Environmental Matters. Available at: www.unece.org/env/pp/documents/cep43e.pdf.

implementation of major infrastructure or extractive industry projects (Colchester and Ferrari 2007). Consistent implementation of FPIC in forest-related decision-making would require significant changes in the way in which governments and corporations interact with communities in many countries (Colchester and Ferrari 2007, 20–1). Not infrequently, forest communities first learn of plans for timber concessions or industrial plantations when the bulldozers arrive.

As a result, there is a strong risk that procedural rights in the formulation and implementation of national climate policies related to forests will not be adequately respected. Significant donor investment in capacity-building for REDD, now focused on building the forest carbon assessment and monitoring capacity of governments, will need to be broadened to include capacity for inclusive decision-making and meaningful participation by less powerful groups. This suggests that REDD interventions will need to be paced and sequenced in accordance with capacity-building achievements, and thus the level of urgency for the latter is high.

At the global level

Inadequate attention to the rights and voices of forest communities within countries is recapitulated at the international level. Advocates for indigenous peoples' rights have pointed out the limited attention that the scientific community, such as the IPCC, has given to the likely impacts of climate change on tropical forest-dependent people (Salick and Byg 2007, 4). Others have observed that the special risks posed by climate change to such communities are not given sufficient attention in negotiations related to mitigation and adaptation options (Macchi 2008, 38).

Indigenous groups have challenged their lack of inclusion in international debates on issues such as REDD (Diaz undated). For example, at COP 13 in Bali in December 2007, activists protested the launch of the World Bank's Forest Carbon Partnership Facility (FCPF) in part due to inadequate consultation with indigenous peoples during the Facility's design (Forest Peoples Program 2008). At that event, the Chair of the United Nations Permanent Forum on Indigenous Issues called for recognition of indigenous rights, consultations with indigenous peoples and representation in the FCPF governance structure (Corpuz-Tauli 2007).

In this context, it is relevant to note that the Aarhus Convention requires signatories to promote its principles in international negotiations related to the environment (UNECE 1998). Thus, most European governments are obligated to incorporate concerns about respect for procedural rights in their negotiations concerning REDD.

Reflections on challenges to justice and equity

Duty bearers and rights holders

The overall discourse about equity and justice in the climate change policy arena has been characterized by a grouping of rich consumers and nations as the perpetrators of injustice (through past and current profligate fossil fuel use, while facing lower vulnerability to the probable impacts of climate change) versus poor people and nations as the victims (in terms of their significantly lower fossil fuel consumption historically and per capita, and higher vulnerability to the probable impacts of climate change). Projections of the impacts of climate change across countries suggest that poor countries will suffer the greater share of damages, despite having historically contributed the least to the problem (Mendelsohn *et al.* 2006, 175). Translated into the language of human rights, the 'duty bearers' and the 'rights holders' related to climate change are discussed as two distinct groups, corresponding to rich countries and poor countries, respectively.

The addition of forest-based emissions to the mix complicates the discourse. Sources of net forest-based emissions are currently concentrated almost exclusively in developing countries, while the net forest cover change in some industrialized countries is marginally positive.[8] This raises the question of the degree to which governments of countries with high forest-based emissions become duty bearers to mitigate the climate change-related impacts of those emissions. Clearly, industrialized countries are responsible for the lion's share of *cumulative* emissions (including those from past deforestation as well as fossil fuel use), and are also implicated as important drivers of current deforestation in developing countries through consumption, trade and investment patterns. Thus, any new 'duties' on the part of forested developing countries would not reduce – nor be fully distinct from – the duties of industrialized countries to reduce emissions.

Indeed, many advocacy groups in both Northern and Southern countries have opposed REDD on the grounds that, by flooding the carbon

[8] The outlier among developing countries is the PRC, which has achieved a significant recent increase in forest cover due to extensive reforestation efforts.

market with cheap carbon emissions reduction credits, REDD will lift pressure on industrialized countries to reduce fossil fuel-based emissions (Hare and Macey 2008). However, recent analyses suggest that catastrophic climate change can be averted only by a 'both/and' solution that includes reduced emissions from both deforestation and fossil fuel use. In addition, the possibility that climate change could itself reduce the ability of forests to sequester carbon implies that reductions of all kinds need to be addressed with urgency to keep this option open.

Another complicating factor is that the underlying causes of deforestation – consumption of forest products that cause forest degradation, and of agricultural and industrial commodities that lead to forest conversion – include both rich and poor people and rich and poor countries alike. For example, the market for palm oil is driven by its use as a domestic cooking oil in developing countries as well as by its use as a 'climate friendly' transport fuel in industrialized countries (Naylor *et al.* 2007). Thus, while debates over agrofuels have been framed as 'food vs. fuel' or 'forest vs. fuel', there is also a potential 'food vs. forest' trade-off that cannot be ignored.

The primary rights holders related to the impacts of climate change – the poor in developing countries – remain the same whether climate change is driven by forest-based or other sources of emissions. But some of those same poor people in developing countries are rights holders who are at most risk from policies intended to mitigate forest-based emissions. Who should be included in the group of duty bearers for reducing forest-based emissions will certainly remain a contentious question. Given the significance of forest-based emissions, and the urgency of reducing emissions overall, *all* actors with potential to affect rates of deforestation – including relevant governments and corporations in both North and South – could be considered to have an obligation to be part of the solution. The question then becomes how such responsibility should be shared (for example, how financed), and how to protect human rights as governments and corporations act on those obligations.

Equity versus efficiency

The prioritization of investments to promote REDD raises a number of thorny equity issues. It is true that there is some scope for 'win–win' opportunities to address forest-based sources of emissions and other objectives related to human rights, such as poverty reduction. For example, payments to rural communities to conserve forests could potentially advance both (Luttrell *et al.* 2007). Unfortunately, there is no neat alignment between need and the most efficient uses of scarce resources to achieve climate protection objectives.

Sources of forest-based emissions are distributed unevenly within and between individual developing countries. Different countries are arrayed along the so-called 'forest transition curve', with some maintaining a significant amount of their original forest cover, others rapidly deforesting and still others with little forest left (Kanninen *et al.* 2007). Efficiency and urgency criteria would dictate concentrating REDD payments in those countries on the 'steep part' of the forest transition curve – such as Indonesia and Brazil – that are responsible for the bulk of current forest-based emissions. But such a narrow targeting leaves out countries that retain significant forests (such as some of those in the Congo Basin) and those that have little forest area left (such as those in South Asia), both with significant numbers of very poor people made further vulnerable by climate change.

Debates over how to incorporate REDD into the global climate protection regime are now grappling with the question of how to achieve equity across countries and regions with different deforestation histories. All else being equal, carbon markets will certainly favour countries and regions with the largest number of inexpensive credits, untempered by concerns about other so-called 'co-benefits', such as poverty reduction or biodiversity conservation. Such objectives will need to be internalized into REDD instruments, or explicitly addressed through complementary initiatives. Developing countries have reason to be nervous that the allocation of development assistance funds will also be skewed by climate protection objectives at the expense of poverty reduction objectives.

Similarly, at a more micro scale, the most effective allocations of funds for REDD will be those that most efficiently affect marginal land use decisions (Wunder 2007). As a result, optimal allocations of REDD funds within countries from a climate protection perspective may offend our sense of justice. For example, if one community – say, migrants to a forest frontier – is rapidly converting forests to a land use that is of lower value than carbon storage (for example, cattle ranching), and another community – say, an indigenous group – has a strong track record and likelihood of continuing to protect its traditional forests, then the most effective allocation of REDD funds would be to the migrant community, to compensate them for foregoing expansion of their ranching activities. Payments to the indigenous community would not 'buy' a reduction in emissions, as the community would be likely to protect the forest anyway. Gender justice advocates have noted that those most likely to benefit from REDD payments would thus be those responsible for the problem, not those – such as women – who have contributed to protecting forests (GCCN 2007, 2).

Trade-offs thus emerge between allocating resources to 'deserving' countries and communities (based on need and merit) versus allocating resources to the countries and communities where emissions reduction potential is highest (in the interest of protecting equally deserving communities from the adverse impacts of climate change). From the perspective of the countries and communities most vulnerable to climate change – such as small island states – there is a need to focus resource allocations on investments that maximize efficiency and effectiveness in reducing emissions rather than other considerations, based on the proposition that the greatest threat to the world's poor is, in fact, climate change. According to a 2007 submission from the Government of Tuvalu to the UNFCCC on REDD:

> The crucial element in the consideration of tropical forest loss under the UNFCCC is the reduction of greenhouse gas emissions at the global level. This must be the principal objective of any action taken by the COP. Other considerations will be important. In fact co-benefits may be possible but these should not outweigh the key principle of reducing emissions at the global level. This principle has significant implications for what can and cannot be achieved with respect to possible actions under the UNFCCC to reduce emissions from deforestation. (UNFCCC 2007c)

Thus, design of REDD regimes at national and international levels will force difficult decisions regarding whether and how to incorporate multiple and potentially conflicting objectives into an instrument designed to reduce emissions. The effectiveness, efficiency and equity of the resulting instrument can be assessed only with respect to those agreed objectives.

Policy implications

> In facing the dangers from climate change there are three options: mitigation, adaptation and suffering. Minimizing the amount of suffering can only be achieved by doing a lot of both mitigation and adaptation. (Holdren 2008)

In the quote above from John Holdren, Director of the Woods Hole Oceanographic Institution, 'suffering' can be used as a proxy for the human rights violations that will surely result from the direct impacts of climate change. Thus, policy initiatives to support mitigation and adaptation are urgently needed. And yet, from the evidence presented in this chapter, it is clear that at least in the forestry sector, human rights violations may be the unintended consequences of mitigation and adaptation efforts themselves. Addressing the linkages among climate change, forests and human rights thus presents potential conflicts among rights and rights holders.

As a result, climate policy needs to include human rights risk as a criterion in the selection among alternative policy options and institutional choice decisions. One way to apply the criterion would be to ask which mitigation or adaptation initiatives significantly preserve or enhance the human rights of the greatest number of people. An alternative method would be to determine which policies pose the lowest risk of significant human rights violations.

Such risk assessment should temper the pace and sequencing of the implementation of alternative policy options. For example, one might argue that, in light of the fraught human rights record of the forestry sector, forest-related emissions reduction schemes should be delayed in favour of less risky interventions related to renewable energy development, at least until more robust institutional safeguards are in place.

But applying such criteria will not be easy in the light of the suffering likely to result from failure to act. For example, a delay in addressing forest-based emissions would almost certainly result in higher emissions overall due to the higher cost of abatement in other sectors, and thus more suffering from climate change. There is also a significant time value of keeping carbon sequestered in trees and soil as long as possible, to buy time for technological advances to support the development of other mitigation options.

In addition, there is the need to avoid the risk of catastrophic dieback of forests as a result of climate change itself. As mentioned above, some models predict that increasing temperatures and drought will lead to the loss of forest cover if global warming is allowed to proceed beyond a certain threshold, thus releasing significant amounts of carbon into the atmosphere in a positive feedback loop (Woods Hole Research Center 2008). In other words, the international community may face the tragic choice between risking human rights violations to address forest-based emissions now, or risking the human rights implications of more catastrophic climate change later.

At a practical level, such choices will be faced in international negotiations regarding the degree to which human rights procedural safeguards should be embedded in global REDD governance and financing regimes.

Governments of forested countries will resist any perceived infringement of their national sovereignty. Others will argue that if REDD mechanisms are too encumbered by such safeguards, finance will flow only to international consultants and not to actual activities on the ground, as is asserted to have happened with the inclusion of afforestation/reforestation measures in the CDM. Achieving the right balance between effectiveness and efficiency in ensuring compliance with standards will not be easy.

At the level of individual REDD investments, such choices will be faced in the selection of specific countries and sites for programmatic interventions. For example, how much national and local institutional capacity is 'enough' to manage risks of unintended consequences for human rights? Consistent implementation of the procedural rights codified in the Aarhus Convention would be one strategy for ensuring that such decisions are made with the participation of key stakeholders.

Weighing trade-offs between the risks of action versus the risks of inaction could make use of the concept of human rights 'thresholds'. Under this approach, policy would require that a minimum acceptable level of protection of individual rights be defined and respected, whether from direct climate change impacts, on the one hand, or adaptation and mitigation efforts, on the other hand (ICHRP 2008). Application of such minimum thresholds would be likely to have the effect of ruling out both the most risky forest-related interventions, as well as the argument for no action.

One clear policy implication is the need for all forest-related mitigation and adaptation initiatives to be accompanied by robust monitoring and early warning systems to flag human rights problems as soon as they appear and force immediate course correction. Proponents of REDD policies and projects and forest-related adaptation interventions need to have policies in place to protect human rights, and ensure that those policies are enforced and subject to both internal self-assessment and external monitoring (Alcorn and Royo 2007, 131–2). For example, the World Bank's new resettlement policy includes a 'process framework' for reviewing human rights issues on an ongoing basis, while the World Wide Fund for Nature (WWF) undertook a self-assessment of its policy on indigenous peoples 10 years after its adoption.

Development assistance related to forests and climate change should be targeted to develop the capacity of duty bearers to meet their obligations, and the capacity of rights holders to claim those rights (UN Interagency Agreement 2003, as quoted in Luttrell and Piron 2005, 7).

Proponents of forest-related responses to climate change must themselves also invest in or be advocates for greater investment in strengthening forest governance more generally. Administrative capacity necessary to implement forest tenure reforms includes the ability to coordinate among different branches of government, sufficient budget to implement land titling, appropriate expertise at local levels and reform of unnecessary regulations that undermine fulfilment of local rights to forest land and resources (Sunderlin *et al.* 2008, 23).

If human rights violations are to be avoided, adoption of a rights-based approach to forest policy and management must accompany, and in many cases precede, implementation of climate-oriented interventions. Such an approach would include, *inter alia*, the training of forestry officials regarding their rights-related responsibilities, accelerated efforts to resolve conflicts over forest land and resources, increased transparency of forest-related data and decision-making and reform of laws, regulations and administrative and judicial mechanisms to recognize and protect forest peoples' rights and forest management systems (Colchester 2007).

Conclusion

The forestry sector illustrates how human rights are put at risk not only by the direct impacts of climate change, but also by policy responses designed to advance adaptation and mitigation objectives. Climate-related interventions risk exacerbating existing weaknesses and inequities that characterize current forest governance regimes. Because of the significance of forest-based emissions, and the possibility that climate change itself will undermine the mitigation potential of forests, the international community faces trade-offs between the human rights risks of forest-related interventions in the short run, and human rights risks of no action in the longer run.

This chapter suggests that such trade-offs be weighed explicitly, and that the risks of forest-related interventions be minimized through human rights safeguard policies, monitoring and assessment. In particular, respect for procedural rights is highlighted as a way of managing risk. In addition, capacity-building efforts should target the ability of duty bearers to guard against human rights violations resulting from changes in forest management, and the ability of rights holders to claim their rights through meaningful participation in climate policy formulation and implementation.

References

Agrawal, A., A. Chhatre and R. Hardin, 'Changing Governance of the World's Forests', *Science*, 320 (2008), 1460, 1462.

Alcorn, J. B. and A. G. Royo, 'Conservation's Engagement With Human Rights: Traction, Slippage or Avoidance', *IUCN Policy Matters*, 15 (2007), 115–39.

Barber, C. and J. Schweithelm, 'Trial by Fire: Forest Fires and Forestry Policy in Indonesia's Era of Crisis and Reform', World Resources Institute, Forest Frontiers Initiative (2000).

Barr, C. *Banking on Sustainability: Structural Adjustment and Forestry Reform in Post-Suharto Indonesia* (Washington, DC: WWF Macroeconomics Program Office and CIFOR, 2001), 140.

Bonan, G., 'Forests and Climate Change: Forcings, Feedbacks, and the Climate Benefits of Forests', *Science* (2008), 320.

Bradshaw, C., N. Sodhi, K. Peh and B. Brook. 'Global Evidence that Deforestation Amplifies Flood Risk and Severity in the Developing World', *Global Change Biology*, 13 (2007), 2379–95.

Bruijnzeel, L. A., A. Van Dijk, M. Van Noordwijk, A. Chappell and J. Schellekens, *Tropical Deforestation, People and Flooding: A Recent Analysis Claiming that Tropical Deforestation Amplifies Flood Risk and Severity Proves Less than Solid, Response to Bradshaw et al.* (World Agroforestry Center, 2007).

Chokkalingam, U., I. Kurniawan and Y. Ruchiat, 'Fire, Livelihoods, and Environmental Change in the Middle Mahakam Peatlands, East Kalimantan', *Ecology and Society*, 10 (2005), 26. Available at: www.ecologyandsociety.org/vol10/iss1/art26/.

Chomitz, K. M., *At Loggerheads? Agricultural Expansion, Poverty Reduction, and Environment in the Tropical Forests* (Washington DC: World Bank, 2007).

Colchester, M., 'Justice in the Forest: Rural Livelihoods and Forest Law Enforcement', *Forest Perspectives, CIFOR*, 3 (2006), 98.

Colchester, M., 'Beyond Tenure: Rights-based Approaches to Peoples and Forests. Some Lessons from the Forest Peoples Program', Presentation to the International Conference on Poverty Reduction in Forests: Tenure, Markets and Policy Reforms, Bangkok, Thailand, 3–7 September 2007.

Colchester, M. *et al.*, 'Procedural Irregularities and Standards Violations in IFC Support for Wilmar Trading', Letter to International Finance Corporation concerning IFC's financing of Wilmar Trading (2007). Available at: www.forestpeoples.org/documents/ifi_igo/ifc_wilmar_fpp_let_jul07_eng.pdf.

Colchester, M. and M. Ferrari, 'Making FPIC Work: Challenges and Prospects for Indigenous Peoples', FPIC Working Papers, *Forest Peoples Program*, June (2007).

Colfer, C. J. P., 'Longhouse of the Tarsier: Changing Landscapes, Gender, and Well Being in Borneo', Monograph No. 10, Borneo Research Council (Maine, 2008). Available at: www.cifor.cgiar.org/Publications/Detail?pid=2689.

Colfer, C. J. P., N. L. Peluso and S. C. Chin, *'Beyond Slash and Burn: Building on Indigenous Management of Borneo's Tropical Rainforests* (Bronx, NY: New York Botanical Garden, 1997).

Corpuz-Tauli, V., *Statement by the Chair of the UN Permanent Forum on Indigenous Issues on the occasion of the Launch of the Forest Carbon Partnership Facility* (Bali, Indonesia, 2007). Available at: www.forestpeoples.org/documents/forest_issues/unpfii_statement_fcpf_dec07_eng.pdf.

Debroux, L., T. Hart, D. Kaimowitz, A. Karsenty and G. Topa, *Forests in Post-Conflict Democratic Republic of Congo: Analysis of a Priority Agenda*, CIFOR, World Bank, CIRARD (2007).

Diaz, E. C., 'Climate Change, Forest Conservation and Indigenous Peoples Rights', *Global Forest Coalition*, Briefing paper (undated).

FAO Corporate Document Repository, 'The Clean Development Mechanism: Promoting Investment Flows From Developed to Developing Countries: The Possibility for Developing Countries to Participate in the Global Carbon Market' (undated), Retrieved on 9 September, 2008. Available at: www.fao.org/DOCREP/005/AC836E/AC836E06.htm.

FAO and CIFOR, 'Forests and Floods: Drowning in Fiction or Thriving on Facts?', *Forest Perspectives* 2, RAP Publication 2005/03 (2005).

Fargione, F., J. Hill, D. Tilman, S. Polasky and P. Hawthorne, 'Land Clearing and Biofuel Carbon Debt', *Sciencexpress*, 7 February 2008.

Forest Peoples Program (FPP), 'Statement by the International Forum of Indigenous Peoples on Climate Change, (IFIPCC, 2007), The 13th session of the Conference of the Parties to the UNFCCC SBSTA 27, agenda item 5/REDD.

Forest Peoples Program (FPP), 2008 Briefing, 'The Forest Carbon Partnership Facility: Facilitating the Weakening of Indigenous People's Rights to Lands and Resources'.

Foti, J., L. de Silva, J. Werksman, L. Shaffer, J. Talbot and H. McGray, 'Voice and Choice: Opening the Door to Environmental Democracy' (World Resources Institute, 2008).

Fujita, Y. and K. Phengsopha, 'The Gap Between Policy and Practice in Laos', in C. Colfer, G. Dahal and D. Capistranto (eds.), *Lessons from Forest Decentralization, Money, Justice and the Quest for Good Governance in Asia-Pacific* (London: CIFOR, Earthscan, 2008), 117–32.

Gender and Climate Change Network (GCCN), 'Women for Climate Justice Position Paper', UNFCCC COP 13, Bali, Indonesia, December 2007. Prepared by GenderCC Network–Women for Climate Justice.

Government of the State of Amazonas and Coordination Committee of the Amazonas Initiative (CC-AI), 'Amazonas Initiative on Climate Change, Forest Conservation and Sustainable Development', Bolsa Floresta Program (2007), 16–17. Available at: www.sds.am.gov.br/dsv/download/img_download/20071226155008Relatorio%202.pdf.

Griffiths, T., 'Seeing "REDD?" "Avoided deforestation" and the Rights of Indigenous Peoples and Local Communities' (Forest Peoples Program, 2007).

Guariguata, M., J. Cornelius, B. Locatelli, C. Forner and G. Sanchez-Azofeifa, 'Mitigation Needs Adaptation: Tropical Forestry and Climate Change', *Mitigation and Adaptation Strategies for Global Change*, 13:8 (2007).

Halvorsen, K., 'A Global Contract', The Ministers of Finance High Level Event on Climate Change, 11 December 2007, Session 3: Common Objectives on Climate Change Financial Issues. Available at: www.regjeringen.no/nb/dep/fin/dep/Finansministeren/taler_artikler/2007/A-Global-Contract-.html?id=493748.

Hare, B. and K. Macey, 'Tropical Deforestation Emission Reduction Mechanism: A Discussion Paper', *Greenpeace* (2008). Available at: www.greenpeace.org/international/press/reports/TDERM-full.

Holdren, J., 'Global Climatic Disruption: Risks and Opportunities', Woods Hole Research Center. See most current PPT version, 20 May 2008.

Hooijer, A., M. Silvius, H. Wosten and S. Page, 'Peat-CO_2 emissions from drained peatlands in SE Asia', Delft Hydraulics Report Q3943 (2006).

Howard, J., A. Downer and M. Turnbull, *Australia to Lead the World: Global Initiative on Forests and Climate*, March (2007), press release. Available at: www.cifor.cgiar.org/publications/pdf_files/media/Howard%20Downer%20Turnbull%20Release.pdf.

Humphreys, D. (ed.), *LogJam: Deforestation and the Crisis of Global Governance* (London: Earthscan, 2006), 302.

Institute for Agriculture and Trade Policy (IATP), *Agrofuels: Opportunity or Danger? A Global Dialogue on U.S. and EU Agrofuels and Agriculture Policies and their Impact on Rural Development in the North and South*, Conference Report, 12–14 December 2007, Berlin, Germany. Available at: www.iatp.org/tradeobservatory/library.cfm?refid=102587.

Intercooperation 2006, 'FORECA-Madagascar: Emission Reduction through Avoided Deforestation in the Context of the UNFCC: Committing Forests as Carbon Reservoir Pilot Project Madagascar', Intercooperation. Retrieved on 10 September 2008, available at www.intercooperation.ch/projects/p152.

International Council on Human Rights Policy (ICHRP), *Climate Change and Human Rights: A Rough Guide* (2008).

IPCC WG II, 2007, 'Summary for Policymakers', in M. L. Parry, O. F. Canziani, J. P. Palutikof, P. J. van der Linden and C. E. Hanson (eds.), *Climate Change 2007: Impacts, Adaptation and Vulnerability. Contribution of Working Group II to the Fourth Assessment Report of the Intergovernmental Panel on Climate Change* (Cambridge University Press, 2007), 7–22.

IPCC WG III, 2007, 'Summary for Policymakers' in B. Metz, O. R. Davidson, P. R. Bosch, R. Dave and L. A. Meyer (eds.), *Climate Change 2007: Mitigation, Contribution of Working Group III to the Fourth Assessment*

Report of the Intergovernmental Panel on Climate Change (Cambridge University Press, 2007).

Jagger, P., 'Negotiating Livelihoods and Sustainability after Uganda's Forest Sector Governance Reform', Pen Project 1, *PEN Newsletters*. Available at: www.cifor.cgiar.org/pen/_ref/events/penews/penews-2-2007.htm.

Kalame, F., M. Idinoba, M. Brockhaus and J. Nkem, 'Forest policies and forest resource flow in Burkina Faso, Ghana and Mali: Conflicting or consistent for adaptation to climate change?', CIFOR, *TroFCCA Brief* (2008).

Kanninen, M., D. Murdiyarso, F. Seymour, A. Angelsen, S. Wunder and L. German, 2007, 'Do Trees Grow on Money? The implications of Deforestation Research for Policies to Promote REDD', CIFOR, *Forest Perspectives*, 4 (2007), 61.

Larson, A. and J. Ribot, 'The Poverty of Forestry Policy: Double Standards on an Uneven Playing Field', Special Feature, *Sustainability Science* (2007).

Laurance, W., S. Bergen, M. Cochrane, P. Fearnside, P. Delamonica, S. d'Angelo, C. Barber and T. Fernandes, 'The Future of the Brazilian Amazon', *Science Magazine* (2001), Available at: http://lba.cptec.inpe.br/publications/Laurance_etal_2001.pdf.

Luttrell, C. and L. Piron, 'Operationalizing Norwegian People's Aid's Rights-Based Approach', Revised Final Report (Overseas Development Institute, 2005).

Luttrell, C., K. Schreckenberg and L. Peskett, 'The Implications of Carbon Financing for Pro-Poor Community Forestry', Forest Policy and Environment Programme (FPEP), *ODI Forestry Briefing*, 14 (2007).

Macchi, M., 'Indigenous and Traditional Peoples and Climate Change', IUCN, Issues paper (2008).

Mayle, F., R. Langstroth, R. Fisher and P. Meir, 'Long-Term Forest–Savannah Dynamics in the Bolivian Amazon: Implications for Conservation', *Philosophical Transactions of the Royal Society B*, 362 (2007), 291–307.

Mendelsohn, R., A. Dinar and L. Williams, 'The Distributional Impact of Climate Change on Rich and Poor Countries', *Environment and Development Economics*, 11 (2006), 159–78.

Menzies, N., *Our Forest, Your Ecosystem, Their Timber. Communities, Conservation, and the State in Community-Based Forest Management* (New York: Columbia University Press, 2007), 264.

Merrill Lynch, 'Merrill Lynch & Carbon Conservation Sign First Commercially Financed Avoided Deforestation Agreement', Merrill Lynch, Press Release 2008. Available at: www.ml.com/?id=7695_7696_8149_88278_95339_96307.

Millennium Ecosystem Assessment, *Ecosystems and Human Well-Being: Synthesis* (Washington, DC: Island Press, 2005).

Naylor, R., A. Liska, M. Burke, W. Falcon, J. Gaskell, S. Rozelle and K. Cassman, 'The Ripple Effect: Biofuels, Food Security, and the Environment', *Environment*, 49 (2007) 30–43.

PEACE 2007, 'Indonesia and Climate Charge: Current Status and Policies', Available at: siteresources.worldbank.org/INTINDONESIA/Resources/Environment/ClimateChange_Full_EN.pdf.

Peluso, N. L., 'Coercing Conservation? The Politics of State Resource Control', *Global Environmental Change*, 32 (1993), 199–218.

Reuters, 'Indonesia's Aceh, Papua Pledge to Protect Forests', 27 April 2007. Retrieved on 15 July, 2008. Available at: unfcccbali.org/unfccc/index2.php?option=com_content&task=view&id=22&pop=1&page=420&Itemid=41.

Salick, J. and A. Byg, 'Indigenous Peoples and Climate Change', *Tyndall Centre Publication* (Oxford: Tyndall Centre for Climate Change Research, 2007).

Searchinger, T., R. Heimlich, R. A. Houghton, F. Dong, A. Elobeid, J. Fabiosa, S. Tokgoz, D. Hayes and T. Yu, 'Use of U.S. Croplands for Biofuels Increases Greenhouse Gases Through Emissions from Land Use Change', *Sciencexpress* (2008).

Seymour, F., 'Sustaining the Environment at the World Bank', *WRI Policy Note*, September (Washington DC: WRI, 2006).

Seymour, F., 'Conservation, Displacement, and Conservation', in Michael Cernea and Hari Mohan Mathur (eds.), *Can Compensation Prevent Impoverishment?* (New Delhi: Oxford University Press, 2008).

Shackleton, S., S. Kaschula, W. Twine, L. Hunter, C. Holding-Anyonge and L. Petheram, 'Forests as Safety Nets for Mitigating the Impacts of HIV/AIDS in Southern Africa', *CIFOR Livelihood Brief 4* (CIFOR, 2006), 4.

Sheil, D., A. Casson, E. Meijaard, M. van Nordwijk, J. Gaskell, J. Sunderland-Groves, K. Wertz and M. Kanninen, 'The Impacts and Opportunities of Oil Palm in Southeast Asia: What do we Know and What do we Need to Know?', Occasional paper No. 51 (CIFOR, Bogor, Indonesia, 2009).

Sheil, D., R. K. Puri, I. Basuki, M. van Heist, Saefuddin, Rukmiyati, M. A. Sardjono, I. Samsoedin, K. D. Sidiyasa, Chrisandini E. Permana, E. M. Angi, F. Gatzweiler, B. Johnson and A. Wijaya, *Exploring Biological Diversity, Environment and Local People's Perspectives in Forest Landscapes: Methods for a Multidisciplinary Landscape Assessment* (Bogor, Indonesia: CIFOR, 2001), 93.

Stern, Nicholas, *The Economics of Climate Change: The Stern Review* (2007).

Sunderlin, W., 'Effects of Crisis and Political Change, 1997–1999', in C. J. P. Colfer and I. A. P. Resosudarmo (eds.), *Which Way Forward? People, Forests and Policymaking in Indonesia* (Washington, DC: Resources for the Future/Center for International Forestry Research, 2002), 246–76.

Sunderlin, W., A. Angelsen, B. Belcher, P. Burgers, R. Nasi, L. Santoso and S. Wunder, 'Livelihoods, Forests and Conservation in Developing Countries: An Overview', *World Development*, 33 (2005), 1383–402.

Sunderlin, W., J. Hatcher and M. Liddle, *From Exclusion to Ownership? Challenges and Opportunities in Advancing Forest Tenure Reform* (Washington, DC: Rights and Resources Initiative, July 2008).

Tacconi, L. (ed.), *Illegal Logging: Law Enforcement, Livelihoods, and the Timber Trade* (London: Earthscan, 2007).

Thomas, D., 'A Brief Review of Upland Agricultural Development in the Context of Livelihoods, Watersheds and Governance for Area-based Development Projects in the Lao PDR', ICRAF Report for International Fund for Agricultural Development (IFAD), September (2003).

Thomas, D. and C. Twyman, 'Equity and Justice in Climate Change Adaptation Amongst Natural-resource-dependent Societies', *Global Environmental Change*, 15 (2005), 115–24.

TroFCCA 2008, CIFOR – Tropical Forests and Climate Change Adaptation, Central America. Available at: www.cifor.cgiar.org/trofcca/_ref/america/index.htm.

UNECE 1998, 'Convention on Access to Information, Public Participation in Decision-Making and Access to Justice in Environmental Matters', 25 June, 1998 in Aarhus, Denmark.

United Nations Development Program (UNDP), United Nations Environment Program, World Bank and World Resources Institute, *The Wealth of the Poor-Managing Ecosystems to Fight Poverty* (Washington DC, 2005).

UNFCCC 2005, 11th Session, Montreal, 28 November to 9 December 2005, Item 6, 'Reducing Emissions from Deforestation in Developing Countries: Approaches to Stimulate Action'. Available at: http://unfccc.int/resource/docs/2005/cop11/eng/misc01.pdf.

UNFCCC 2007a, Bali, Indonesia, 3–15 December 2007. 1/CP.13. Bali Action Plan. Available at: http://unfccc.int/resource/docs/2007/cop13/eng/06a01.pdf#page=8.

UNFCCC 2007b, Bali, Indonesia, 3–15 December 2007. 2/CP.13. Reducing Emissions from Deforestation in Developing Countries: Approaches to Stimulate Action. Available at: http://unfccc.int/resource/docs/2007/cop13/eng/06a01.pdf#page=8.

UNFCCC 2007c, Subsidiary Body for Scientific and Technological Advice, 27th Session, Bali, Indonesia, 3–11 December 2007, 'Reducing Emission From Deforestation in Developing Countries: Approaches to Stimulate Action', Government of Tuvalu proposal.

Wells, A., C. Luttrell, D. Brown and N. Bird, 'Public Goods and Private Rights: The Illegal Logging Debate and the Rights of the Poor', *Forestry Briefing* (Overseas Development Institute, 2006).

White, A. and A. Martin, *Who Owns the World's Forests?* (Washington, DC: Forest Trends, 2002), 30.

Wollenberg, E., B. Belcher, D. Sheil, S. Dewi and M. Moelino, *Why are Forest Areas Relevant to Reducing Poverty in Indonesia?, Governance Brief* (CIFOR, 2004).

Woods Hole Research Center (WHRC), South America, 'Fire and Savannization', (2008). Available at: www.whrc.org/southamerica/fire_savann/index.htm.

World Bank (2004), *Sustaining Forests: A Development Strategy* (Washington, DC: World Bank, 2004). Available at: siteresources.worldbank.org/INTFORESTS/Resources/SustainingForests.pdf.

World Bank (2006), 'Sustaining Economic Growth, Rural Livelihoods, and Environmental Benefits: Strategic Options for Forest Assistance in Indonesia' (2006).

World Bank (2007), 'Forest Carbon Partnership Facility Launched at Bali Climate Meeting', Press release No. 2008/142/SDN (2007).

Wright, T., 'Merrill Lynch: Turning Trees into Money', *Wall Street Journal* (2008), accessed on 15 July 2008.

Wunder, S., 'The Efficiency of Payments for Environmental Services in Tropical Conservation', *Conservation Biology*, 21:1 (2007), 48–58.

8

Climate change and the right to the highest attainable standard of health

PAUL HUNT* AND RAJAT KHOSLA**

The health of individuals, communities and populations requires more than medical care. Equally important are the environmental, social, cultural, economic, political and other conditions that make people need medical care in the first place.[1]

For this reason, the right to the highest attainable standard of health is an inclusive right extending not only to timely and appropriate medical care but also to the underlying determinants of health, such as access to safe water and adequate sanitation, an adequate supply of safe food, nutrition and housing, healthy occupational and environmental conditions, access to health-related education and information, including on sexual and reproductive health and freedom from discrimination.[2] In short, the right to the highest attainable standard of health encompasses both medicine and public health. The International Covenant on Economic, Social and Cultural Rights (ICESCR) and the Convention on the Rights of the Child (CRC), for example, are clear that the right to health extends beyond medical care.[3]

Given the massive public health challenge posed by climate change, especially in the developing world, there is an urgent need for a global

* Professor of Law, University of Essex, UK; Adjunct Professor, University of Waikato, New Zealand; United Nations Special Rapporteur on the right to the highest attainable standard of health (2002–8). We would like to thank Kate Raworth for her comments on a draft of this chapter.

** Senior Research Officer, Human Rights Centre, University of Essex, UK.

1 R. Beaglehole, 'Overview and Framework', in R. Detels (ed.), *Oxford Textbook of Public Health*, 4th edn. (Oxford University Press, 2002).

2 UN Committee on the Economic, Social and Cultural Rights (CESCR), General Comment No. 14: The right to the highest attainable standard of health, UN Doc. E/C.12/2000/4 (2000), para. 11.

3 Article 12 of the International Covenant on Economic, Social and Cultural Rights, and Article 24 of the Convention on the Rights of the Child.

partnership aimed at establishing an effective, integrated environmental regime capable of ensuring healthy environmental conditions for all. Here we signal some dimensions of the contribution that the right to the highest attainable standard of health can make as the international community seeks to respond to this challenge. Climate change represents an extremely grave risk to the health of individuals, communities and populations, especially those living in poverty in developing countries. Crucially, states have an obligation, arising from the right to the highest attainable standard of health, to take reasonable steps to slow down and reverse climate change.

In this chapter, we briefly examine the contours and content of the right to the highest attainable standard of health, explore some of the linkages between climate change and health, outline the right-to-health analytical framework, apply some of this framework to climate change and close with some brief concluding remarks. Our survey provides no more than an introduction to a wide range of complex issues that call for more detailed examination.

The right to the highest attainable standard of health

At the international level, the right to the highest attainable standard of health was first articulated in the Constitution of the World Health Organization (WHO) 1946.[4] Subsequently, the right to the highest attainable standard of health was codified in numerous legally binding international and regional human rights treaties.[5] (The full name of the right is 'the right of everyone to the enjoyment of the highest attainable standard of physical and mental health'. For convenience, this is often shortened to the 'right to the highest attainable standard of health' or the 'right to health'.) Adopted by the United Nations Committee on Economic, Social and Cultural Rights, General Comment 14 provides an authoritative interpretation of the right to the highest attainable standard of health.[6] This substantive instrument confirms that the right to health not only encompasses access to health care, but also the underlying determinants of health, such as a healthy environment. The right has a preoccupation with disadvantaged groups, participation

[4] See Preamble, Constitution of the World Health Organization, 1946.

[5] P. Hunt, *Report of the Special Rapporteur on the Right of Everyone to the Enjoyment of Physical and Mental Health*, UN Doc. E/CN.4/2003/58 (2003).

[6] CESCR, General Comment 14.

and accountability. Moreover, it places a responsibility on high income countries to help developing countries to deliver the right to health to their people. The right requires an effective, inclusive health system of good quality. Although General Comment 14 leaves many questions unanswered, it remains groundbreaking and marks the moment when the right to health ceased to be a slogan and became an important instrument for all health policymakers and practitioners.

The right is also enshrined in numerous national constitutions: over 100 constitutional provisions include the right to health or health-related rights.[7] Moreover, in some jurisdictions constitutional provisions on the right to the highest attainable standard of health have generated significant jurisprudence, for example, the Ecuadorian case of *Mendoza and others* v. *Ministry of Public Health and the Director of the HIV-AIDS National Programme*.[8] In this case, when a public hospital withdrew antiretroviral therapy from the applicants who were living with HIV/AIDS, the Court held that the Ministry of Health had violated the applicants' fundamental rights to life and health and required that remedial measures be taken.

At root, the right to the highest attainable standard of health consists of globally legitimized standards; out of these standards derive legal obligations, and these obligations demand effective mechanisms of accountability. In principle, the combined effect of these three dimensions – standards, obligations and accountability – is to empower vulnerable individuals and disadvantaged communities.

While the right to the highest attainable standard of health is a powerful campaigning and advocacy tool, it also has normative depth and something constructive and concise to say to policymakers. The right to health can help to ensure that health policies devote particular attention to the vulnerable and disadvantaged, enhance community participation, ensure that health interventions strengthen health systems and so on. If integrated into national and international health policy-making, the right to health can help to establish policies that are robust, sustainable, equitable and meaningful to those living in poverty.[9]

7 E. Kinney and B. Clark, 'Provisions for Health and Health Care in the Constitutions of the Countries of the World', *Cornell International Law Journal*, 37 (2004), 285–305.

8 Resolution No. 0749-2003-RA, 28 January 2004.

9 P. Hunt, *Report of the Special Rapporteur on the Right of Everyone to the Enjoyment of Physical and Mental Health*, UN Doc. A/HRC/4/28 (2007).

The right to health brings with it the crucial requirement of accessible, transparent and effective mechanisms of monitoring and accountability.[10] Those with right-to-health responsibilities must be held to account in relation to the discharge of their duties, with a view to identifying successes and difficulties; so far as necessary, policy and other adjustments can then be made. While states are permitted to choose different forms of accountability, all accountability mechanisms must be accessible, transparent and effective.

Accountability mechanisms fall into various categories. For example, it may be possible to pursue a complaint and seek a legally binding decision in the courts if the right to health or another relevant right (for example, freedom from discrimination) is enshrined in the Constitution or other statute.[11] A national human rights commission or independent ombudsperson who is independent from the government may be empowered to make inquiries into complaints and to issue authoritative recommendations. Political accountability mechanisms may also exist, such as parliamentary scrutiny committees.

One way to advance the right to health is to ensure that it is integrated in all relevant policy-making processes, including those for poverty reduction and international development.[12] A policy approach also demands vigilant accountability and monitoring mechanisms including, but not limited to, some of those previously mentioned. It may include use of publicly available human rights impact assessments to anticipate the likely impact of a proposed policy upon the right to health. A policy approach also requires the use of indicators and benchmarks to measure whether or not policies advance the right to health over time.[13]

At the international level, if a state is a party to a treaty that enshrines the right to health, accountability mechanisms commonly fall into three groups. First, the state may be obliged to submit periodic reports to an independent body responsible for monitoring compliance with the treaty, such as the United Nations Committee on Economic, Social and Cultural Rights. This Committee reviews the reports received from

[10] P. Hunt, *Report of the Special Rapporteur on the Right of Everyone to the Enjoyment of Physical and Mental Health*, UN Doc. A/63/263 (2008).

[11] P. Hunt, *Report of the Special Rapporteur on the Right of Everyone to the Enjoyment of Physical and Mental Health*, UN Doc. A/HRC/4/28 (2007), ch. 3.

[12] P. Hunt, 'Using All the Tools at Our Disposal: Poverty Reduction and the Right to the Highest Attainable Standard of Health', *Development Outreach* (2006), 18–20.

[13] P. Hunt, *Report of the Special Rapporteur on the Right of Everyone to the Enjoyment of Physical and Mental Health*, UN Doc. E/CN.4/2006/48 (2006).

states, and 'shadow reports' received from non-governmental organizations (NGOs), before issuing observations about the states' compliance with their obligations. NGOs have a vital role to play in preparing accurate and authoritative 'shadow reports'.[14] Second, in relation to some, but not all, treaties, individuals are entitled to submit complaints to the relevant treaty body after first pursuing the matter at the domestic level.[15] Third, some treaties establish an inquiry mechanism authorizing the relevant treaty body to investigate and report on gross or systematic violations of a particular right.[16]

Even if a state has not ratified a treaty enshrining the right to health, it may still be possible to access some accountability mechanisms at the international level. For example, Special Rapporteurs are independent experts appointed by the United Nations Human Rights Council in relation to a particular country or a theme, including the right to health. A Special Rapporteur may ask a state to respond to complaints of human rights violations that he or she receives from individuals, irrespective of whether that state is a party to one or more human rights treaties.[17]

Impact of climate change on health

The climate is changing and we are experiencing more extreme weather patterns. The extent of climate change, and also its impact, varies across countries and the world. Some areas, such as the Arctic, are warming more and faster than others. Some coastal regions will be affected by rising sea levels, while some regions will be more affected by extreme weather events, such as droughts, storms and floods. Good health depends on access to safe drinking water, sufficient food, secure shelter and favourable social conditions, all of which will be affected by a changing climate. Climate change is likely to have wide-ranging impacts on human health, resulting in significant loss of life.[18]

[14] M. Craven, *The International Covenant on Economic, Social and Cultural Rights: A Perspective on Its Development* (Oxford: Clarendon Press, 1995).

[15] See, for instance, First Optional Protocol to the International Covenant on Civil and Political Rights, 1966.

[16] Optional Protocol (1999) of the Convention on the Elimination of All Forms of Discrimination against Women, 1979.

[17] For information on United Nations Special Rapporteurs, see www2.ohchr.org/english/issues/health/right/index.htm (accessed September 2008).

[18] Second Assessment Report of the Intergovernmental Panel on Climate Change, *Synthesis of Scientific-Technical Information Relevant to Interpreting Article 2 of UNFCCC* (1995), 35.

McMichael observes that climate change is not merely another addition to the list of environmental health hazards, each warranting separate epidemiological study and risk management. A complex global environmental hazard, climate change is intrinsically different from exposure to radiation or a specific toxic chemical. The overall risk to health is more than the aggregation of specific disease risks due to particular climatic factors.[19]

Climate change is likely to impact human health directly and indirectly. The direct health impacts include those due to changes in exposure to thermal extremes, both hot and cold, increases in extreme weather events, such as floods, cyclones, storms, droughts and increased production of certain air pollutants and aeroallergens, such as spores and moulds which result in severe health repercussions.[20] Additionally, it is likely that climate change will impact human health indirectly, such as by affecting regional food productivity. It will affect the transmission of many infectious diseases, especially water, food and vector-borne diseases. Excessive rainfall and runoff can lead to large numbers of microorganisms entering drinking water, resulting in outbreaks of water-borne diseases.[21] In the longer term, these indirect impacts are likely to have greater magnitude than more direct impacts.

The Intergovernmental Panel on Climate Change (IPCC) confirms that climate change is likely to affect the health status of millions of people, particularly those with low adaptive capacity, through a combination of factors, such as increase in malnutrition and consequent disorders, with implications for child growth and development, increased deaths, disease and injury due to heat waves, floods, storms, fires and droughts, increased burden of diarrhoeal disease, increased frequency of cardio-respiratory diseases due to higher concentrations of ground level ozone and the altered spatial distribution of some infectious disease vectors.[22] Changing temperatures will cause some infectious diseases to spread into new areas, so extending the range of diseases. Further, while

[19] A. McMichael, 'Climate Change and Risk to Health', *British Medical Journal*, 329 (2004), 1416–17.

[20] A. McMichael, 'Global Environmental Changes, Climate Change and Human Health', in Kelly Lee and Jeff Collin (eds.), *Global Change and Health* (Berkshire: Open University Press, 2005), 140.

[21] McMichael, 'Global Environmental Changes', 144.

[22] Contribution of Working Group II to the Fourth Assessment Report of the Intergovernmental Panel on Climate Change, *Impacts, Adaptation and Vulnerability* (Cambridge University Press, 2007), 8.

some impacts will occur soon, others will depend on a succession of changes in natural systems, and may occur incrementally over a longer period.

The UNFCCC's Investment and Financial Flows report of 2007 estimates that for atmospheric concentrations of 550 parts of CO_2 equivalent per million by volume (ppmv), which is highly likely to be reached and surpassed, the projected number of additional cases of diarrhoeal diseases (in 2030 as compared with 1990) as a result of climate change will be 113 million per year, the number of additional cases of malnutrition will be 3.1 million cases per year and the number of additional cases of malaria will be 17.4 million per year. Estimates for these diseases are even higher for atmospheric concentrations of 750 ppmv.[23]

Warmer and wetter conditions resulting from climate change are increasing the range and season of vectors, such as mosquitoes and tsetse flies, which spread diseases such as malaria, dengue and yellow fever and encephalitis. Climate change will adversely affect the world's hydrological cycle and result in more droughts and floods. Drought poses serious threats to health. As clean water sources evaporate, people resort to more polluted alternatives that may lead to epidemics of water-borne diseases. Likewise, floods not only increase the risk of drowning and destruction of crops, they also spread disease by extending the range of vectors and by washing agricultural pollutants into drinking water supplies.

The impact of climate change on human health will not be evenly distributed around the world. Developing country populations, particularly in small island states, arid and high mountain zones and densely populated coastal areas are particularly vulnerable. The Human Development Report (HDR) 2007 estimates that major killer diseases could expand their coverage. For example, an additional 220–400 million people could be exposed over the next hundred years to malaria, a disease that already claims around 1 million lives annually.[24] Exposure rates for sub-Saharan Africa, which accounts for around 90 per cent of malarial deaths, are projected to increase by 16–28 per cent.[25] Dengue fever is already in evidence at higher altitudes than has previously been the case,

[23] UNFCCC, *Investment and Financial Flows to Address Climate Change* (Bonn: UNFCCC, 2007).

[24] United Nations Development Programme (UNDP), *Human Development Report: 2007* (New York: UNDP, 2007), 19.

[25] UNDP, *Human Development Report: 2007*, 45.

especially in Latin America and parts of East Asia. Climate change could further expand the reach of these diseases.

Despite these disturbing trends, the international community has not yet confronted the health threats posed by climate change. The failure of the international community to take the health impact of climate change seriously endangers the lives of millions of people across the world.

The poor and other marginalized groups are most affected by climate change. By impacting their health and livelihoods, climate change imposes a twin threat on people living in poverty. An increase in morbidity (prevalence of illness) and mortality leads to a reduction in earning capacity, creating a vicious cycle of poverty and ill-health that is devastating among the poorest. Poverty renders women and men ill-equipped to protect themselves and their children from diseases or to seek treatment for illness. Poor health and an impaired ability to work, compounded by high health costs, deepens poverty.

Moreover, health systems in developing countries are not equipped to tackle the threat posed by climate change. Whereas the rich countries are already preparing their public health systems to deal with future climate shocks, developing countries, which will experience greater effects of climate change because of high levels of poverty, weak public health systems and the geographical location of some of these countries, which render them more susceptible to extremes of climate, are not prepared.[26] Climate change superimposes new demands on already weak health systems.

The *Stern Review* emphasises that climate change will amplify health disparities between rich and poor parts of the world. The WHO estimates that climate change since the 1970s has already been responsible for 150,000 deaths each year through increasing incidence of diarrhoea, malaria and malnutrition predominantly in Africa and other developing regions.[27] A 1°C increase in global temperatures above pre-industrial levels could, according to WHO, double deaths from climate change to at least 300,000.[28] At higher temperatures, death rates will increase sharply, for example, with millions more people dying from malnutrition each year.[29]

[26] UNDP, *Human Development Report: 2007.*

[27] A. McMichael *et al.*, 'Global Climate Change', in M. J. Ezzati *et al.* (eds.), *Comparative Quantification of Health Risks: Global and Regional Burden of Disease Due to Selected Major Risk Factors* (Geneva: World Health Organisation, 2004), 1543–649.

[28] McMichael, 'Global Climate Change'.

[29] International Council on Human Rights Policy, *Climate Change and Human Rights: A Rough Guide* (Geneva: ICHRP, 2008), 100.

The right-to-health analytical framework

In recent years the Committee on Economic, Social and Cultural Rights (CESCR), WHO, civil society organizations, academics and many others,[30] have developed a way of 'unpacking' or analysing the right to health with a view to making it easier to understand and apply in practice to health-related policies, programmes and projects. For his part, the Special Rapporteur on the right to the highest attainable standard of health has applied and refined this analytical framework in his country-level and other reports.[31] Importantly, the framework has general application to all aspects of the right to the highest attainable standard of health, including the underlying determinants of health, such as a healthy environment.

While the analytical framework is discussed in detail elsewhere,[32] its key elements may be very briefly outlined as follows.

National and international human rights laws, norms and standards: the laws, norms and standards relevant to the particular issue, programme or policy must be identified.[33]

Resource constraints and progressive realisation: international human rights law recognizes that the realization of the right to health is subject to resource availability. Thus, what is required of a developed state today is of a higher standard than what is required of a developing state. However, a state is obliged – whatever its resource constraints and level of economic development – to realize progressively the right to the highest attainable standard of health.[34]

Obligations of immediate effect: despite resource constraints and progressive realization, the right to health also gives rise to some obligations of immediate effect, such as the duty to avoid discrimination.[35]

[30] See for instance, WHO, *Human Rights, Health and Poverty Reduction Strategies* (Geneva: World Health Organisation, 2005); Physicians for Human Rights, *Deadly Delays: Maternal Mortality in Peru* (Boston: Physicians for Human Rights, 2007).

[31] See for instance, P. Hunt, *Report of the Special Rapporteur on the Right of Everyone to the Enjoyment of Physical and Mental Health*, UN Doc. A/61/338 (2006); see also P. Hunt, *Report of the Special Rapporteur on the Right of Everyone to the Enjoyment of Physical and Mental Health on Mission to Uganda*, UN Doc. E/CN.4/2006/48/Add.2 (2006).

[32] See, P. Hunt, *Report of the Special Rapporteur on the Right of Everyone to the Enjoyment of Physical and Mental Health*, UN Doc. E/CN.4/2005/51 (2005).

[33] P. Hunt, *Report of the Special Rapporteur on the Right of Everyone to the Enjoyment of Physical and Mental Health*, UN Doc. E/CN.4/2003/58 (2003).

[34] ICESCR, Article 2(1).

[35] CESCR, General Comment 14, para. 43.

These are obligations without which the right would be deprived of its raison d'être and as such they are not subject to progressive realization, even in the presence of resource constraints.[36]

Freedoms and entitlements: the right to health includes both freedoms (for example, freedom from discrimination) and entitlements (for example, the provision of a system of health protection that includes minimum essential levels of water and sanitation). For the most part, freedoms do not have budgetary implications, while entitlements do.

Available, accessible, acceptable and good quality: all health services, goods and facilities should comply with each of these four requirements; in other words, a particular health service must be available, accessible (for example, affordable and physically accessible), culturally acceptable and of good quality.[37] There is a similarity between this right to health analysis and the public health requirements set out in the Declaration of Alma-Ata (1978): geographical accessibility; financial accessibility; cultural accessibility; and functional accessibility.[38]

Respect, protect, fulfil: states have duties to respect, protect and fulfil the right to the highest attainable standard of health, as explained and used by CESCR and the Committee on the Elimination of Discrimination Against Women (CEDAW).[39]

Non-discrimination, equality and vulnerability: because of their crucial importance, the analytical framework demands that special attention be given to issues of non-discrimination, equality and vulnerability in relation to all elements of the right to the highest attainable standard of health.

Active and informed participation: the right to health requires that there be an opportunity for individuals and groups to participate actively and in an informed manner in health policy-making processes that affect them.[40]

International assistance and cooperation: in line with obligations envisaged in the United Nations Charter and some human rights treaties,[41] developing countries have a responsibility to seek international

[36] CESCR, General Comment 3, para. 10. [37] CESCR, General Comment 14, para. 12.

[38] WHO, A Joint Report by the Director-General of the WHO and the Executive Director of UNICEF presented at the International Conference on Primary Health Care (1978), Alma-Ata (Geneva: WHO, 1978).

[39] See, CESCR, General Comment 14, para. 33; CEDAW, General Recommendation 24, para. 13.

[40] CESCR, General Comment 14, para. 54.

[41] United Nations Charter, 1945; International Covenant on Economic, Social and Cultural Rights, 1966; Convention on the Rights of the Child, 1990.

assistance and cooperation, while developed states have some responsibilities toward the realization of the right to health in developing countries.[42]

Monitoring and accountability: transparent, effective and accessible monitoring and accountability mechanisms, including redress, are among the most crucial characteristics of the right to the highest attainable standard of health.[43]

By way of illustration we next briefly apply some of the elements of this framework to climate change.

Climate change and the right to health

Progressive realization and obligations of immediate effect

The right to the highest attainable standard of health, including the underlying determinants of health, such as healthy environmental conditions, are subject to progressive realization and resource availability.[44] Put simply, progressive realization means that a state is required to be doing better in two years time than it is doing today, while resource availability means that what is required of a developed state is of a higher standard than what is required of a developing state.

Progressive realization has a number of important implications. For example, states need appropriate indicators and benchmarks so they know whether or not they are progressively realizing the right to health.[45] Also, progressive realization gives rise to 'a strong presumption that retrogressive measures taken in relation to the right to health are impermissible'.[46] As the United Nations Committee on Economic, Social and Cultural Rights continues:

> If any deliberately retrogressive measures are taken, the State party has the burden of proving that they have been introduced after the most careful consideration of all alternatives and that they are duly justified by

[42] P. Hunt, *Report of the Special Rapporteur on the Right of Everyone to the Enjoyment of Physical and Mental Health*, A/HRC/7/11/Add.2.

[43] H. Potts, *Accountability and the Right to the Highest Attainable Standard of Health* (Essex: Human Rights Centre, 2008).

[44] ICESCR, Article 2(1).

[45] For a human rights-based approach to health indicators, see the UN Special Rapporteur's report E/CN.4/2006/48.

[46] CESCR, General Comment 14, para. 32.

> reference to the totality of the rights provided for in the Covenant in the context of the full use of the State party's maximum available resources.[47]

In short, there is a rebuttable presumption that acts or omissions worsening environmental conditions and health are inconsistent with the right to the highest attainable standard of health.

However, progressive realization also has an important qualification: the right to health includes some core obligations of immediate effect that are not subject to progressive realization.[48] These are obligations without which the right is deprived of its raison d'être.[49] In its General Comment 14, the Committee lists some obligations that are 'at the very least' minimum core obligations arising from the right to the highest attainable standard of health.[50] These obligations include, for example, access to health facilities, goods and services on a non-discriminatory basis, especially for vulnerable or marginalized groups, access to the minimum essential food which is nutritionally adequate and safe, to ensure freedom from hunger to everyone, access to basic shelter, housing and sanitation and an adequate supply of safe and potable water. The wording is clear: this is not an exhaustive list. According to the Committee, it may revise and update its General Comments.[51] In other words, in the light of contemporary developments, new obligations may be added to the minimum core obligations listed by the Committee in General Comment 14.

As we have seen, climate change represents an extremely grave risk to the health of individuals, communities and populations, especially those living in poverty in developing countries. In our view, states have a core obligation, arising from the right to the highest attainable standard of health, to take reasonable steps to slow down and reverse climate change. For example, states have a core obligation to take reasonable steps to stabilize greenhouse gas concentrations in the atmosphere at a level that prevents dangerous anthropogenic interference with the climate system.[52]

Here we focus on the national dimension of the right to health. But this fundamental human right also has an international dimension. Later in

[47] CESCR, General Comment 14 para. 22. [48] CESCR, General Comment 14, para. 43.

[49] CESCR, General Comment 3, para. 10. [50] CESCR, General Comment 14, para. 43.

[51] United Nations Economic and Social Council, *Committee on Economic, Social and Cultural Rights, Report of the Twentieth and Twenty-First Sessions* (26 April–14 May 1999, 15 November–3 December 1999) UN ESCOR, Supp. No. 2, [51] UN Doc. E/2000/22-E/C.12/1999/11.

[52] United Nations Framework Convention on Climate Change (UNFCCC), 1992, Article 2.

this section, we outline this international dimension especially developed states' human rights responsibility of international assistance and cooperation in health.

Particular vulnerability of those who are already disadvantaged

Climate change tends to aggravate existing vulnerabilities. As the IPCC puts it: 'Overall, climate change is projected to increase threats to human health, particularly in lower income populations, predominantly within tropical/subtropical countries.'[53] According to United Nations Secretary General Ban Ki-moon, the terrible irony for many developing countries is that, while they have contributed the least to the process of climate change, they are the ones most at risk from its consequences, and least able to cope.[54] In short, those living in poverty are disproportionately affected by the adverse impact of global warming.

Arising from the concepts of non-discrimination and equality, international human rights law has a preoccupation with vulnerability and disadvantage. States have an obligation to take measures in favour of disadvantaged individuals, communities and populations.[55] Thus, they have a human rights duty, arising from non-discrimination and equality, to take reasonable steps to stop and reverse climate change. This resonates with the position taken by the IPCC that climate change must be stopped from reaching a level that is dangerous to the most vulnerable.[56]

Participation

The active and informed participation of individuals and communities in health policy-making that affects them is an important feature of the right to the highest attainable standard of health.[57] States must make a special effort to facilitate the active and informed participation of

[53] IPCC, *Third Assessment Report of the Intergovernmental Panel on Climate Change* (2001), 9.

[54] United Nations Secretary General, Ban Ki-moon, 'Address to High-Level Event on Climate Change', 24 September 2007.

[55] United Nations Office of the High Commissioner for Human Rights, *Principles and Guidelines for a Human Rights Approach to Poverty Reduction Strategies* (2006).

[56] IPCC, *Fourth Assessment Report of the Intergovernmental Panel on Climate Change* (Cambridge University Press, 2007).

[57] See, P. Hunt, *Report of the Special Rapporteur on the Right of Everyone to the Enjoyment of Physical and Mental Health*, UN Doc. A/59/422, 2004; see also H. Potts, *Participation and the Right to the Highest Attainable Standard of Health* (Essex: Human Rights Centre, 2008).

individuals and communities that have traditionally been excluded or marginalized. Even though those living in poverty are disproportionately affected by the adverse affects of global warming, they are invariably excluded from relevant policy discussions. While it might not be necessary for those living in poverty to take part in all the technical deliberations that underlie policy formulation, they are entitled to participate in the policy-making that will guide such deliberations. In other words, states have a human rights responsibility to take measures that facilitate, in all relevant policy-making, the active and informed participation of those affected by climate change, including those living in poverty.

Poor and marginalized people, for example, have a right to be involved in national adaptation initiatives. Such initiatives should be community-based, with the community's needs and interests at the heart of planning and implementation. Women, indigenous people and ethnic minorities in particular must be supported as participants, since their needs and interests are often marginalized within communities. Further, countries that are home to many of the world's most climate-vulnerable people must be able to participate in target setting for cutting global emissions.

Participation is not only right as a matter of principle; it also makes sense as a matter of practice. Individuals, communities and populations usually have a very keen sense of their environmental needs and priorities.

International assistance and cooperation

States have an obligation to take steps individually, and through international assistance and cooperation, towards the realization of the right to the highest attainable standard of health.[58] This obligation gives rise to the human rights responsibility of international assistance and cooperation in health.[59] Subject to resource availability and progressive realization, developed countries have a human rights responsibility to provide financial and technical assistance to supplement the resources of developing countries, with a view to ensuring that everyone has access to healthy environmental conditions.

[58] ICESCR, Articles 2(1) and 12.

[59] For a detailed discussion on the human rights responsibility of international assistance and cooperation in health, see Paul Hunt, *Report of the Special Rapporteur on the Right of Everyone to the Enjoyment of Physical and Mental Health on Missions to World Bank and the International Monetary Fund and Uganda*, UN Doc. A/HRC/7/11/Add.2 (2008).

The right to the highest attainable standard of health gives rise to three layers of obligations on states: to *respect*, *protect* and *fulfil*. In the context of international assistance and cooperation in health, states must ensure that their actions *respect* the right to health in other countries. The duty to respect obliges states, for example, to refrain from activities that harm the composition of the global atmosphere or arbitrarily interfere with healthy environmental conditions. States must also, so far as possible, *protect* against third parties undermining the right to health in other countries. For example, states must take effective steps to ensure that private companies operating in their jurisdictions do not compromise emission limitation and reduction commitments.[60] Subject to resource availability and progressive realization, states' obligations to *fulfil* the right to health include responsibilities to facilitate access to essential health facilities and services, especially for those in poorer countries.[61] Also, the duty to fulfil would appear to place responsibility on high income countries to enable adaptation to unavoidable climate impacts in low income states.

However, enabling adaptation is not enough. The HDR 2007 argues that the world's poor cannot be left to sink or swim, with their own resources, while high income countries protect their citizens behind climate defence fortifications.[62] The human rights responsibility of international assistance and cooperation should not be narrowly understood as only a duty to mitigate the damage caused by climate change. States have a human rights responsibility to ensure that their policies do not obstruct, but support, the realization of the right to the highest attainable standard of health in other countries.[63] They have a duty to work actively toward an equitable international order that is conducive to the realization of the right to the highest attainable standard of health,[64] including healthy environmental conditions.

Importantly, developed countries must honour their commitments regarding overseas development assistance. Over the years there has been a massive under-investment in developmental assistance for health, which has left millions of people vulnerable.[65] Many developed countries

[60] See Article 3 of the Kyoto Protocol to the United Nations Framework Convention on Climate Change (1998).

[61] CESCR, General Comment 14, para. 39. [62] *Human Development Report: 2007*, 7.

[63] CESCR, General Comment 14, paras. 38–42.

[64] CESCR, General Comment 14, para. 38.

[65] WHO, *Report of the Commission on Macroeconomics and Health* (Geneva: World Health Organization, 2001).

still give far less than 0.7 per cent of their national income as foreign aid.[66] Most developed countries are now diverting additional foreign aid into climate change funding, while counting it toward their 0.7 per cent aid targets.[67] Countries must live up to their development and climate change commitments so as to ensure the realization of human rights, including the right to the highest attainable standard of health.

Accountability

Human rights empower individuals and communities by granting them entitlements and placing legal obligations on others. Critically, rights and obligations demand accountability: unless supported by a system of accountability they can become no more than window dressing. Accordingly, a human rights – or right to health – approach emphasizes obligations and requires that all duty holders be held to account for their conduct.[68]

All too often, accountability is used to mean blame and punishment.[69] But this narrow understanding of the term is much too limited. For example, a right to health accountability mechanism establishes which health policies and institutions are working and which are not, and why, with the objective of improving the realization of the right to health for all. Such an accountability device has to be effective, transparent and accessible.[70]

Accountability comes in many forms. At the international level, human rights treaty bodies such as CESCR, CRC or the Human Rights Committee provide an embryonic form of accountability, while at the national level a health commissioner or ombudsperson may provide a degree of accountability. The Ugandan National Human Rights Institution, for example, has established a unit to monitor the realization of the right to health in

[66] United Nations Millennium Project, *Investing in Development: A Practical Plan to Achieve the Millennium Development Goals* (London: Earthscan, 2005).

[67] Oxfam, *Climate Wrongs and Human Rights* (Oxford: Oxfam, 2008), 20.

[68] P. Hunt, *Report of the Special Rapporteur on the Right of Everyone to the Enjoyment of Physical and Mental Health*, UN Doc. A/59/422 (2004).

[69] See L. P. Freedman, 'Human Rights, Constructive Accountability and Maternal Mortality in the Dominican Republic: A Commentary', *International Journal of Gynaecology and Obstetrics*, 82 (2003), 111–14.

[70] The accountability device should clarify who has the responsibility to do what – and whether they have done it. If they have not done it, the device should explore why not, with a view to ensuring that it is properly done next time. For more on accountability, see A/59/422, para. 38.

Uganda.[71] A democratically elected local health council is another type of accountability mechanism. Administrative arrangements, such as publicly available health impact assessments, may also enhance accountability. In relation to issues as complex as climate change a range of national and international accountability mechanisms is required; the form and mix of devices will vary from one state to another. Each accountability device must be adequately resourced.

Human rights, including the right to health, can strengthen the weak accountability mechanisms presently associated with climate change. Provided that they are adequately briefed and resourced, for example, existing human rights accountability mechanisms can consider the adequacy of what states are doing in relation to climate change. The examination by a human rights treaty body of a state's periodic report could consider the dimensions of climate change falling within the treaty body's mandate. On country missions, United Nations Special Rapporteurs could explore the dimensions of climate change relevant to their mandates. At the country level, a national human rights institution could establish a monitoring and accountability unit that considers climate change and human rights.

The international community – and other actors – will have to identify appropriate, effective, transparent and accessible accountability mechanisms in relation to climate change. If it does not, then attempts to combat climate change are unlikely to be effective.

There is a long-standing perception among developing countries that accountability arrangements are imbalanced and mainly applicable to them, while developed countries escape accountability when failing to fulfil their international pledges and commitments that are of particular importance to developing countries.[72] There is no doubt that national and international accountability mechanisms in relation to developed countries' human rights responsibility of international assistance and cooperation remain weak. There is a lack of national or international accountability mechanisms that would give adequate attention to the impact of a developed country's policies on climate change and the right to the highest attainable standard of health. This state of affairs is

[71] P. Hunt, *Report of the Special Rapporteur on the Right of Everyone to the Enjoyment of Physical and Mental Health*, UN Doc. A/HRC/7/11/Add.2, (2007), para. 71.

[72] J. Vandenmoortele, K. Malhotra and J. A. Lim, *Is MDG 8 on track as a Global Deal for Human Development?* (United Nations Development Programme, Bureau for Development Policy, 2003).

unacceptable because human rights require effective, transparent and accessible accountability mechanisms in relation to all human rights responsibilities and all actors, including high income states' responsibility of international assistance and cooperation in health.

Accountability mechanisms are urgently needed to address the issue of climate change. This is crucial, not only because of the magnitude and gravity of the issue, but also because there are differentiated responsibilities for climate change. All states have some responsibility for climate change, but the responsibilities are not the same. Suitable mechanisms are needed to monitor, and hold to account, regarding these differentiated responsibilities.

The design of appropriate, independent accountability mechanisms demands creativity and imagination. Often associated with accountability, lawyers must be willing to understand the distinctive challenges of climate change, and learn from the rich experience of experts working in the fields of the environment and health.

Conclusion

The right to the highest attainable standard of health requires states to adopt and implement laws, plans, policies, programmes and projects that tackle the adverse effects of climate change. The right also demands measures that stop and reverse climate change.

The vital importance of addressing climate change must be reflected in national budgets and international assistance and cooperation. Large-scale public awareness health campaigns are needed to provide information relating to the adverse effects of climate change. Effective, transparent and accessible monitoring and accountability mechanisms must be established. They may take the form of a national human rights institution, health ombudsperson, environmental regulator, or other body. The mechanism should have the responsibility to monitor, and hold all relevant public and private actors to account, in relation to national and international policies bearing upon climate change.

Climate change demands close collaboration between those working in international human rights and those specializing in the environment. Environmental law, policy and research provide indispensable analysis, standards, insights, expertise and information, especially on complex technical issues. As argued in this volume (and this chapter), international human rights law and practice also have a substantive contribution to make in relation to climate change.

Because it explicitly includes healthy environmental conditions,[73] the right to the highest attainable standard of health serves as a bridge between the domains of the environment and human rights. The right to health provides a bridge in another sense, too: international human rights law is primarily designed to promote and protect the rights of individuals, yet the right to health encompasses both medical care and public health. This means that, unlike some other components of international human rights, the right to health inescapably encompasses the entitlements and interests of individuals, communities and populations. For its part, climate change impacts upon individuals, communities and populations. Thus, here is another distinctive connection between climate change and the right to the highest attainable standard of health.

These are some of the issues motivating the United Nations Special Rapporteur on the right to the highest attainable standard of health to call upon the United Nations Human Rights Council to urgently study the impacts of climate change on human rights, a recommendation that the Council acknowledges in its resolution establishing such a study.[74]

[73] See ICESCR, Article 12(2); CESCR, General Comment 14, para. 11.

[74] See P. Hunt, *Report of the Special Rapporteur on the Right of Everyone to the Enjoyment of Physical and Mental Health*, UN Doc. A/62/214 (2007), para. 107(j) and UN Human Rights Council Resolution No. A/HRC/7/L.21/Rev.1 (2008), 3.

9

Human rights and vulnerability to climate change

JON BARNETT[*]

This chapter offers a sys`tematic examination of the connections between vulnerability to climate change and human rights. Vulnerability is a key concept in climate change research and policy. It is defined here as the degree to which people and the things that they value are susceptible to damage arising from climate change (adapted from Schneider *et al.* 2007).

There is at present only a very small body of academic scholarship that explicitly addresses the risks that climate change poses to human rights, and this is largely from theoretical (Sachs 2006) and legal perspectives (Juss 1998, Posner 2007, Sinden 2007). The topic has not been considered in any detail by the global environmental change research community as yet, although there has been some mention of climate impacts as a human rights issue (see, for example, Adger 2004, Barnett and Adger 2003), and there are points of commonality in entitlements-based approaches to vulnerability (for example, Adger and Kelly 1999, Bohle *et al.* 1994) and in the research on global environmental change and human security (see, for example, Barnett *et al.* 2007, O'Brien 2006).

The most substantial body of existing research on rights and climate change concerns the issue of mitigation of greenhouse gases (GHGs), the bulk of which is about rights-based approaches to reducing emissions. This research is for the most part concerned with *property rights* (things that can most often be exchanged), as distinct from *human rights* (which are most often inalienable) (Shue 1999). The modest points of overlap between property and human rights arises in this research from the deployment of the ethical principle of an equal per capita or group right to emit for reasons of fairness, or an allocation of a minimum right for the purposes of the human right of subsistence (Shue 1993).

* Australian Research Council Fellow at the University of Melbourne.

This issue is well canvassed elsewhere, and so is not discussed in this chapter (see, for example, Baer *et al.* 2000, Brown *et al.* 2006, Fankhauser 1995, Grubb 1995, Meyer 2000, Muller 2001, Paterson 2001, Shue 1993, Toth 1999).

The human rights of concern in this paper are those that are specified in existing legal texts (that is, the human rights that are, rather than those that could be). The chapter explains the ways in which human rights violations increase vulnerability to climate change. It does this both through a general explanation of the way in which human rights violations affect the factors that give rise to vulnerability, and through three short cases where human rights violations, to varying degrees, increase vulnerability to climate change.

Human rights violations as drivers of vulnerability to climate change

Vulnerability to climate change is generally seen to be the function of three factors (Alwang *et al.* 2001, Cutter 1996, Eakin and Luers 2006). The first of these is the degree to which an entity is exposed to a climate risk; for example, a house located on a low-lying coast in an area prone to storm surges is more at risk from flooding than one that is in a more elevated and inland location, and the house on the coast is more at risk if the probability of a storm surge increases due to climate change. The second factor is the susceptibility of the entity that is exposed to a climate risk to damage when the event happens; for example, a low-lying house on a flood-prone coast is more susceptible to damage than one which is elevated on stilts above the floodline. The final factor is the capacity of the entity exposed to a climate risk which may be damaging (a hazard) to avoid or recover from that hazard in order to prevent unrecoverable loss; for example, households that move away from a flood-prone area in response to increased flood risk, elevate their buildings above the floodline, or have insurance that will pay for losses are less likely to experience irrecoverable loss from storm surge than those which cannot move, cannot elevate and do not have insurance. In climate change research this third factor is called 'adaptation'.

Human rights violations are powerful drivers of vulnerability to climate change because they can influence each of these three factors. They influence exposure to risk inasmuch as those whose standards of living are inadequate (ICESCR, Article 1) and those who have been displaced for fear of, or as a consequence of, cruel and inhuman treatment (ICCPR, Article 7) are often the most exposed because they tend to live where land is cheap or free because it is exposed to climatic extremes such as flood, drought and landslides, or is unproductive (see de Waal 1991, Wisner 2001). For example, pastoral populations are highly vulnerable to hunger in times of drought when there are restrictions placed on their freedom of movement (ICCPR, Article 12) (see Franke and Chasin 1980, HRW 1992); vulnerability to the increased spread of vector-borne diseases, such as malaria, is highest among those whose basic rights, including the right to health (ICESCR, Article 12) are already not upheld (Bates *et al.* 2004, Winch 1998); thousands of small farmers died during Hurricane Mitch in Central America in 1998 after agricultural developments displaced them onto steep and unstable slopes that collapsed under conditions of high rainfall (Wisner 2001).

The susceptibility of an individual or group of people exposed to climate change damage is in many ways influenced by the degree to which they enjoy human rights. There is a close (but by no means perfect) association between groups that tend to be most dependent on natural resources that are sensitive to climate – for example, farmers who grow in soils that are not irrigated, or fishers who rely on fish such as tuna that move with warm water currents – and losses of food and income during extreme climatic events. Resource dependence often means a lack of diversity in sources of income and materials such that when the supply of the resource that is the basis for livelihoods and community stability fails then communities have few, if any, options to sustain income and materials (Adger 1999). In these situations enjoyment of basic human rights, such as social security, including social insurance (ICESCR, Article 9) and liberty of movement (ICCPR, Article 12), can assist communities to access alternative strategies to sustain their livelihoods and maintain their communities. In contrast, communities in which there is no welfare system and mobility is restricted are more susceptible to suffering caused by climatic variability and change.

The degree to which an individual or group of people enjoys human rights is strongly associated with their capacity to adapt to climate change. Research into disasters and environmental change has consistently shown that severe adverse outcomes (hunger, disease, poverty, mortality) are not evenly distributed among exposed populations (see, for example, Blaikie *et al.* 1994, Leichenko and O'Brien 2008, Pelling 2003). The difference between those who lose and those who remain unaffected (or who benefit) can be explained by the differences in the entitlements of people to various resources that can assist in avoiding risk in advance and in recovering from a crisis (see Adger 2006). These resources include income, natural capital and welfare support from the state and civil society.

However, people's entitlements are restricted or denied for many reasons, including poor or no access to health care and education for reasons of gender, race and sexual and political preference; unequal exchanges of raw materials and goods in markets; and restrictions on labour rights to reduce the bargaining power of labour in markets. Famines, for example, may be understood in part as the product of failures to protect both material and human rights in that democracy and freedom of speech (and the media) reduces famine risk because it creates incentives for the state to intervene with social security packages to avoid hunger (because failure to act can result in eviction from office) (de Waal 1991, Sen 1981). Hence, the right to freedom of expression (ICCPR, Article 19), the right to vote (ICCPR, Article 25) and the right to social security (ICESCR, Article 9) are all important in reducing vulnerability to the kinds of adverse effects likely due to climate change.

In the following sections, I will further clarify and illustrate these links between human rights violations and vulnerability to climate change through a brief examination of the vulnerability of three groups of people: farming households in the upland areas of Timor-Leste (drawing on Barnett 2006, Barnett *et al.* 2007 and CAVR 2005); farmers in the north China plain (drawing on Barnett *et al.* 2006, Webber *et al.* 2006, Webber *et al.* 2008); and people living on low-lying atolls in the South Pacific (drawing on Barnett 2005, Barnett and Adger 2003 and Mortreux and Barnett 2009). The first two cases detail the denial of different sets of rights, which in turn produces different kinds of vulnerability; in the third contrasting case, climate change vulnerability is not fundamentally related to human rights failures.

Food security in Timor-Leste

Timor-Leste is likely to experience significant changes in its climate in the future. The country's climate is influenced by the Asian monsoon system, giving it a 4- to 6-month wet season beginning in December, and a longer dry season. Annual rainfall is particularly low along the northern coast. Timor-Leste's climate is highly variable, and in El Niño years annual rainfall is up to 50 per cent less than average in many places, and the wet season is delayed by two to three months (Barnett *et al.* 2007). Climate projections suggest that El Niño events are likely to become more intense. They do not show how much rainfall may change in the wet season, but large decreases in rainfall in the early part of the dry season in the north are projected. Increases in the intensity of extreme short-term precipitation events are expected but changes in the frequency of droughts and floods (resulting from sustained high rainfall) are uncertain, although the hydrological cycle is expected to become more vigorous. In the absence of considerable changes to reduce the sensitivity of food systems to changes in climate, these projected changes in climate seem likely to impact on food security in Timor-Leste.

Farming households in the upland areas of Timor-Leste are particularly susceptible to increased food insecurity arising from these changes in climate. They are highly dependent on climate-sensitive agricultural systems for their food security. In the upland areas of Timor-Leste rural people are extremely dependent on household production of maize for their food security. Yet crop productivity is low, stagnant and falls in times of drought since soils are thin and maize is not an irrigated crop. When droughts occur – as they already do during El Niño events, and will increasingly do so because of climate change – farmers have few options with which to maintain food security. They have insufficient income to buy food, they cannot borrow money since credit schemes are not available and land tenure uncertainties mean that they have no capital to serve as collateral, they cannot find work due to high unemployment and the cost (and opportunity costs) of travel to urban centres and they cannot depend on emergency relief assistance. The result is that in dry periods farming households eat one meal per day for extended periods, there is increasing malnutrition particularly among children, and increasing illness and mortality.

Abuses against the Timorese by the Indonesian armed forces (1974–99) are the most powerful causes of the sensitivity of food security in upland areas in Timor-Leste to climate change, and of the limited capacity of households to adapt to climate change to maintain (let alone improve) food security. These abuses have been many and varied, and include over 18,000 unlawful killings and over 80,000 deaths due to hunger, displacement of over half of households, widespread detention and torture, extensive sexual violence and widespread and often extreme violations of children's rights (CAVR 2005). While these practices may have to some extent impacted on labour productivity in rural areas today, it is the violations of economic and social rights between 1974–99 that have directly caused the current situation of vulnerability.

During this period of Indonesian occupation, for example, education was used as a propaganda tool rather than to empower people, and this has restricted opportunities for people to earn higher incomes, learn new skills and access new information and technologies (Millo and Barnett

2004). The diversion of capital works by the Indonesian armed forces for military-related purposes has led to a stock of economic and social infrastructure that is totally insufficient to facilitate rural development. Resettlement and the imposition of the Indonesian land tenure system has created much uncertainty about land tenure, which in turn constrains the investment of capital that might increase productivity and denies farmers the use of land as collateral against loans. The destruction of farming tools, livestock and, in some cases, irrigation infrastructure further impeded rural production and the accumulation of surpluses. Farming households were denied opportunities to accumulate income from the most widely grown cash crop, coffee, as all surpluses were expropriated by the Indonesian forces, and the crop itself was not maintained (see Nevins 2003). The use of defoliants, extraction of sandalwood and other resource extraction activities created widespread land use change with associated problems of biodiversity loss and soil erosion (Aditjondro 1994, Bouma and Kobryn 2004).

The combined effect of these and other human rights violations has been massive decapitalisation of rural areas and arrested rural and human development, to the extent that rural people in Timor-Leste are the poorest people in one of the world's poorest nations. In the absence of these violations, rural areas would arguably now be more developed, with, for example, higher levels of literacy, more diversified income sources, better transport, communications and irrigation infrastructure, higher levels of capital investment on-farm, clearer property rights regimes, higher levels of income, more diversified cropping patterns and clearer agricultural policy settings. All of these developments would mean that food security would be less dependent on climate-sensitive production of maize, and households would have more options to adapt in order to maintain food security. Because of these human rights violations, farming households in the upland areas of Timor-Leste are highly vulnerable to hunger due to climate change.

Irrigation and farmers on the north China plain

The north China plain produces 60 per cent of the PRC's wheat and 40 per cent of its corn, even though it only has 22 per cent of its cultivated land and 4 per cent of its water resources (Yang and Zehnder 2001). Farmers are dependent on rainfall, groundwater and irrigation channels to water their crops, which are in most cases maize and wheat. Surface water predominantly comes from the Yellow River or one of its tributaries. A significant source of the river's run-off is water supplied by spring melting of glaciers. In addition, the climate of the Northern PRC is dominated by the Asian monsoon, with most rain falling in the summer period. Rainfall and glacial melting combine so that 60 per cent of annual run-off occurs in the months of July to October. In the last fifty years the river has experienced significant decreases in run-off, almost entirely due to increasing extraction of water rather than changes in climate (Cai 2008, Changming and Shifeng 2002). From 1992 onwards there have been increasing periods where the river ceased to flow to the sea. The problem peaked in 1997 when the river bed was dry for 687 km upstream from the river mouth for 227 days, and ceased to flow to the sea for 330 days (Fu and Chen 2006).

Climate change is likely to alter the temporal distribution of rainfall, with more intense rainfall in winter and decreased and less predictable rainfall in summer. In the longer term this will coincide with decreased river flows in spring as glaciers contract and so release less water (although increased spring river flows are likely as glaciers progressively melt). So the principal impacts of climate change on water resources in the Yellow River basin will be decreased run-off and, therefore, less supply of irrigation water during the summer growing season (Cruz *et al.* 2007, Fu and Chen 2006). This change, superimposed on the trend toward increasing extraction of surface and groundwater, suggests that in the future there will be increasingly critical shortages of water for irrigation in the north China plain, exposing farming households to considerable risk of decreased income.

It has been estimated that between 1972 and 1996 the drying of the lower Yellow River caused a decline in production of some 9.86 billion kg of cereals and RMB12.2 billion worth of lost agricultural output, with annual losses in production in the order of RMB1.6 billion per annum in the 1990s (Changming and Shifeng 2002, Fu and Chen 2006). Data from Shandong province shows that grain production fell markedly when the river ran dry in 1997, and that it was a decline in maize production that accounted for the overall decline in grain production, because maize needs to be irrigated in summer when the river did not flow. Nevertheless, the drying of the river did not have any effect on growth in the value of agricultural production from Shandong, which increased unabated throughout the late 1990s, principally because of increasing production of vegetables which sell for much higher prices than grains and which are irrigated from groundwater.

The impact of the 1997 drying of the river on farming households varied, depending on the availability of off-farm work, groundwater and local rainfall. Most farmers increased their use of groundwater to compensate for declining supplies of river water, and given that rainfall was good in that year, while yields fell there was still a maize crop that was sold. There were no reports of crop switching during or immediately after 1997. Off-farm work and remittances from children working in cities helped to hedge against the minor losses in income arising from lower yields on farm. Because of these adaptation strategies, household income does not appear to have been significantly affected by the drying up of the river.

Farmers have options to adapt to shortages in irrigation water: they can switch crops, they can find off-farm employment, they may be able to ask for remittance income from family members working in cities and the state can implement a range of measures to protect farmers. Chinese farmers have more capacity to adapt under today's more liberal market systems than they did in the past, and in this sense the expansion of economic freedoms and gains made in the achievement of economic and social rights since 1978 has very significantly increased the capacity of farmers in the north China plain to adapt to the likely reductions in the availability of irrigation water arising from climate change.

However, farmers cannot control the availability of surface water in times of drought. Today, as in 1997, when river water becomes scarce the needs of cities and industries are prioritised above the needs of farmers. In response to the 1997 drying episode, from 1999 onwards the Chinese state, via the Yellow River Conservancy Commission (YRCC), remotely controls all the gates that release water from the Yellow River into irrigation districts. The gates can be opened only by the YRCC. When there is enough water in the river, the YRCC will open the gates, but when flow is low it will not open any gates to irrigation districts so as to allow sufficient water to flow to the sea. In this situation, farmers simply do not get surface water, even though they may have requested it via a (poorly functioning) system of requests emanating from village governments up through township and provincial governments through to the YRCC. Farmers adapt through the above mentioned strategies, but the sustainability of the strategies depends critically on growth in off-farm employment and replenishment of groundwater, and the supply of both jobs and groundwater in the future may be less secure than it has been in the past decade as much employment was in construction as part of state-driven investments in infrastructure, and groundwater levels are falling rapidly.

Farmers have no formal right to water in the PRC. The 2002 water law states that all rights to surface and groundwater belong to the state. Even the right to access water is a *de facto* rather than *de jure* arrangement (Lohmar *et al.* 2003). Indeed, it is not yet recognised that people have a *right to participate* in water resources management (Liu 2005). People in the PRC also have no right to vote to change the leadership of the state, so farmers have no effective means of recourse against the water policies and management practices that disadvantage them. Decision makers have no incentive to ensure supply since there is no electoral punishment for not doing so. Thus, farmers can choose only to accept management decisions and adapt as best they can, migrate (but this is restricted somewhat by the *hukuo* system which regulates movement), or resist – and it is notable that there have been some significant examples of organised resistance to water management decisions in rural areas of the PRC in recent years (see Jun *et al.* 2005). So the capacity of farmers to adapt to drought in the PRC remains considerably less than it could be because of a lack of a right to participate in water resource management and to vote. Therefore, while ongoing expansion of economic freedoms increases the capacity of farmers to adapt to climate-induced water stress, that capacity is nevertheless still constrained by a lack of political rights.

South Pacific atolls

The previous two examples suggest that enhancing human rights is an important means of reducing the sensitivity of people to climate impacts and enhancing their capacity to adapt. However, in some circumstances vulnerability is largely a function of exposure to climate risks, and the scope for reducing vulnerability by enhancing human rights is limited. The Pacific atoll states of Kiribati, Tuvalu and the Marshall Islands are all states where human rights are largely upheld – especially political liberties, and there is a good range of economic freedoms if not so many opportunities. For example, all are categorised by Freedom House (2007) as 'free' countries, and although Kiribati and Tuvalu are least developed countries, this lack of development is due entirely to low factor endowments rather than a lack of liberties or a history of rampant resource extraction under colonial rule (Bertram and Watters 1984).

These atoll countries are highly vulnerable to climate change because of their high ratio of coastline to land area, low elevation, poor soils, limited water resources, fragile artisanal fisheries, migratory deep-water fish stocks (largely tuna), sensitivity to coral bleaching, high population densities, limited economic resources, economic marginalisation due to isolation and generally low levels of human resource development. A 2°C rise in temperature is likely to cause annual coral bleaching (Donner *et al.* 2005), and changing rainfall patterns, more intense extreme events and sea level rise, such that local food production from the land and sea declines, water scarcity increases and coasts erode to the point that the islands may cease to be able to sustain existing numbers of people, and in the longer term may be subsumed (Barnett and Adger 2003).

It is the case that larger amounts of public and private income would expand adaptive options and prolong the ability of people to inhabit their atoll islands by, for example, allowing for progressive elevation of infrastructure, improved water storage and distribution systems (including desalination), removal of hazardous wastes from the islands and imports of food. Technically, then, it may be possible for these islands to sustain current levels of population indefinitely, although the costs of such adjustments are prohibitive and the costs to quality of life would be so high that such adjustments might be better understood as transmuted climate impacts rather than adaptations *per se*. Similarly, migration away from islands to avoid exposure to climate risks will also result in psychosocial impacts that might be better understood as climate impacts rather than adaptations (see Mortreux and Barnett 2009). Further, migration creates new sets of questions about rights fulfilment in new locations.

So, there are limits to adaptation on atolls: there is no land to which atoll people can retreat – short of international emigration – should there be a significant and rapid rise in sea level, and the costs to amenity and lifestyle arising from expensive engineering solutions would also be high. Given the barriers to adaptation, in the worst case scenario the carrying capacity of atolls may be significantly or completely reduced (either progressively or suddenly) by climate change, with impacts including mortality and morbidity, stagnant if not negative economic growth and, means permitting, increased emigration.

People living on atolls are vulnerable principally because of the characteristics of the ecosystem in which they live and not because of social vulnerability created by human rights violations. Unlike most vulnerable communities, the vulnerability of atoll populations cannot be reduced merely by the enhancement of human rights at the level of the state alone. Only the reduction of GHG emissions can address its root causes. So, while increased human rights fulfilment can reduce vulnerability to climate change in many cases, it is clearly neither a sole nor sufficient solution to the numerous problems raised by climate change.

Conclusions

There are good empirical and legal bases to link climate change and human rights. Climate change poses risks to established human rights and the social processes that make people vulnerable to climate change often entail the violation of human rights. There is, therefore, a multiplication of human rights injustices caused by climate change: those whose rights have been most violated are most often those whose rights are most vulnerable to climate change and who are the least responsible for the emission of GHGs. This understanding has led a number of authors to suggest a rights-based approach to reducing vulnerability to climate change (see Adger 2004, Brooks *et al.* 2005, Dow *et al.* 2006 and Yamin *et al.* 2005). Avoiding human rights violations arising from climate change demands a renewed commitment to human rights to reduce vulnerability to climate change (that is, human rights fulfilment reinforces adaptation), but, as the case of the atoll countries suggests, it also requires reductions in GHG emissions to reduce changes in climate *per se*.

An implication of this discussion is that in places where there are gross violations of human rights – like those occurring in conflict zones such as Darfur, or in highly repressive systems such as North Korea – specific actions to promote adaptation may be far less effective in reducing vulnerability than actions to promote the kinds of political freedoms and social opportunities necessary for people to take actions themselves. In other words, action to protect political, social and economic rights may be a far more important adaptive strategy than locally-based actions that address specific climate impacts. This also implies that there are limits to what the United Nations Framework Convention on Climate Change can in, and of, itself do to achieve its ultimate objective of preventing dangerous climate change: in some cases the largest gains in reducing vulnerability may come from eliminating the most severe human rights violations through strengthening and developing the application and enforcement of human rights law. This implies strong complementarity between what may otherwise appear to be the disparate policy issues of climate change and human rights.

References

Adger, W. N., 'Social Vulnerability to Climate Change and Extremes in Coastal Vietnam', *World Development*, 27 (1999), 249–69.

'The Right to Keep Cold', *Environment and Planning A*, 36 (2004), 1711–15.

'Vulnerability', *Global Environmental Change*, 16 (2006), 268–81.

Adger, W. N. and P. M. Kelly, 'Social Vulnerability to Climate Change and the Architecture of Entitlements', *Mitigation and Adaptation Strategies for Global Change*, 4 (1999), 253–66.

Adger, W. N., J. Paavola, M. Mace and S. Huq (eds.), *Fairness in Adaptation to Climate Change* (Cambridge, MA: MIT Press, 2006).

Aditjondro, G., *East Timor: An Indonesian Intellectual Speaks Out* (Deakin, ACT: Australian Council for Overseas Aid, 1994).

Alwang, J., P. Siegel and S. Jorgensen, *Vulnerability: A View From Different Disciplines. Social protection discussion paper No. 0115* (Washington: World Bank, 2001).

Baer, P., J. Harte, B. Haya, A. Herzog, J. Holdren, N. Hultman, D. Kammen, R. Norgaard and L. Raymond, 'Climate Change: Equity and Greenhouse Gas Responsibility', *Science*, 289 (2000), 2287.

Barnett, J., 'Titanic States? Impacts and Responses to Climate Change in the Pacific Islands', *Journal of International Affairs*, 59 (2005), 203–19.

'Climate Change, Insecurity and Justice', in W. Adger, J. Paavola, M. Mace and S. Huq (eds.), *Fairness in Adaptation to Climate Change* (Cambridge, MA: MIT Press, 2006), 115–29.

Barnett, J. and W. N. Adger, 'Climate Dangers and Atoll Countries', *Climatic Change*, 61 (2003), 321–37.

Barnett, J., M. Webber, M. Wang, B. Finlayson and D. Dickinson, 'Ten Key Questions about the Human Dimensions of Water in the Yellow River Basin', *Environmental Management*, 38 (2006), 179–88.

Barnett, J., S. Dessai and R. Jones, 'Vulnerability to Climate Variability and Change in East Timor', *Ambio*, 36 (2007), 372–8.

Barnett, J., R. Matthew and K. O'Brien, 'Global Environmental Change and Human Security' (2007), in H. Brauch, J. Grin, C. Mesjasz, P. Dunay, N. Chadha, B. Chourou, U. Oswald-Spring, P. Liotta and P. Kameri-Mbote (eds.), *Globalisation and Environmental Challenges: Reconceptualising Security in the 21st Century*. (Berlin: Springer-Verlag, 2007).

Bates, I., C. Fenton, J. Gruber, D. Lalloo, A. Lara, S. Squire, S. Theobald, R. Thomson and R. Tolhurst, 'Vulnerability to Malaria, Tuberculosis, and HIV/AIDS Infection and Disease. Part 1: Determinants Operating at Individual and Household Level', *The Lancet Infectious Diseases*, 4 (2004), 267–77.

Bertram, G and R. Watters, *New Zealand and its Small Island Neighbours* (Wellington: Institute of Policy Studies, 1984).

Blaikie, P., T. Cannon, I. Davis and B. Wisner, *At Risk: Natural Hazards, People's Vulnerability, and Disasters* (London: Routledge, 1994).

Bohle, H., T. Downing and M. Watts, 'Climate Change and Social Vulnerability: Toward a Sociology and Geography of Food Insecurity', *Global Environmental Change*, 4 (1994), 37–48.

Bouma, G. and H. Kobryn, 'Vegetation Cover Change in East Timor, 1989–1999', *Natural Resources Forum*, 28 (2004), 1–12.

Brooks, N., N. Adger and M. Kelly, 'The Determinants of Vulnerability and Adaptive Capacity at the National Level and the Implications for Adaptation', *Global Environmental Change*, 15 (2005), 151–63.

Brown, D., M. Tuana, M. Averill, P. Baer, R. Born, C. E. Lessa Brandão, M. T. S. Cabral, R. Frodeman, C. Hogenhuis, T. Heyd, J. Lemons, R. McKinstry, M. Lutes, B. Mueller, J. D. Gonzalez Miguez, M. Munasinghe, M. S. Muylaert de Araujo, C. Nobre, K. Ott, J. Paavola, C. Pires de Campos, L. Pinguelli Rosa, J. Rosales, A. Rose, E. Wells and L. Westra, *White Paper on the Ethical Dimensions of Climate Change* (College Park, PA: Rock Ethics Institute, Penn State University, 2006).

Cai, X., 'Water Stress, Water Transfer and Social Equity in Northern China – Implications for Policy Reforms', *Journal of Environmental Management*, 87 (2008), 14–25.

Changming, L. and Shifeng, Z., 'Drying Up of the Yellow River: Its Impacts and Counter-Measures', *Mitigation and Adaptation Strategies for Global Change*, 7 (2002), 203–14.

Commission for Reception, Truth, and Reconciliation Timor-Leste (CAVR), *Chega: The Report of the Commission for Reception, Truth, and Reconciliation Timor-Leste: Executive Summary* (Dili: CAVR, 2005).

Cruz, R. V., H. Harasawa, M. Lal, S. Wu, Y. Anokhin, B. Punsalmaa, Y. Honda, M. Jafari, C. Li, N. Huu Ninh, 'Asia' in: M. L. Parry, O. F. Canziani, J. P. Palutikof, P. J. van der Linden and C. E. Hanson (eds.), *Climate Change 2007: Impacts, Adaptation and Vulnerability. Contribution of Working Group II to the Fourth Assessment Report of the Intergovernmental Panel on Climate Change* (Cambridge University Press, 2007), 469–506.

Cutter, S., 'Vulnerability to Environmental Hazards', *Progress in Human Geography*, 20 (1996), 529–39.

de Waal, A., 'Famine and Human Rights', *Development in Practice*, 1 (1991), 77–83.

Donner, S., W. Skirving, C. Little, M. Oppenheimer and O. Hoegh-Guldberg, 'Global Assessment of Coral Bleaching and Required Rates of Adaptation under Climate Change', *Global Change Biology*, 11 (2005), 2251–65.

Dow, K., R. Kasperson and M. Bohn, 'Exploring the Social Justice Implications of Adaptation and Vulnerability' in W. N. Adger, J. Paavola, M. Mace and S. Huq (eds.), *Fairness in Adaptation to Climate Change* (Cambridge MA: MIT Press, 2006), 79–96.

Eakin, H. and A. Luers, 'Assessing the Vulnerability of Social-ecological Systems', *Annual Review of Environment and Resources*, 31 (2006), 365–94.

Fankhauser, S., *Valuing Climate Change: The Economics of the Greenhouse* (London: Earthscan, 1995).

Franke, R. and B. Chasin, *Seeds of Famine* (New Jersey: Rowman and Allanheld, 1980).

Freedom House, *Freedom in the World 2007* (New York: Freedom House, 2007).

Fu, G. and S. Chen, 'Water Crisis in the Yellow River: Facts, Reasons, Impacts, and Countermeasures', *Water Practice and Technology* 1:2 (2006), doi10.2166/wpt.2006.0028, IWA Publishing.

Grubb, M., 'Seeking Fair Weather: Ethics and the International Debate on Climate Change', *International Affairs*, 71 (1995), 463–96.

Human Rights Watch (HRW), *Indivisible Human Rights: The Relationship of Political and Civil Rights to Survival, Subsistence and Poverty* (New York: Human Rights Watch, 1992).

Jun, M., W. Xuejun and Y. Xiobo, 'Water Conflict Resolution in China', *China Environment Series*, 7 (2005), 99–103.

Juss, S., 'Global Environmental Change: Health and the Challenge for Human Rights', *Indiana Journal of Global Legal Studies*, 5 (1998), 121–78.

Leichenko, R. and K. O'Brien, *Double Exposure: Global Environmental Change in an Era of Globalization* (New York: Oxford University Press, 2008).

Liu, Y., 'On Establishment of People's Right to Participate in Water Resources Management' in YRCC (ed.), *Proceedings of the 2nd International Yellow River Forum*, vol. I (Zhengzhou: Yellow River Conservancy Press, 2005), 341–6.

Lohmar, B., J. X. Wang, S. Rozelle, J. K. Huang and D. Dawe, *China's Agricultural Water Policy Reforms: Increasing Investment, Resolving Conflicts, and Revising Incentives* (Washington DC: Market and Trade Economics Division, Economic Research Service, US Department of Agriculture, 2003).

Meyer, A., *Contraction & Convergence: The Global Solution to Climate Change* (Totnes: Green Books for the Schumacher Society, 2000).

Millo, Y. and J. Barnett, 'Educational Development in East Timor', *International Journal of Educational Development*, 24 (2004), 721–37.

Mortreux, C. and J. Barnett, 'Climate Change, Migration and Adaptation in Funafuti, Tuvalu', *Global Environmental Change*, 19:1 (2009), 105–12.

Muller, B., 'Varieties of Distributive Justice in Climate Change', *Climatic Change*, 48 (2001), 273–88.

Nevins, J., 'Restitution Over Coffee: Truth, Reconciliation and Environmental Violence in East Timor', *Political Geography*, 22 (2003), 677–701.

O'Brien, K., 'Are We Missing the Point? Global Environmental Change as an Issue of Human Security', *Global Environmental Change*, 16 (2006), 1–3.

Paterson, M., 'Principles of Justice in the Context of Climate Change' in U. Luterbacher and D. Sprinz (eds.), *International Relations and Global Climate Change* (Cambridge, MA: MIT Press, 2001), 119–26.

Pelling, M., *The Vulnerability of Cities: Natural Disaster and Social Resilience* (London: Earthscan, 2003).

Posner, E., 'Climate Change and International Human Rights Litigation: A Critical Appraisal', *University of Pennsylvania Law Review*, 155 (2007), 1925–45.

Sachs, W., 'Climate Change and Human Rights', *The Pontifical Academy of Sciences*, 106 (2006), 349–57.

Schneider, S., S. Semenov, A. Patwardhan, I. Burton, C. Magadza, M. Oppenheimer, A. Pittock, A. Rahman, J. Smith, A. Suarez and F. Yamin, 'Assessing Key Vulnerabilities and the Risk from Climate Change' in M. Parry, O. Canziani, J. Palutikof, P. van der Linden and C. Hanson (eds.), *Climate Change 2007: Impacts, Adaptation and Vulnerability, Contribution of Working Group II to the Fourth Assessment Report of the Intergovernmental Panel on Climate Change* (Cambridge University Press, 2007), 779–810.

Sen, A., *Poverty and Famines: An Essay on Entitlement and Deprivation.* (Oxford: Clarendon, 1981).

Shue, H., 'Subsistence Emissions and Luxury Emissions', *Law and Policy*, 15 (1993), 39–59.

'Bequeathing Hazards: Security Rights and Property Rights of Future Humans' in M. Dore and T. Mount (eds.), *Global Environmental Economics: Equity and the Limits to Markets* (Oxford: Blackwell, 1999), 38–53.

Sinden, A., *Climate Change and Human Rights*, Social Science Research Network Working Paper, 1 May 2007. Available at SSRN: http://ssrn.com/abstract=984266.

Toth, F. (ed.), *Fair Weather? Equity Concerns in Climate Change* (London: Earthscan, 1999).

Webber, M., J. Barnett, B. Finlayson and M. Wang, 'Managing the Yellow River: Questions of Borders, Boundaries and Access', *Transforming Cultures*, 1 (2006), 114–35.

Webber, M., J. Barnett, M. Wang and B. Finlayson, 'The Yellow River in Transition', *Environmental Science and Policy*, 11:5 (2008), 422–9.

Winch, P., 'Social and Cultural Response to Emerging Vector-borne Diseases', *Journal of Vector Ecology*, 23 (1998), 47–53.

Wisner, B., 'Risk and the Neoliberal State: Why Post-Mitch Lessons Didn't Reduce El Salvador's Earthquake Losses', *Disasters*, 25 (2001), 251–68.

Yamin, F., A. Rahman and S. Huq, 'Vulnerability, Adaptation and Climate Disasters: A Conceptual Overview', *IDS Bulletin*, 36 (2005), 1–14.

Yang, H. and Y. Zehnder, 'China's Regional Water Scarcity and Implications for Grain Supply and Trade', *Environment and Planning A*, 33 (2001), 79–95.

10

Climate change, evolution of disasters and inequality

JOHN C. MUTTER AND KYE MESA BARNARD*

So devastating were the outcomes of Cyclone Nargis for the people of the Irrawaddy Delta in Myanmar in May 2008, and so inadequate and perverse were the responses of their government, that numerous calls were made for international intervention by invoking the Responsibility to Protect (R2P) doctrine.[1] The foundations for R2P actions lie in 'specific legal obligations under human rights and human protection declarations, covenants and treaties, international humanitarian law and national law'.[2] If taken by invoking R2P, an action to intervene in a disaster setting on behalf of those who have been affected would, therefore, be undertaken in the name of human rights; an action in which the international community assumes the role of duty bearer because a given sovereign state has failed to fulfil or is unable to meet its primary duty to provide adequate protection.

Disasters have three phases: the evolution of vulnerability through social processes preceding the disaster; the disaster singularity; and the recovery that follows. It is well established, though not always set out in sharp relief or adhered to in practice, that human rights principles of fair and equal treatment for all who suffer underpin approaches to the care of disaster victims.[3] That is, established human rights norms have been drawn upon to govern the treatment of those impacted in the second and

* Columbia University, New York.

1 Evans, Gareth, 'Facing up to our Responsibilities', *The Guardian*, 12 May 2008, available at: http://commentisfree.guardian.co.uk/gareth_evans/2008/05/facing_up_to_our_responsbilities.html.

2 G. Evans and M. Sahnoun (co-chairs), *The Responsibility to Protect*, Report of the International Commission on Intervention and State Sovereignty (Ottawa: IDRC Books, 2001).

3 Brookings–Bern project on Internal Displacement, Human Rights and Natural Disasters Pilot Report (2008).

third phases of a natural disaster. While the nature and intensity of a natural extreme event is a physical force that strongly shapes the outcome for all affected human populations, the social and political landscapes impacted by an extreme event create pre-conditions that have at least as forceful a role in determining outcomes. Where social inequalities precede the arrival of a storm they will shape the outcomes after the storm passes.

Human agency is always deeply complicit in the transformation of natural extremes into disasters[4] – the events are not merely 'acts of God'. It is, therefore, appropriate to ask whether the deaths, injuries and subsequent disruption to education, health, work and general welfare that result from a natural disaster should be seen as a failure to ensure the right to life, personal security and livelihood on the part of sovereign states and other actors. Can the tragedy of massive human losses in natural disasters be seen as a failure on the part of a government to afford its people basic rights that would secure their lives and a failure of international norms such as the Responsibility to Protect; failures that are located in the first phase of a disaster?

Climate change aligns our conception of disasters and their effects along a new axis. Because it is widely conceded that climate change will increase the incidence, strength and distribution of extreme events and their attendant consequences for a growing number of people, and since that increase is widely understood to be caused by human agency of a known and deliberate form through greenhouse gas (GHG) emissions, a new set of *ex ante* questions arises concerning disasters and human rights.[5] These questions focus more on the evolution of vulnerable pre-conditions of disaster; the first, rather than the second or third phase of a disaster. We suggest that government action or inaction in securing human rights for its people directly impacts natural disaster inequalities by deeply modulating the risks they face.

This contribution begins with a short introduction of the science underlying climate change and the impact on the frequency and intensity of extreme meteorological events. After a brief description of the changing nature and path of such events, we present the cases of Hurricane

[4] J. C. Mutter, 'Pre-conditions of Disaster: Premonitions of Tragedy', *Social Research: An International Quarterly of the Social Sciences*, 75:3 (Fall 2008), 691–724.

[5] The recognized source for current information on climate change is the UN Intergovernmental Panel on Climate Change (IPCC), *Fourth Assessment Report: Climate Change* (2007).

Katrina in New Orleans and Cyclone Nargis in the Irrawaddy Delta to illustrate the role of rights in these respective tragedies. We then examine the ways in which human agency determines whether such events become human disasters. This requires a consideration of the relationship between economic development and disaster outcomes in the context of human rights attainment. The discussion leads into an analysis of various forms of vulnerability resulting from failures to protect human rights and the consequences manifested during and following disasters. The chapter concludes with a brief discussion of some implications for policy.

The nature of climate change and some consequences

The primary driver of modern climate change is the perturbation of the natural atmospheric greenhouse effect by addition of the products of fossil fuel combustion, primarily carbon dioxide, resulting largely from transport-related emissions and energy production in power plants. The basic phenomenon is, in principle, fairly simple to understand. While the Earth receives radiation from the sun, not all the solar energy received at the top of the atmosphere passes through. Everyone has had the experience of sunning themselves at a beach and having a cloud come in front of the Sun, causing shivers until the cloud moves and warmth returns. This occurs because water vapour in the atmosphere (the cloud) has reflected some of the incoming energy back into space. Greenhouse gases such as carbon dioxide and methane, though not visible, have a similar filtering effect. Solar radiation that reaches Earth warms the surface, which then re-radiates heat back into the atmosphere and beyond. The atmosphere presents a partial barrier to the outgoing radiation, just as it did for the incoming solar radiation. However, the atmosphere presents a much stronger barrier to long wavelength infrared energy radiated from the heated Earth than it does to shorter wavelength visible radiation from the Sun, so the atmosphere retains a greater proportion of heat radiated from Earth. Clouds again provide a familiar effect – very clear nights are typically much colder than cloudy nights because the cloud cover acts to retain the warmth radiated by the Earth. Additions of GHGs, therefore, lead to an overall warming of the planet. The industrialized countries and emerging countries such as China, India and Brazil far exceed developing countries in their emissions, but because the atmosphere is very well mixed the warming effect is global.

Considerable uncertainty attends projections of total average temperature increase at a given time in the future. One uncertainty arises from needing to predict the future composition of the atmosphere. To do this scenarios are constructed based on how emissions of important gases might evolve and these are driven by rates of economic growth, particularly in Brazil, China and India that have historically had low emissions but are rapidly increasing emissions as their economies grow. To make forecasts of future temperature conditions, a suite of emission scenarios is used as input to global climate models that calculate a range of projected temperatures for each scenario. There are several computer models and each treats the Earth a little differently by placing different values on the effects of various GHG components of the atmosphere[6] and by treating critical feedbacks differently.[7] The total uncertainty is the sum of those uncertainties associated with the scenario projections and computer model uncertainty.

Regional variations and impact disparities

Despite the uncertainties associated with scenarios and models, a number of outcomes are very clear. For example, the greatest temperature increases will most likely occur in high northern latitudes because of the known positive feedback of ice albedo.[8] These areas are today showing the most convincing signs that warming is taking place. Relatively few people live at very high latitudes[9] simply because of the difficulty such climates present; those small populations are typically indigenous people whose livelihoods depend critically on the mastery of the present harsh conditions. The changing conditions will likely significantly impact on these traditional livelihoods and means of survival. In contrast, rising temperature may benefit non-indigenous populations as, for instance,

[6] The effects of aerosols in the atmosphere is currently quite poorly understood with some models presuming they will have a net cooling effect, others assuming the effect is positive (aerosols are small airborne particles, such as soot and sulphur dioxide, released from fossil fuel or biomass burning. They are also released from natural sources, such as volcanic eruptions or dust storms).

[7] One source of uncertainty is the role of clouds in a warmer world as we can expect there to be a greater amount of cloud cover because the added warmth will cause increased evaporation, but clouds both cool the Earth by reflecting sunlight in the daytime and warm the Earth by retaining heat at night and the net effect is not well known at present.

[8] The melting of ice that usually reflects incoming solar radiation exposes land that in turn absorbs radiation and heats the Earth. IPCC, *Climate Change 2001*, 7.5.2 'Sea Ice', available at: www.grida.no/climate/ipcc_tar/wg1/295.htm.

[9] The gridded population map of the world (available at: http://sedac.ciesin.columbia.edu/gpw/) shows that population density is very low above 60° North and essentially zero above 60° South.

agriculture that is typically managed by non-native groups may become more productive due to longer growing seasons. Shipping routes may open up across the Arctic in the traditional hunting grounds of indigenous people, facilitating trade that almost solely benefits others. Changes in northern regions impact the world as a whole through the melting of grounded ice sheets that raise sea level and alteration of the thermohaline circulation that controls major ocean circulation patterns including the Gulf Stream.

The low latitude tropics and sub-tropics, with the exception of desert regions, are often host to very high population densities and the highest concentrations of extreme poverty.[10] They are the only parts of the world where droughts still lead to famines that cause significant mortality,[11] not only from direct starvation, but also from illness associated with malnourishment. Historically, it is very possible that drought-induced food shortages combined with cruel, bias-based or malfeasant government actions have given rise to famines that have caused the greatest mortality of all disasters.[12] The poorest people in these regions live close to the edge of survival and often depend on rain-fed agriculture for basic food production. They are, therefore, highly vulnerable to even small perturbations in ambient conditions[13] that can have cascading feedbacks that propagate very small shifts into catastrophic outcomes. Although model projections suggest that a warmer world is also likely to be wetter due to increases in evaporation, all projections show a worsening of conditions in the tropics, including potentially devastating impacts on already stressed agricultural production. In the very poorest regions of the world, few, if any, groups will gain from climate change.

In a warmer world, a more energetic atmosphere and oceans will most likely cause an increase in the power of cyclones and hurricanes and change their spatial distribution.[14] These intense storms do not occur in

[10] The twenty-two countries that make up the lowest ranking in the UNDP's Human Development Index are in sub-Saharan Africa.

[11] UNDP, *Reducing Disaster Risk: A Challenge for Development* (2004).

[12] M. Davis, *Late Victorian Holocausts: El Niño Famines and the Making of the Third World* (Verso, 2002).

[13] This statement pertains to human welfare and not to other species of animals which might well suffer the greatest impacts in higher latitudes; the fate of the polar bear populations being frequently discussed in this regard.

[14] M. Bister and Kerry A. Emanuel, 'Dissipative Heating and Hurricane Intensity', *Meteorology and Atmospheric Physics*, 65:3–4 (1998), 233; Lars R. Schade and Kerry A. Emanuel, 'The Ocean's Effects on the Intensity of Tropical Cyclones: Results from a Simple Coupled Atmosphere–Ocean Model', *Journal of the Atmospheric Sciences*, 56:4

the immediate vicinity of the equator and are more common in the northern than the southern hemisphere. Because conditions found in the tropics will expand toward the poles in a warmer world, storm tracks that have been associated with the Caribbean and US Gulf coast will move northward, and similar northward shifts will occur in the Pacific. Areas that previously were free from, or experienced very few, hurricanes and cyclones will start to experience them in greater numbers. While the total number of storms on an annual basis may not change in the future, there is reason to believe that changing conditions will lead to a greater proportion of very intense storms.

Hurricane Katrina and Cyclone Nargis: some consideration of inequalities and human rights

Many factors, physical, institutional and social govern the outcome of natural extremes for impacted populations. Broadly, on the global stage it is well established that the poorest people are at greatest risk. Thus, the global inequality of well-being that can be viewed as lack of attainment of basic human rights can be thought of as a principal determinant of disaster risk. We focus here on two contrasting cases in which extreme events impacted populations characterized by deprivation, and ask what role pre-existing inequalities played in the harsh outcomes, and if viewing them through the lens of human rights might add clarity to our understanding.

Death in New Orleans

Hurricane Katrina made its deadly entry into history on 29 August 2005 in one of the poorest sections of the United States.[15] Louisiana suffered

(1999), 642; Kerry A. Emanuel, 'Thermodynamic Control of Hurricane Intensity', *Nature*, 401 (1999), 665; Stanley B. Goldenberg *et al.*, 'The Recent Increase in Atlantic Hurricane Activity: Causes and Implications', *Science*, 293 (5529) (2001), 474; Thomas R. Knutson and Robert E. Tuleya, 'Impact of CO_2-induced Warming on Simulated Hurricane Intensity and Precipitation: Sensitivity to the Choice of Climate Model and Convective Parameterization', *Journal of Climate*, 17:8 (2004), 3477; P. J. Webster *et al.*, 'Changes in Tropical Cyclone Number, Duration, and Intensity in a Warming Environment', *Science*, 309 (5742) (2005), 1844; M. E. Mann and Kerry A. Emanuel, 'Atlantic Hurricane Trends Linked to Climate Change', *Eos*, 87:24 (2006); Jun Yoshimura, Masato Sugi and Akira Noda, 'Influence of Greenhouse Warming on Tropical Cyclone Frequency', *Journal of the Meteorological Society of Japan*, 84:2 (2006), Yoshimura *et al.*, 2006, IPCC, 2007.

[15] US Census Bureau, available at: www.census.gov/hhes/www/poverty/poverty.html.

the greatest numbers of deaths, Mississippi the second. Mississippi is the poorest state, the first in poverty among the elderly and the lowest in life expectancies. Louisiana is the third lowest in these same indicators. While the average US poverty rate is 12.5 per cent, the seven southern states have rates that exceed 15 per cent, around double that of the wealthiest states. There is less regional variation for poverty among African Americans, which is, on average, double that of whites. Before Katrina the Greater New Orleans area had the nation's second highest rate of urban 'concentrated poverty',[16] correlating quite closely with concentrations of African Americans. There are neighbourhoods in New Orleans in which the population is close to 100 per cent African American and where one in four people are classified as extremely poor. New Orleans was also on track to be the nation's most violent city, with homicide deaths at approximately ten times the national average. As Katrina approached, many poor people in New Orleans were living with physical and social risks that were quite unlike those of any other US city.

New Orleans is also the largest city vulnerable to hurricanes located on the Gulf coast,[17] and is geographically centred in the low lying Mississippi River Delta. The river repeatedly floods and an extensive system of levees controls the flow of water, protecting people and property. However, in New Orleans, residents also need protection from Lake Pontchartrain and from a series of man-made canals, several of which were created for commercial purposes – the Gulf Inter-coastal Waterway (GIWW), the Mississippi River Gulf Outlet (MR-GO) and the Industrial Canal. The latter was built in 1918 and connects Lake Pontchartrain to the Mississippi River and connects to the GIWW and MR-GO. MR-GO provides a shorter route for shipping to the Gulf of Mexico. The 17th Street Canal and London Avenue Canals, although originally built for a different purpose, today serve to assist in draining low lying areas of New Orleans into Lake Pontchartrain. Much of the city lies below sea level and is subject to flooding during heavy but not storm-related rains. Breaches in the commercial and drainage canals, not those holding back the river

[16] Alan Berube and Bruce Katz, *Katrina's Window: Confronting Concentrated Poverty Across America* (The Brookings Institution Metropolitan Program, October 2005); L. S. Kathryn, G. Pettit and Thomas Kingsley, *Concentrated Poverty: A Change in Course* (Urban Institute, May 2003), available at: www.urban.org/url.cfm?ID=310790.

[17] Tampa Saint-Petersburg is quite a bit larger but is not in the direct path of typical hurricane tracks. Houston is larger still but is considerably inland.

or the lake, were responsible for the main flooding in New Orleans that took so many lives.[18]

Inundation of the city began the day before the storm made landfall and continued throughout the day of the storm and the days following.[19] First to be hit were eastern wards of St. Bernard Parish and the now infamous Lower Ninth Ward of Orleans Parish, the poorest neighbourhood in the city that experienced the greatest depth of flooding as it lies at one of the lowest points beneath sea level. Homes there were destroyed due to their proximity to major breaches in the Industrial Canal caused by the pressure of the storm surge that forced water along MR-GO and the GIWW and into the Industrial Canal. The break created a tsunami-like surge of water that swept houses off their foundations. By 0900 on 29 August the entire eastern part of the city was flooded, and a city so starkly divided on race and social class was also divided into flooded and un-flooded areas. As the storm continued, levees of the 17th Street and London Avenue Canals were breached, flooding relatively wealthy areas of Lakeview and Gentilly, both of which lie as far below lake level as the Lower Ninth Ward and experienced significant but less housing damage due, in general, to stronger and generally newer housing stock.

Virtually no one who stayed in a dwelling typical of the poorer parts of the city and faced the water raging out from canal breaks could possibly have survived. Away from the breaks, flood water arrived much less violently and escape was possible by moving to higher levels in a house. Few houses were completely submerged, so escape to a rooftop generally meant survival, perhaps after many arduous hours waiting for help. Very few parts of the city were truly safe. People died in poor and wealthy areas even if they were flooded by a modest amount; some died in places with no flooding at all.[20] Causes of death are numerous,[21] with the leading causes being drowning, injury and trauma, but many people died from existing conditions such as heart disease, which were exacerbated by the trauma of the event. The most

[18] A very comprehensive analysis of levee breaches is available at: www.freerepublic.com/focus/f-news/1517817/posts.

[19] An excellent animation that tracks the flooding hour-by-hour was created by staff of the New Orleans newspaper the *Times Picayune* from numerous sources and is available at: www.nola.com/katrina/graphics/flashflood.swf.

[20] Joan Brunkard, Gonza Namulanda and Raoult Ratard, *Hurricane Katrina Deaths, Louisiana, 2005* (Disaster Medicine and Public Health Preparedness, August 2008).

[21] Dr Raoult Ratard, Louisiana State Epidemiologist, personal communication.

common location where deceased victims were found was in their homes. Despite the raucous news coverage only one person is known to have died in the Superdome.

No consensus exists on the true total mortality from Hurricane Katrina. The Louisiana Family Assistance Center (LFAC) gives the figure at 1,464 for Louisiana residents, including deaths that occurred out of state, and a total of 224 deaths in Mississippi. Total deaths from so-called direct and indirect effects is estimated to be 1,833.[22] By contrast the Brunkard *et al.* study lists only 988 deaths. Neither of these estimates include possible mortality among those who remain missing or the potentially very large number of people who died in the weeks and months subsequent to the event through loss of continuity of health care or the exacerbation of prior medical conditions. It may also underestimate the violent deaths associated with looting and other criminal behaviour. Excess mortality rates in the months following Katrina are very difficult to determine because the number who remained in the city is largely unknown (and remains uncertain today). One estimate[23] suggests an excess mortality rate of 47 per cent in New Orleans. This estimate is contested[24] because it is based in large part on obituaries that are known to be unreliable sources of mortality data. A conservative estimate of mortality might be 3,000[25] and, as such it is quite anomalous. Hurricane Andrew, the first named storm of 1992 and equally intense as Katrina devastated south Florida but took only sixty-five lives. Commonly, Atlantic hurricanes claim a very small number of lives in the United States. Katrina's death toll is substantially larger than any American natural disaster since the Galveston Flood of 1900 that may have taken as many as 8,000 lives.[26]

[22] Richard D. Knabb, Jamie R. Rhome and Daniel P Brown, *Tropical Cyclone Report: Hurricane Katrina* (National Hurricane Center, updated August 2006).

[23] K. Stephens, D. Grew, K. Chin, *et al.*, 'Excess Mortality in the Aftermath of Hurricane Katrina: A Preliminary Report', *Disaster Medicine and Public Health Preparedness*, 1:1 (2007), 15.

[24] Joanna Eavey and Raoult Ratard, 'Post-Katrina Mortality in the Greater New Orleans Area, Louisiana', *Journal of the Louisiana State Medical Association*, 160:5 (2008), 267–72.

[25] In one effort to determine the true mortality Mutter has constructed a web-based tool in which people can volunteer names of those they consider to be direct or indirect victims of the hurricane even if the deaths occurred many months after the event, available at: www.katrinalist.columbia.edu.

[26] The Galveston disaster predates the National Hurricane Center, modern telecommunications and the automobile so it is hardly surprising that the death toll was so high.

Many factors contributed to the high death toll in New Orleans relative to other hurricanes. Physical factors differ – the storm surge associated with Hurricane Andrew, for instance, was not as large as for Katrina and there was little flooding and drowning deaths were few. But even Brunkard *et al.*'s figures give several hundred non-drowning deaths in New Orleans, and we suggest that physical factors alone have insufficient explanatory power to account for the high victim toll from Katrina. We suggest that Katrina's death toll exceeds the norm because the people of New Orleans lived lives well below the norm, lacking the education to understand the gravity of their situation, lacking trust in a government that oversaw their lives and lacking the means to escape even if they understood the risks. Lack of vehicle access is high in New Orleans and correlates strongly with concentrations of poverty and racially with concentrations of African American residents.[27] These factors are no doubt reinforcing ones as New Orleans lacks an extensive and efficient public transportation system such as the New York subway system that provides accessible and reasonably cheap transportation to employment. Many people had also experienced previous evacuations in which they incurred considerable expenses that were never compensated, so may have made what was for them an economically reasonable judgment based on past experience.

Fully 64 per cent of deceased victims whose remains were received at the Saint Gabriel morgue established by FEMA's Disaster Mortuary Operational Response Team (DMORT) were over 65 years old in a population in which that age group represented only about 15 per cent of the existing population. Mortality rates for those over 75 were even higher. The elderly mostly lived alone and in isolation, perhaps little aware of the dangers they faced and with no ability to seek safety even if they did. Mortality rates were higher among older females than older males. Many elderly people died in their homes[28] or in nursing homes, unwilling or unable to leave if they lived alone or fearing the consequences of leaving. Death among elderly whites exceeded those of elderly African Americans because there were a greater number of elderly whites; their life expectancies considerably exceeding those of blacks.

[27] US Census Bureau, *United States Census* 2000, available at: www.census.gov/main/www/cen2000.html.

[28] In Harrison County, Mississippi, the coroner Mr Gary Hargrave found all but a very small number of deceased victims in their homes; personal communication, June 2008.

In all age categories death *rates* among African Americans exceeds that among whites.

In 2008, New Orleans was effectively and efficiently evacuated in the days before Hurricane Gustav made landfall to the south-east of the city almost three years to the day after Katrina. Both the people of New Orleans and the city government were taught a harsh lesson by Katrina. It must be remembered though that Gustav made landfall well west of New Orleans and the evacuation was made easier because perhaps 200,000 people have not returned since Katrina. Those who remain displaced were the people who had stayed during Katrina, found themselves in the Superdome and were evacuated days after the storm.

Hurricane Katrina, the people of New Orleans and human rights

The US Army Corps of Engineers has acknowledged that the levees designed to protect the citizens of New Orleans were inadequate to withstand the onslaught of Katrina's storm surge. That fixes blame on one specific external group and suggests that an engineering solution could restore safety to the community. But the very existence of the Industrial Canal imposed a risk to the neighbourhoods through which it traversed, and those neighbourhoods are today and were historically the poorest and most segregated in the city. The purpose of the Industrial Canal is very different from the drainage canals designed to ensure safety for the people in the city.[29] They breached too, but the Industrial Canal did nothing but put the adjacent poor communities at risk, while the drainage canals help to make their adjacent wealthier communities safe. Maps showing the locations where deceased victims remains were recovered show a far greater density close to the Industrial Canal than the drainage canals (see Brunkard *et al.*, note 20, above) despite an essentially identical depth of flooding and proximity to levee breaks. Living in the shadow of a levee put one at great risk, but living in the shadow of poverty greatly enhanced that risk.

Though possibly underestimated, both physical and social risks were known before the storm arrived. Most individuals have little capacity to mitigate those risks, particularly the poor and the elderly. In the United

[29] John M. Barry, author of *Rising Tide* has made this point also. See www.johnmbarry.com/bio.htm.

States the responsibility to protect falls, first, locally at the city government level, then to the state if the city is overwhelmed, then nationally if the state cannot cope. Failure of government to protect in Hurricane Katrina occurred at every level of government[30] and seen through the prism of human rights, comprise a wide range of specific rights that were not ensured. These include the inherent right to life (not only secured under international law, but also specifically in the US Bill of Rights), as well as the right to dignity and physical, mental and moral integrity. Specific to disasters, people have a right to have access to or be provided with: (a) essential food and water; (b) basic shelter and housing; and (c) appropriate clothing. Several of these considerations could be captured under the general conditions of the fundamental right to health if one considers a disaster to be a massive public health threat. Additionally, the right to information must be met in order to help to facilitate all the other rights being met. The US Human Rights Network has levelled specific charges against the US government under the Covenant on Civil and Political Rights, Article 6 (right to life) and Article 26 (prohibition against discrimination) in which it claimed that evacuation plans were discriminatory and that humanitarian relief was inadequate and also discriminatory. The American Civil Liberties Union (ACLU) has claimed extensive mistreatment of prisoners at the Orleans Parish Prison during the disaster using rights arguments to support their case.[31]

Tierney[32] and Adler[33] have discussed the role that social inequality played in the outcome of Katrina and provide suggestions as to how considerations of equity might be incorporated into future considerations of disaster management. On a global basis, Kahn[34] has made formal statistical analyses of several factors that contribute to specifically imperilling the poorest in a natural disaster, and his study suggests that open

[30] D. Brinkley, *The Great Deluge: Hurricane Katrina, New Orleans, and the Mississippi Gulf Coast* (New York: William Morrow, HarperCollins, 2006), 686.

[31] American Civil Liberties Union, 'Abandoned and Abused', available at: www.aclu.org/prison/conditions/26198res20060809.html.

[32] Kathleen Tierney, 'Social Inequalities, Hazards and Disasters' in Ronald Daniels, Donald Kettl and Howard Kunreuther (eds.), *On Risk and Disaster: Lessons from Hurricane Katrina* (Philadelphia, PA: University of Pennsylvania Press, 2006).

[33] Matthew D. Adler, 'Equity Analysis and Natural Hazard Policy' in Daniels, Kettl and Kunreuther, *On Risk and Disaster.*

[34] Matthew E. Kahn, 'The Death Toll from Natural Disasters: The Role of Income, Geography and Institutions', *The Review of Economics and Statistics*, 87:2 (May 2005), 271–84.

governance and strong institutions play an important role in shielding people from disaster risks. He includes measures of inequality and government transparency in his analysis, but acknowledges some uncertainty as to why these factors should be important.

We are unable to determine if specific actions or inactions on the part of government discriminated against some people in the disaster that engulfed New Orleans, though such claims have certainly been made and may have merit.[35] Many of the failures to protect can be ascribed to incompetence, but the disproportionate death rates among African Americans of all ages, accents their greater vulnerability. We can say with certainty that those residents of New Orleans who routinely suffered the greatest deprivations and lack of rights attainment in their normal lives experienced the impacts of Katrina most harshly as they struggled to overcome an extreme of nature. Human rights attainment predicts disaster outcomes – those whose rights were most neglected before the storm were most neglected during the storm.

Cyclone Nargis and the people of Myanmar

Like Katrina, Cyclone Nargis made landfall in one of the most impoverished countries in the region – one that has long been notorious for egregious human rights violations. Over one quarter of the people of Myanmar live below the national poverty line and nearly half of the rural population does not have access to safe sanitation facilities.[36] Myanmar is ranked 132 (of 177) on the Human Development Index, and 164 out of 174 when ranked by GDP.[37] It is tied with Somalia as the most corrupt country in the world according to Transparency International.[38] The Global Peace Index ranks Myanmar as 126 out of 140, the lowest in the region.[39] The International Labour Organization has found extensive

[35] As one example, Michael Eric Dyson, *Come Hell Or High Water: Hurricane Katrina and the Color of Disaster* (Basic Civitas Books, 2006).

[36] Asian Development Bank, *Asian Development Bank and Myanmar: 2008 Fact Sheet*; United Nations Economic and Social Council, Commission on Human Rights, 'Situation of Human Rights in Myanmar, Report of the Special Rapporteur' (ECOSOC Report, 2006), 1–29.

[37] See UNDP, *Statistics of the Human Development Report*, available at: http://hdr.undp.org/en/statistics.

[38] See Transparency International, *Corruptions Perceptions Index 2007*, available at: www.transparency.org/policy_research/surveys_indices/cpi/2007.

[39] See Vision of Humanity, *Global Peace Index Rankings*, available at: www.visionofhumanity.org/gpi/results/rankings.php.

evidence of forced labour.[40] The junta engages in massive, severe violations of children's rights, contravening their ratification of the UN Convention on the Rights of the Child. Among the gravest violations are the government's recruitment and use of children in their own armed forces, and failure to address the recruitment of underage children by non-state armed forces.[41] The government is accused of serious human rights violations in the conduct of military operations against ethnic groups, including forced labour, indiscriminate killing and rape.[42] These conflicts have led to tens of thousands fleeing the country and over half a million internally displaced people without protection from their government or other armed forces.[43] Such movement arises both because people flee conditions of conflict and because of state-sponsored land and crop confiscation and forced relocations.[44]

In 2005, the Global Fund to Fight AIDS, Tuberculosis and Malaria announced that it could no longer continue its operations in Myanmar due to government restrictions on its operations.[45] The junta also restricted access of the International Committee of the Red Cross (ICRC) to prisoners , leading the ICRC to close several of its local offices.[46] The government also created barriers to the distribution of food aid from the World Food Program (WFP) in 2005 at a time when the WFP was supporting several hundred thousand hungry people in two regions primarily inhabited by ethnic minorities.[47] To underscore the callousness of the junta, on 10 May 2008, only eight days after Cyclone Nargis, the government went ahead with a referendum for most of the country, followed by voting in the hardest hit regions two weeks later. The government claims that over 92 per cent of eligible voters showed up to approve the referendum.[48]

[40] International Labour Organization, 'Forced Labour in Myanmar (Burma)' (Geneva, 2 July 1998). Available at: www.ilo.org/public/english/standards/relm/gb/docs/gb273/myanmar.htm.

[41] UN Security Council Report of the Secretary-General on Children and Armed Conflict in Myanmar, 2007. Available at: www.un.org/children/conflict/english/myanmar.html.

[42] ECOSOC Report, 2006.

[43] Human Rights Watch, 'Burma Events of 2007', *World Report* 2008, available at: http://hrw.org/englishwr2k8/docs/2008/01/31/burma17601.htm.

[44] ECOSOC Report, 2006.

[45] The Global Fund to Fight AIDS, Tuberculosis and Malaria, 'Update on Closure of Myanmar Grants', 30 August 2007. Available at: www.theglobalfund.org/en/media_center/press/an_070830.asp.

[46] UN Security Council Report of the Secretary-General on Children and Armed Conflict in Myanmar, 2007.

[47] ECOSOC Report, 2006.

[48] *The New York Times*, 'When it Comes to Politics, Burmese Say, Government is All Too Helpful', *The New York Times* (28 May 2008).

Cyclone Nargis left more than 132,000 people dead or missing, over a million homeless and over 2 million needing relief.[49] The storm track hit highly populated regions and the bulk of the wetland rice-producing land in the country.[50] The government severely restricted foreign aid and assistance and continued to curtail aid efforts, limit foreign involvement and require that government officials accompany all foreign aid workers even when they did allow aid three weeks after the cyclone struck. Although the feared public health disaster did not occur and there were no apparent disease outbreaks or famine, many thousands remained without aid, housing and other relief.[51] In cases where survivors did receive health care and limited assistance, it occurred through efforts of individuals and local civil society groups rather than government support.[52] Beyond preventing foreign assistance, the Myanmar government imposed greater restrictions on its people's already limited access to communication sources and detained community members and high profile activists attempting to provide aid to cyclone survivors.[53] Some victims of the cyclone who were fortunate enough to survive and find refuge in nearby monasteries or other temporary camps were already being evicted from their new shelters by the government, just one month after the cyclone.[54]

The extensive human rights abuses both before and after the cyclone meant that the people of Myanmar were not only ill-prepared to buffer the immediate effects of the cyclone but also extremely limited in their ability to seek and secure support from various sources, including their government, during a time of great need. The government's rejection of food aid, loss of health care assistance and rights-inspired sanctions in years prior to the cyclone helped to weaken vital health systems and infrastructure further debilitated by decades of corruption and malfeasance. The resulting vulnerability of the population was intensified not only by negligent refusal to address state failures but also violent repression of opposition and active rights violations. When Cyclone Nargis struck, the pre-conditions for disaster were well established; the

[49] *The New York Times*, 'When it Comes to Politics', 2008.

[50] See University of Texas Libraries, available at: www.lib.utexas.edu/maps/thematic.html.

[51] *The New York Times*, 'Burmese Endure in Spite of Junta, Aid Workers Say', *The New York Times* (18 June 2008).

[52] Washington Post Foreign Service, 'Frustrated Burmese Organize Aid Forays', *Washington Post* (21 June 2008).

[53] A. H. Tun, 'Myanmar Junta Slams Citizens over Cyclone Report' (Reuters, 6 June 2008).

[54] *The New York Times*, 'Myanmar Junta Begins Evicting Cyclone Victims from Shelters', *The New York Times* (7 June 2008).

government's disregard for its people's safety and security following the event served only to ensure their continued suffering.

While death tolls of 100,000 or more from major cyclones in Asia are not unknown, they are *not* inevitable. In neighbouring Bangladesh, an even poorer, much more densely populated country, the Category 4 Cyclone Sidr resulted in less than 4,000 deaths, orders of magnitude less than previous cyclones.[55] The government evacuated 40 per cent of coastal inhabitants in the 36 hours before the storm, saving countless lives. While Bangladesh may not be a human rights beacon, the government's concerted efforts to protect its people from the cyclone's devastating effects reveal the capacity for even very poor countries to experience outcomes very different from those in Myanmar and leave that government no reasonable excuse for its inaction. Bangladesh can be credited with recognizing its responsibility to protect its citizens; a responsibility clearly unrecognized by the leaders of Myanmar. There are, of course, vast differences between these two countries in every aspect of governance, economic structure and rate of progress. They are similar only in that they continue to be host to a very large number of very poor people at high levels of risk. Bangladesh's recent actions have demonstrated that even modest protective measures can save many lives even in the poorest settings. Poverty alone cannot be blamed for high disaster mortality.

Taken together, the cases of Hurricane Katrina and Cyclone Nargis, along with the brief comparison to Cyclone Sidr, reveal a critical qualifier to the relationship between poverty and disaster outcomes. Government behaviour and willingness to actively protect its people and their rights can be a more significant determinant of disaster outcomes than national economic conditions. While extreme poverty can constrain the ability of a government to ensure human rights, it does not preclude the government from having the capacity to protect its people during an extreme event. Conversely, great wealth at a national level does not guarantee minimal disaster consequences, particularly if the government fails to adequately ensure its people's rights.

Predicting natural extremes; anticipating disasters for people: Katrina and Nargis in a global context

Does an image of natural disasters refracted through the lens of human rights provide a clarity from which we might learn how to avert the harm

[55] US Department of State: Humanitarian Information Unit, 'Bangladesh: Cycle of Vulnerability' (December 2007).

that increasing disasters will bring to people in a world where climate is changing and extreme events are expected to increase? In trying to answer this question we must first appreciate that there are at least two distinct parts to the problem of predicting disasters. The first is the prediction of the physical event itself: when and where cyclones and other natural extremes will occur and how severe they will be. The second involves predicting the consequences of an extreme event for the affected human societies: whether or not it will become a disaster and for whom. For instance, could we know the outcome of a cyclone's landfall in terms of mortality, economic impact, changes in settlement patterns and future consequences for the affected population, even if we knew in advance exactly when and where it would occur and how strong it would be? Could we know whether a society will shrug off a storm and recover quickly or be weakened and impaired to the degree that every subsequent storm would result in a progressively worse outcome? Could societies actually get stronger by managing recovery in such a way that they build more efficient infrastructure for commerce and become more resilient to the impacts of subsequent events? The fundamental question is: how can we know what turns an extreme of nature into a disaster for humans? And for our discussion: can considerations of human rights attainment of impacted populations provide insight into the prediction problem and guidance on how to reduce harm?

Rights protection during a disaster and specific, constructed vulnerabilities

Many articles within the Universal Declaration of Human Rights and subsequent covenants can be applied to the needs that arise in the course of disaster relief and recovery.[56] The Inter-Agency Standing Committee (IASC) has produced very specific rights-based guidelines for actions taken during disaster events.[57] In addition to a protection framework, three elements of humanitarian assistance are identified: (1) ensuring non-discriminatory relief; (2) informing and consulting disaster affected individuals; and (3) monitoring rights protection. Those suffering from

[56] G. Kent, 'The Human Right to Disaster Mitigation and Relief', *Environmental Hazards*, 3 (2001), 137–8.

[57] Inter-Agency Standing Committee, 'Human Rights and Natural Disasters: Operational Guidelines and Field Manual on Human Rights Protection in Situations of Natural Disaster' (2008). This resulted from a collaboration between the Inter-Agency Standing Committee and the Brookings–Bern Project on Internal Displacement.

the devastation of a disaster are considered to be individuals with extensive, recognized rights including the protection of life and health and the provision of the basic necessities of life (food, clothing, etc.) that are fundamental economic, social and cultural (ESC) rights. ESC rights also include the right to education, property and possessions, housing, livelihood and work. Disaster victims may also require protection of civil and political (CP) rights, including the right to documentation, free movement and return, communication with family members and knowledge of family member whereabouts and fates, assembly and religion and electoral rights.

The right to life-saving information via providing early warning and assisting in finding aid after disasters is crucial. This, of course, requires that the government can provide, or at least secure, sufficient funds for such early warning systems and aid efforts. In addition to receiving information to access essential services, victims also have the right to the information necessary to help them contact friends and family, locate missing loved ones and bury their dead.[58] After Hurricane Katrina, for instance, the Louisiana Family Assistance Center (LFAC) provided this service.[59] No such agency existed prior to Katrina to inform people of their potential vulnerabilities or suggest means of protection. It is well known that in 2004, FEMA contracted the enactment of a simulation exercise – Hurricane Pam – carried out for New Orleans that predicted even greater devastation than actually occurred in Katrina. If lessons were learnt from that simulation, they were not passed on to those who would face Katrina less than a year later.

Another broad-ranging component of information rights entails the right to sufficient and appropriate information necessary for individuals to make informed decisions. This involves information regarding the availability and location of aid and also for participation in decisions about livelihoods, governance, children's education and even physical aid distribution. Adequate access to information typically requires an existing, well-functioning infrastructure and open and free communication system (social or technical). These systems might not be sufficient if they have been constrained through government efforts to control

[58] See IASC (2008) and Article IXX, Global Campaign for Free Expression, 'Humanitarian Disasters and Information Rights: Legal and Ethical Standards on Freedom of Expression in the Context of Disaster Response' (London, 2005).

[59] Louisiana Department of Health and Hospitals, *Hurricane Katrina*, available at: www.dhh.louisiana.gov/offices/?ID=192.

information and restrict rights. The extraordinary level of information and communications restrictions found in Myanmar prior to Cyclone Nargis significantly impeded the ability of people to prepare for the storm and recover once it had passed.

Often it is children and the elderly, particularly the frail elderly, who face acute risks in disaster situations and require targeted assistance. The reasons are perfectly obvious: older persons are vulnerable to severe or lethal injuries due to limited mobility and frail health that also make it difficult for them, for instance, to escape rising waters in a flood. They often require services to be brought to them since they are unable to easily access available resources and face difficulties or absolute barriers in participating in livelihood activities.[60] The absence of these services and protections in New Orleans revealed the extent to which being frail and older can increase one's likelihood of dying in a disaster. Children, if separated from their families, may be physically or sexually abused, exploited or trafficked and recruited into armed groups.[61] If children face some or all of these risks on a daily basis as part of their normal lives, a natural disaster will increase those risks, particularly when the rule of law deteriorates or displacement erodes social barriers and increases opportunities for undetected exploitation or abuse. The rights of children, the elderly and women, in fact, exemplify the way in which disasters threaten basic rights and can erode gains made in rights attainment.

The enhanced risks to people with disabilities or poor physical or mental health are quite similar to those of the elderly. In addition to increased susceptibility to illness for people living with HIV/AIDS resulting from poorer sanitation conditions, a disaster-induced breakdown of social norms leading to elevated sexual violence, exploitation and prostitution can increase HIV infection rates. Community perceptions, historical prejudices and institutional discrimination can also all contribute to the creation of acute vulnerabilities for internally displaced persons, refugees, indigenous peoples and ethnic minorities. These diverse groups typically experience systematic discrimination before disasters and face significant barriers to receiving aid in the aftermath,[62] as highlighted in

[60] HelpAge, 'Older People in Disasters and Humanitarian Crises: Guidelines for Best Practice' (London, 2003), 1–27.

[61] The International Save the Children Alliance, 'Child Protection in Emergencies: Priorities, Principles and Practices' (Stockholm, 2007), 1–80.

[62] IASC (2008).

both Katrina and Nargis. In this way the level of human rights attainment prior to an extreme event significantly determines the outcome of the disaster for these groups.

There are clearly disparities in the extent to which different individuals and groups of individuals suffer from disaster. In one conceptualization,[63] disaster vulnerability is a public bad and vulnerability reduction an impure public good because not all people benefit equally. In order to ensure that humanitarian assistance compensates for these disparities rather than exacerbating them, relevant actors must be aware of the population affected and their specific needs. As the International Federation of Red Cross and Red Crescent Societies notes, 'one-size-fits-all relief planning' cannot address the social realities and avoid discriminatory provision of services.[64] The human rights-based approach supports non-discriminatory practices by recognizing the specific, differentiated rights of individuals based on given circumstances. Women, for instance, comprised 90 per cent of victims of the 1991 cyclone in Bangladesh[65] and typically face much greater risks in and after disasters. Neumayer and Plümper[66] show that disaster experience lowers the life expectancy of women more than men and that the greater the strength of the disaster, the greater the effect on life expectancy. They also reveal that women of higher socio-economic status experience smaller effects. Their conclusion is that it was the '*socially constructed gender-specific vulnerability of females built into everyday socioeconomic patterns that led to the relatively higher female mortality*' (original emphasis). This statement may be a metaphor that applies to all socially marginalized groups and those who lack attainment of basic rights.

[63] J. Boyce, 'Let Them Eat Risk? Wealth, Rights and Disaster Vulnerability', *Disasters*, 24:3 (2007), 254–61.

[64] International Federation of Red Cross and Red Crescent Societies, 'World Disaster Report, 2007' (Geneva, 2007), 1–244.

[65] L. Aguilar, 'Climate Change and Disaster Mitigation' (Switzerland: IUCN, 2004); Oxfam, 'The Tsunami's Impact on Women' (2005), available at: www.oxfam.org/en/files/bn050326_tsunami_women; Eric Neumayer and Thomas Plümper, 'The Unequal Burden of War: The Effect of Armed Conflict on the Gender Gap in Life Expectancy', *International Organization*, 60 (2006), 723–54; Gender and Disaster Network, available at: www.gdnonline.org.

[66] Eric Neumayer and Thomas Plümper, 'The Unequal Burden of War: The Effect of Armed Conflict on the Gender Gap in Life Expectancy', *International Organization*, 60 (2006), 723–54.

The influence of economic and social pre-conditions; ex ante *and* ex post

Recent UNDP[67] and World Bank[68] reports show that disasters in poor countries take many more lives in both absolute *and* relative terms (that is, more people killed in total and more people killed as a proportion of population) than those that occur in wealthier nations. For instance, while Niger and Norway have approximately equal population density, disaster severity as measured in terms of annual mortality is more than an order of magnitude greater for those in Niger. Citizens of New Zealand and Morocco have roughly equal levels of exposure to flooding events, yet the mortality risk for Moroccans is two orders of magnitude higher than that of New Zealanders – New Zealand ranks 19th in the UNDP's Human Development Index[69] (HDI) and Morocco ranks 125th. Flood mortality for people in Botswana and Germany is roughly the same even though Germany's exposure level is three orders of magnitude greater. Botswana's HDI ranking is 132nd, while Germany is presently 20th. The examples of Katrina and Nargis show that the poor are vulnerable both globally and within a rich country.

A human rights-based approach to poverty locates their situation as, 'a human condition characterized by sustained or chronic deprivation of the resources, capabilities, choices, security and power necessary for the enjoyment of an adequate standard of living and other civil, cultural, economic, political and social rights'.[70] We need to examine whether such a redefinition of poverty is helpful in understanding the vulnerability of certain groups or whether it merely takes a different perspective on a well-understood problem.

Major disasters that completely level cities have long been seen as opportunities for urban renewal and general improvement.[71] Arguments are constructed based on the economist Joseph Schumpeter's concept of

[67] United Nations Development Programme (UNDP), 'Disaster Risk Reduction: A Challenge for Development' (2004).

[68] M. Dilley, R. S. Chen, U. Deichmann, A. Lerner-Lam and M. Arnold, 'Natural Disaster Hotspots: A Global Risk Analysis' (Washington, DC: World Bank, 2005).

[69] United Nations Development Programme (UNDP), *Human Development Report 2006, Beyond Scarcity: Power, Poverty and the Global Water Crisis* (New York: Palgrave Macmillan, 2006).

[70] United Nations Economic and Social Council, Poverty and the International Covenant on Economic, Social and Cultural Rights, 10 May 2001, E/C.12/2001/10.

[71] Kevin Rozario, *The Culture of Calamity: Disaster and the Making of Modern America* (University of Chicago Press, 2007).

a 'gale of creative destruction' that, setting the tragedy of mortality aside, disasters are important positive forces in economic progress. Old, inefficient productive capital assets may be destroyed but replaced with newer more efficient assets, and cities rebuilt to improve commerce. Schumpeter's view, as is the case with most perspectives from economics, considers aggregate conditions and applies only where market economies are functioning well. Many individuals may suffer despite an aggregate economic benefit and inequalities may increase as the wealthy cope and potentially benefit while the poorer cannot.

Economic impacts of disasters may be difficult to predict from the strength of the physical event or death toll. We know now that the Chinese economy will not suffer a setback from the magnitude 7.9 Sichuan earthquake on 12 May 2008, despite the massive death toll of over 60,000 people, including an unusually large number of children.[72] The earthquake area is remote and largely agricultural, contributing relatively little to the national economy. Similarly, the US economy did not register that at least 2,000 people lost their lives in Hurricane Katrina because those who died or were displaced were not major contributors to the economy. Economic recovery in New Orleans is centred around the casino and tourist industry because it is the most easily regenerated and does not rely on a local base. Recovery is stratified in a way that the pre-existing conditions might predict – vast areas that were characterized by chronic concentrated poverty remain largely uninhabited, while the wealthier areas have revived.

In essence, loss of life and loss to an economy need not be directly related. The dwellings of the poor are typically of little value and are seldom insured and hence unassessed. Squatter dwellings in peri-urban slums may technically be considered to have no value, with the state or other actors owning the land they occupy. Thus, loss of property by the poor may not be accounted for in the way that economic losses are typically assessed. Similarly, the very poor quite often live outside the formal cash or tax-based economy of a country and are either not considered to be contributors to the economy in a formal sense or may be unknown. If they access social services but do not contribute taxes, they may even be considered a net liability. Hence, their loss as humans may not translate directly to an economic loss. This contributes to an apparent inverse relationship of disaster losses: while rich countries often

[72] Keith Bradsher, *The New York Times*, 4 June 2008, International Report.

incur small loss of life and large financial losses, poor countries typically experience large loss of life and apparently small financial losses.

Such a conclusion is very misleading. What is most important for the economic calculus is the loss relative to the size and strength of the economy and the distribution of the losses within the economy.[73] Disasters act as economic shocks, not unlike the shock of a major currency devaluation or hyperinflation. These shocks are systemic because they perturb almost the entire economy so that very few people, perhaps only the richest who might have assets outside the country, can weather the impact. So the poorest inhabitants suffer disproportionately.[74] Additionally, geographically small countries like island states in the Caribbean and states in Central America are at greater risk because a cyclone or drought can impact very large sections of the country. For the very poorest countries, those caught in poverty traps, disasters may have little real economic impact because the economy already has a small productive base and cannot lose a great deal – they literally have nothing to lose except people's lives.

Very rich countries can probably buffer even the most extreme events (though the modern world has yet to experience a major disaster in a mega-city that is the economic centre of a regional economy, such as Tokyo). In contrast, countries in between these limits, especially the emerging market economies (EMEs), may be the most vulnerable because they rely on a few crucial industries that are rarely sited with any consideration of possible disasters. Disaster-induced economic shocks for countries in this latter category can cause such severe setbacks through loss of productive capital assets that the economy can be thrown into a poverty trap situation. The disaster serves to trigger the descent but is not the trapping mechanism holding the economy in stagnation. Disasters cause very large and instantaneous loss of productive capital that often has to be re-established with meagre national savings; donor contributions seldom covering more than 50 per cent of the recovery costs. This means that capital becomes restricted for schools, hospitals and other requirements for development and countries slide back into economic traps in which the capital needed for improvement is unavailable and decline follows.

[73] J. C. Mutter, 'Pre-conditions of Disaster: Premonitions of Tragedy', *Social Research: An International Quarterly of the Social Sciences*, 75:3 (Fall 2008), 691–724.

[74] N. Lustig, 'Crises and the Poor: Socially Responsible Macroeconomics', *Economia* (Fall 2000).

Differing from an economics perspective then, the human rights focus argues that natural disasters first pose greater threats to those individuals who have yet to attain a reasonable standard of rights, and will erode those rights they have attained, making them more vulnerable still when they experience the inevitable next extreme of nature. This erosion of rights and enlargement of inequalities may well occur despite a general improvement in average conditions as measured by standard economic metrics.

Human rights and the evolution of disasters

Sovereign states bear the responsibility to protect their people. Ensuring human rights provides those protections. Climate change will cause the intensity of meteorological extremes to increase and hence increase the need to protect people against the dangers of these events. Those dangers arise during the disaster itself, and the norms of human rights provide clear guidance as to the equitable treatment of people in disaster situations. But no level of attention to human rights principles at the time of a disaster can fully compensate for the uneven exposure and vulnerability to threat that characterizes disaster situations globally in all countries rich and poor.

The threat posed by a natural disaster is far more than the threat of the moment, the terror experienced when the Earth moves underfoot or when people face the hurricane's gale and advancing storm surge. It is incubated through a history of discrimination that may involve either active denial or inability to ensure fundamental rights. The threats and responses to major disasters are rehearsed repeatedly by those most deprived of basic rights in every daily challenge they face, from obtaining sufficient nourishment, to educating their children, to finding work and a safe reasonable place to live. And it is not only the poor who face greater risks. Clearly the elderly, particularly the frail, isolated and immobile elderly were at hugely disproportionate risk in New Orleans and women often bear an uneven burden. But the vulnerability of these groups is well known to their governments. States that fail to protect their people from deprivation and discrimination in the years preceding a disaster are unlikely to abandon those actions during and after a disaster and become more protective and ensure even treatment. Both Cyclone Nargis and Hurricane Katrina suggest that years of discrimination and unequal treatment portend tragedy for rights-deprived groups, propelling their situation from tragedy to nightmare.

One of the most pressing challenges of our time is to understand how the global climate will change and how those changes will impact people around the world. The physical effects of increasing disasters, for instance, sea level rise and changing agricultural conditions, will directly impact people. People may also be at risk from the actions taken by their governments to adapt to and/or mitigate climate change. It is critical that we come quickly to an understanding of who could be harmed most by these changes and responses, both because they are living in vulnerable locales that do not afford protection and because their governments do not provide them human rights protections.

Human rights attainment may be an unrecognized but robust way of anticipating who is at greatest personal risk. A human rights approach, focusing on the individual condition rather than the aggregate benefit may also provide a better way to understand the nature of risks. Looking ahead, decisions will need to be made concerning approaches to protections from extremes and adaptations to slower changes. It is very unlikely that all countries in the world will be able to take adaptation measures to fully protect all of their people. Some, with a history of discrimination, may take the opportunity to protect only favoured groups. If we cannot protect everyone, how will we decide who we will protect? The current international framework of climate change negotiations does not include consideration of human rights in adaptation planning, opening the potential for tragically different outcomes for different groups. Along with economic analyses a human rights lens must be focused on these considerations to provide the clearest image to guide decisions toward equitable human prospects in a world re-arranged by climate change.

Conclusion

11

Conceiving justice: articulating common causes in distinct regimes

STEPHEN HUMPHREYS*

It is often said that climate change confronts the present generation with a challenge greater than any previous. If so, the global nature of the challenge is perhaps timely, in that it comes at a time when the 'present generation' might itself be thought of, really for the first time, as 'global'. But if several decades of the phenomenon still generally called 'globalisation' have helped to create a global audience for this global challenge, it is far less clear that this same event has provided the 'global' public policy tools with which the challenge can be managed. Certainly, the 'promotion' of human rights and the rule of law around the world in recent decades has not aimed at, or led to, the empowerment of governments to address pressing social and economic rights; if anything the reverse has been the case. Insofar as development policy has concerned itself with law and rights, the emphasis has been firmly on economic freedoms – property rights, contract, investors' rights, banking and commercial law – and judicial enforcement. There has been some attention to civil and political rights, notably gender equality and press freedom, but relatively little to rights to food, water, health, shelter and so on. At the same time, there is an at least plausible case – perhaps best presented in the writing of Thomas Pogge – that the persistence and exacerbation of global poverty over recent decades is itself a human rights violation, largely sustained by international actors and systematised through international law.[1] In this chapter, I first look briefly at Pogge's thesis and its potential application to climate change, particularly through Simon

* Research Director, International Council on Human Rights Policy. See the Acknowledgements for the background to the present chapter and a list of those who have contributed comments to some sections of it. Special thanks to Robert Archer for valuable editorial and substantive suggestions.

[1] See Pogge (2005) and (2002).

Caney's conceptual frame of human rights as 'thresholds'. I then examine one issue from each of the two disciplines viewed from the perspective of the other (emissions trading and participatory rights), before finishing with a brief discussion of cross-regime interaction.

Pogge's claim is not especially extravagant; that is, he does not speak of a positive duty, such as redistribution of global wealth in order to mitigate poverty, but merely of a negative duty; that is, not to make matters worse for the poor. On Pogge's account, however, developed countries ('the West') are not attaining even this modest moral duty, given a 'global institutional order [that] reflects the shared interests of the governments, corporations and citizens of the affluent countries more than the interest in global poverty avoidance, insofar as these interests conflict'.[2] Pogge points in particular to the international trade regime, which, he says, 'favor[s] the affluent countries by allowing them to continue protecting their markets through quotas, tariffs, anti-dumping duties, export credits and subsidies to domestic producers in ways that poor countries are not permitted, or cannot afford, to match'.[3] So if global income inequality and absolute global poverty have been growing (as they have) in recent decades, despite exponential global economic growth overall, the fault lies partly with an international institutional regime that skews the benefits of growth sharply toward wealthier countries. According to Pogge, the common argument that the introduction of the trade regime may have lessened overall deaths from poverty below the number that would have died had there been no such regime does not indicate adequate execution of the negative duty to respect human rights. The proper baseline is not a world without any trade regime at all, but one with a trade regime that accounts properly for the negative duty not to harm.[4] As an example of such a baseline, Pogge cites Article 28 of the Universal Declaration of Human Rights: 'Everyone is entitled to a social and international order in which the rights and freedoms set forth in this Declaration can be fully realized.' As some such international order is in any case constructed, Article 28 offers a standard below which it presumptively should not fall – a standard that, Pogge shows, is not currently met.

[2] Pogge (2005), 7. [3] Pogge (2005), 7.

[4] Pogge discusses in detail other common arguments raised against his thesis – in particular that local factors are primarily to blame for poverty and that the existence of examples of success in some countries necessarily indicates that global institutions must benefit all countries. See Pogge (2002), 6–11.

Pogge's observations transpose with ease to the climate change phenomenon as well as to the construction of an international regime with which to address it. As to climate change itself, the violation of a negative (moral) duty is, if anything, even starker in the case of greenhouse gas (GHG) emissions that contribute to affluent lifestyles in some parts of the world but result directly in misery and hardship in others. A negative duty not to violate basic human rights presumably calls at a minimum for an urgent and stringent mitigation regime. Beyond this, however, some climate change impacts are now being felt and others cannot be halted due to the extent of historical and current emissions and the time lag between emissions and their effects on the climate. In the most vulnerable places – Arctic regions, for example, Saharan Africa and some small island states such as Tuvalu – the consequences are already being experienced as threats to human rights. To recognise this fact and to view it as ethically, and even legally, relevant – and perhaps, in some sense, actionable – requires factoring in human rights consequences into current and future climate change scenarios. For example, a UN study on threatened island communities in sinking territories observed that such individuals will find themselves in the unprecedented situation of being citizens of a state that no longer has territory and being *de facto* refugees outside any existing Convention definition of the term.[5]

Pogge's general position is also clearly relevant to the construction of global policy measures to address climate change. Any regime that fundamentally reorganises and determines the access of countries and individuals to highly efficient energy resources, as a successful climate regime must, will have profound redistributive consequences, determining the long-term access of many hundreds of millions of people to some basic public goods, including human rights protections. There is clearly no way that such a regime can be constructed without these consequences, so presumably the primary obligation is to design one that will have the least negative impact on the most vulnerable. Again UDHR, Article 28's just 'social and international order' comes to mind. And here Pogge's thesis is again applicable in reminding us of the shortcomings of the current regime to address even the ordinary

[5] For an informed discussion, see E/CN.4/Sub.2/2005/28, Expanded working paper by Françoise Hampson on the human rights situation of indigenous peoples in states and other territories threatened with extinction for environmental reasons (16 June 2005). A total of just under half a million individuals are likely to be affected, from the islands of Tuvalu, Nauru, Kiribati, Maldives and the Bahamas. *Ibid.*, para. 25.

incidence of poverty in the world, which, as we have seen elsewhere in this book, tends to coincide with great vulnerability to climate change impacts. How can the same architecture be expected to cope with the further expected ramping up of vulnerability and insecurity that climate change will bring to those same persons and communities? The international trade regime is not the sole source of institutional concern from this perspective. Widespread promotion of deregulation has also had the cumulative effect of disempowering governments and inoculating publics against broad regulatory actions of a kind that might deal swiftly and cleanly with a problem of the scale and public interest weight of climate change.[6]

Against this backdrop, in the following section I pick up a number of themes raised in earlier chapters and flesh them out in more detail. Three topics are broached, each of which presents the disciplinary distinction between climate change and human rights in a different light: first, emissions trading, a subject that easily escapes human rights analysis, given that its effects are largely indirect; second, 'process rights', an area where human rights and environmental law already coincide to a degree; and finally, the possible application of human rights 'thresholds' to climate change policy.[7] The chapter then ends with a short discussion of the potential 'defragmentation' of international law in this context.

Emissions trading and the clean development mechanism

An international market in emissions reductions (also known as 'carbon trading') is likely to be at the centre of any future global mitigation regime.[8] As emissions trading is potentially both the most far-reaching mechanism, the most speculative in its potential outcomes and the least independent in terms of its broader effects, its human rights impacts are not easy to assess. Indeed, there is no obvious hook for a human rights analysis of an emissions market. The following section nevertheless flags some broad concerns particularly with regard to market access and the allocation of rights to emit, both discussed below. Before turning to these issues, the following paragraphs briefly describe the evolving regime.

[6] For recent examples of 'rule of law promotion', see contributions to Trubek and Santos (2006), Carothers (2006) and also Humphreys (2008).

[7] Other mitigation options would also bear closer scrutiny from this perspective: the imposition of carbon taxes and the possibility of recourse to nuclear power in particular.

[8] See IPCC, *Climate Change 2007*, AR4, WG III, ch. 13.

Under an emissions trading regime, mandatory national emissions reductions are converted into tradable commodities. The principle is to achieve cuts as cheaply as possible by allowing those who are best placed to make cuts the freedom to do so, and permitting others, for whom cuts are too expensive, to buy them instead. Companies that can make cuts cheaply can sell their excess reductions (which amount to rights to emit) to those for whom it is cheaper to buy these rights to emit than it is to achieve reduced emissions. Again, differences in the relative costs of making reductions make it attractive to include developing countries within the regime. According to the *Stern Review*:

> The ability to trade obligations across borders would improve efficiency by ensuring that deployment takes place where it is cheapest to do so. The benefits from this may be significant where there are major differences between countries in, for instance, the availability of a natural resource such as sunshine, or in lower labour or other costs.[9]

These benefits are already built into the mitigation regime. Companies from the wealthy countries listed in Annex I to the UNFCCC are not constrained to make cuts solely at home. Through the clean development mechanism (CDM), companies that reduce emissions cheaply in non-Annex I (developing) countries, relative to what would 'otherwise' have taken place, can then trade those reductions (known as certified emission reductions) (CERs) on the emissions market. By substituting cleaner technologies for dirtier ones, the CDM also aims to facilitate lower carbon development paths in poorer countries. At present, a very limited number of CDM CERs can be traded on the main existing market, the EU's emissions trading system (ETS). (Other markets are also being tried and tested.) In the longer term, the trajectory is toward an increasingly global market for fungible emissions reductions, resulting in a global carbon price.

The emissions market claims other benefits. It is expected to spur technological innovation, particularly among institutions and states for whom it will be particularly expensive to achieve future targets. Research into and development of clean technologies ought, in principle, to become a better investment for companies over time than repeatedly buying and using rights to emit.[10] Trading also promises the likelihood of a global price on carbon, which is one way to require companies and

[9] *Stern Review* (2006), Part VI, ch. 24, p. 529.

[10] It is far from clear that these incentives are, in fact, built into the system as currently designed. For informed discussion, see Lohmann (2006), 104–21 and 175–86.

other carbon users to internalise the social cost of producing GHG emissions. (Trading is not the only way of establishing a carbon price, of course: carbon taxes or simple fines would have a similar effect.) If efficiency is a guiding principle, the optimal carbon price should be global – carbon emissions should cost the same everywhere.[11] Trading too, therefore, ought ideally to function globally.[12] The CDM is a step toward creating such a global system.

Backing up the emissions trading regime is the hard mathematics of long-term stabilisation. If average global temperature rises are to be kept below a 'dangerous' threshold, *total* global emissions need to have fallen by at least 50 to 85 per cent from 2000 levels by 2050, which on most accounts means that the heavily polluting OECD countries will need to have reduced emissions by 80 to 90 per cent. This is well known. Less often discussed, however, is the arresting fact that, even if this ambitious target is achieved, developing countries will probably have collectively to have cut their emissions by 30 to 60 per cent (having peaked around 2025).[13] In other words, in just a couple of decades *no* country will be in a position to *increase* GHG emissions, not even those that today lack the resources necessary for basic public goods, such as food security, clean drinking water and access to basic health services. The dilemma is well captured in a report recently published by the Stockholm Environmental Institute (SEI) and EcoEquity.[14]

> If we are to [achieve a] plausibly precautionary global pathway, the South's emissions must leave their projected path almost immediately, and be dropping precipitously by 2025. And even [under] optimistic assumptions about both equity and economic growth, many people in the South would still be struggling against poverty when its emissions had to begin this steep decline. Moreover, the less stringent pathways – despite their substantially higher risks of catastrophic climate change – provide only another few extra years of emissions growth.

The problem is that the market now in design, focused as it is on easing cuts in rich countries, has little to say about the long-term development needs of poorer countries (beyond vague promises of cleaner technologies

[11] *Stern Review* (2006), Executive Summary, xviii ('Economic efficiency points to the advantages of a common global carbon price: emissions reductions will then take place wherever they are cheapest').

[12] In principle, a carbon tax too could achieve a global price if there was international harmonisation. For discussion see *Stern Review* (2006), Part VI, 470. See also Stiglitz (2006).

[13] See Baer *et al.* (2007), 23–4.

[14] Baer *et al.* (2007).

and adaptation funding that has been slow to materialise), even though the effective privatisation of GHGs must determine, or at least strongly influence, the eventual global distribution of development capacity. As the SEI report points out, the enormous challenge facing poorer countries can be met only by taking actions to ensure that by 2025 even the poorest countries will be on transition paths to low carbon economies while continuing a sustained development drive all the while. Future fulfilment of developmental needs will thus depend on the intermediate steps taken by then to ensure that such countries can maximise their limited carbon use or have access to inexpensive low carbon technologies. So far, no such plan exists. Indeed, for many countries, especially those with urgent or emergency needs, the immediate benefits of *selling on* carbon capacity may outweigh the more substantial benefits of utilising them, particularly if the technology to do so is inefficient or lacking. Poor countries might then be stuck selling off their development potential to rich countries. It is against this background that questions must be raised about the longer-term human rights implications of the emissions trading regime as currently designed.

Further dilemmas will arise regarding the allocation of emission rights today and equitable access to them in future. A country's emission cap is the basis for determining its allocation of emission rights. At present only Annex I country parties to the Kyoto Protocol (developed countries) have emission caps. These have mostly been passed on to national private actors, amounting in practice to rights to emit, which can then be traded between them.[15] Although developing countries do not have caps, they can nevertheless gain access to the emissions market at present by co-funding CDM projects and selling the resulting reductions on the emissions market. Furthermore, as CDM projects are not subject to the caps taken on by Annex I countries, they will generally represent a net *increase* in global emissions that cannot be properly factored into the

[15] The principal technique for allocation has so far been 'grandfathering'. Rights to emit are assigned to both countries and, down the line, companies according to the amount they already emit. The companies in question are frequently multinational, so that the national cap they inherit does not represent an absolute limit to their global emissions capacity. Companies may trade internally, and may negotiate CDM deals directly. So although the ETS, the most advanced market, is limited to the EU, transnational actors running enterprises outside the EU are somewhat free to decide where and how they make and sell emissions cuts. As other trading regimes emerge (in specific US states or elsewhere), the rights become increasingly fungible, particularly for ubiquitous actors with a presence in many markets.

global account.[16] In order to be functional and reliable over time, then, the emissions market will eventually require the long-term accession of all relevant actors to a system of caps, including CDM host countries.

For middle income developing countries (Brazil, the PRC, India and some countries of east and southeast Asia), this scenario need not pose a significant problem and, indeed, presents an opportunity. They may be able to defer taking on caps (and so trading directly in an international emissions market) until they have reached the point where their economies can transition successfully into a low carbon system. This option is unlikely to be available for many countries, however, especially not the poorest. So although it may seem wise (as well as just) for such countries not to accept caps at present (indeed, they have fought hard to avoid doing so), over time the number of allocations available can only fall, and will do so increasingly sharply. When poor countries enter the game, fewer allocation rights will be available and they will be more expensive.[17]

Market mechanisms are, of course, premised on discrimination: they discriminate against those who cannot afford to pay, which in this case will mean the very same countries that have not historically used the global carbon dump and have most need of its benefits. By mid-century, fewer emission rights will be available; they are likely to be in high demand and scarce; and wealthy countries will be pushing hard for concessions as well as buying up available capacity. A high price will nominally suit countries that cannot, in any case, use up all their allocation.[18] But, in fact, there are few *ex ante* guarantees that the exchange of emission capacity for cash will be a good deal for populations who thereby stand increasingly to lose their right to use the atmosphere to this end. For one, if the revenues from sales go to governments, broad-

[16] Arguably, since in the long run emissions from CDM projects must be set against global targets, CERs derived from CDM projects amount to a free gift to rich countries from poor countries. Early discussions as to whether they might be 'banked' by host countries came to nothing.

[17] In one version, the 'contraction and convergence' model (see Introduction to the present volume) proposed allocating emission rights (caps) to developing countries at levels significantly higher than present usage. This would have provided an extra source of income, year-on-year that could have been earmarked for technological investment up until convergence. It was not adopted.

[18] In principle, emissions allocations will be renewable: targets will be fixed within a period, following which new caps will be set and new allocations made. However, with each step, caps will become tighter than before. For poorer actors, selling may appear profitable even if gains are short term. Banking will be difficult and purchasing may be impossible.

based benefits will depend on redistributive disposition, administrative and technical capacity and incorrupt good offices of the state, which many of the countries in question currently lack. Equally or more likely, however, such revenues will accrue to private actors, local or foreign, who may simply repatriate or reinvest or otherwise send the money out of the country, and are unlikely in any case to invest it in the public interest. The emissions market may effectively place future development capacity in private, and largely transnational, hands.

All told, regardless of negotiating position, it is extremely unlikely that many of the poorest countries will ever be able to rely on carbon-based energy to fuel the kind of living standards that would ensure even basic rights guarantees for all. They will, therefore, have to rely on the appearance of affordable, clean technological fixes. This in turn presumes not only that rich countries will invest seriously in R&D, but also that they will encourage and facilitate the transfer of new technologies, and prevent intellectual property barriers to their access. Yet, technology transfer has not received the volume of attention devoted to emissions trading. As things stand, the available store of emission allocations is likely to run down before the poorest countries reach a sustainable and clean level of development. Without robust and detailed policies of technology transfer and adaptation, their development and policy options will steadily shrink, with deleterious effects on basic rights. From a rights perspective, therefore, it will be important to assess the extent to which potential impacts on rights are taken into account when trading regimes are implemented. Where the social consequences are likely to be inequitable, as suggested above, they should presumably be revised.

The above paragraphs are speculative. They name several long-term equity concerns that emissions trading raises, and suggest that the systems now being devised need to be assessed rigorously in terms of social and developmental consequences, also viewable as human rights concerns. In theory, of course, a portion of any country's emission rights might be considered inalienable, or emissions rights might be reallocated to the least developed countries. The adoption of a market mechanism, however, presumes against (if it does not actually preclude) these solutions. That is likely to penalise the very countries that are already most vulnerable to climate change impacts.

A core question raised by the emissions market, then, is whether it will put carbon-based development out of the reach of certain countries, without making any alternative readily available. From this perspective, the human rights impacts of emissions trading can be properly assessed only in the context of other elements of the climate regime, now and in future. These include: robust adaptation policies and programmes; research and development of new technologies and affirmative approaches to technology transfer; the relaxation of access barriers (to intellectual property, for example) and so on. These policy vectors do not depend upon market-based climate change solutions, and are ordinarily viewed in isolation, but they need not be.

Participation and information

The previous section looked at an area of climate change policy that appears distant or wholly distinct from the ordinary concerns of human rights law, but that, on closer analysis, clearly has significant human rights implications. The following section, by contrast, looks at one of the few clear areas of cross-fertilisation between human rights and environmental law. Of the menu of internationally recognised human rights, those to information and public participation have the longest history and most secure place within environmental law.[19] The first significant source is the 1992 Rio Declaration on Environment and Development, proclaimed by the UN General Assembly at the same time as the UNFCCC was opened for signature, which includes the following principle:

> Environmental issues are best handled with the participation of all concerned citizens … At the national level, each individual shall have appropriate access to information concerning the environment that is held by public authorities, including information on hazardous materials and activities in their communities, and the opportunity to participate in decision-making processes. States shall facilitate and encourage public awareness and participation by making information widely available. Effective access to judicial and administrative proceedings, including redress and remedy, shall be provided. (Rio Declaration, Principle 10)

Rio Principle 10 received its fullest expression to date in the 1998 Aarhus Convention on Access to Information, Public Participation in Decision-Making and Access to Justice in Environmental Matters. This treaty establishes comprehensive and binding standards in each of the three procedural areas of its title.[20] Although Aarhus has a regional (European) basis, it is

[19] For a full account of the history and relevant legal background, see Shelton (2001), 194–213 and 218–25.

[20] Full text online at: www.unece.org/env/pp/documents/cep43e.pdf. The right to information also appears in the Rotterdam Convention on the Prior Informed Consent Procedure for Certain Hazardous Chemicals and Pesticides in International Trade

open for signature to all states, and provides standards that might usefully be drawn upon at international level and by states everywhere faced with the peculiar hazards posed by climate change. Article 7 of the Aarhus Convention concerns public participation:

> Each Party shall make appropriate practical and/or other provisions for the public to participate during the preparation of plans and programmes relating to the environment, within a transparent and fair framework, having provided the necessary information to the public.

The *human right* to information, as commonly codified in national legislation, might not at first glance appear to be of much help when faced with these challenges, because it only affirms the general public's right to receive *on request* information *already held* by public authorities. This could not be applied to advance adaptation policy in countries where relevant information has not been collected. However, the Aarhus Convention goes beyond the usual wording of (domestic) freedom of information acts, requiring that states actively compile periodic reports on environmental risks, update them systematically and make them available to the public proactively (Aarhus, Article 5). On its own, this stipulation is still insufficient to resolve the information dilemmas facing developing countries in the context of climate change. An Aarhus equivalent for African states, for example, would not overcome the resource and capacity gaps that stymie good information gathering in many countries. Nor would it necessarily strengthen the hand of developing country delegates (or civil society groups, journalists or others) at larger negotiating tables. However, the Aarhus Convention goes still further. Article 3(7) states that:

> Each Party shall promote the application of the principles of this Convention in international environmental decision-making processes and within the framework of international organizations in matters relating to the environment.

This principle is supposedly binding on Aarhus parties, including all EU member states (except Ireland). It must presumably also apply to the international adaptation and mitigation negotiations in which these countries are involved, and inform the obligations outlined in the UNFCCC and reiterated and expanded in the Bali road map.[21] It is perhaps surprising,

(PIC Convention); EC Directive on Combating Air Pollution from Industrial Plants, 84/360/EEC, OJL 188 of 16 July 1984; Directive 90/313/EEC of 27 June 1990 on the Freedom of Access to Information on the Environment, OJL 158 of 23 June 1990. Also WHO European Charter on the Environment and Health. See Shelton (2001), 200–3.

[21] Decision -/CP.13, Bali Action Plan (Advance Unedited Version), Article 1(c)(i).

given the essential role that information and participation must play in elaborating adaptation policies, that Aarhus, Article 3(7) appears not to have been invoked or insisted upon in those negotiations. The Aarhus principles recognise that information gathering involves choices and decisions about resource distribution and capacity, about what and how much to gather at what cost, and about who should receive it and how it may be used. How, for example, might the application of Aarhus affect the respective roles of national and international actors in information gathering, analysis and dissemination for adaptation? Given the recurrent impasse in international adaptation negotiations, such questions merit further investigation and advocacy.

Climate change negotiations have long suffered from complications of process and participation, rooted in systemic inequalities.[22] Resource-poor countries in need of adaptation funding often can afford only a few delegates at climate negotiations, where wealthy countries can field hundreds. As a result, those present face inevitable difficulties, given the complexities and interdependence of the many themes under negotiation at a given time. In these circumstances, diplomats from poor countries are unlikely to influence outcomes effectively even when the states they represent have a clear interest.

Not surprisingly, the situation has fuelled distrust, even acrimony, between the parties.[23] Donor countries have been criticised for apparently seeking to avoid a participatory process through the UNFCCC to fund the immense adaptation needs of developing countries. The core funds of the Global Environmental Facility (GEF) – the World Bank-hosted clearing house for funding related to the principal environmental treaties including the UNFCCC – have been subject to criteria that, while sensible for mitigation activities, have little to do with adaptation. Projects must contribute to 'global environmental benefits' to qualify, whereas adaptation actions will necessarily bring primarily local (rather than global) benefits. Projects are funded on an 'incremental costs' basis, which makes sense for mitigation activities, where the GHG emissions of a given project can ordinarily be reduced at an additional cost, but not for adaptation, where costs are likely to be wholly new, just like the causes they address.[24] As the World Resources

[22] See generally Mace (2005). [23] McGray *et al.* (2007), 33.

[24] The GEF eventually introduced the notion of 'additional' to replace 'incremental' costs – the idea being that funding 'additional' to ODA would be made available to 'add' extra 'climate adaptive' components onto existing development projects. McGray *et al.* (2007), 33. On the persistence of both 'global environmental funding' and 'incremental costs', see Mace (2005), 226–8, 244, 335–6. In addition to these criteria a Resource Allocation

Institute has remarked, this suggests that adaptation is somehow parasitic upon existing development planning, ignoring the unique character of adaptive responses to climate change. 'The lack of [adaptation] implementation', they remark, is 'largely driven by the conflicts in funding procedures [and] has further exacerbated tensions in the international climate negotiations.'[25]

A root cause of tension is the governance structure of the World Bank itself. The Bank acts as a trustee for the GEF as well as one of its three implementing agencies (the others are UNEP and UNDP) and, in addition, frequently co-finances GEF-supported projects. As a result the Bank wields enormous influence over the GEF. However, decision making at the Bank's Board is weighted in favour of its largest shareholders by financial contribution, and this tends to minimise or exclude recipient countries from the outset.[26] In this context, the emphasis in Aarhus on public participation is again relevant, particularly when read in conjunction with the Article 3(7) requirement regarding international negotiations. For as long as the participation of some relevant actors can be curtailed by channelling decisions through the World Bank, donor countries will continue to exercise undue influence and many recipient countries will continue to perceive consultation as merely formal. Since European countries are under a binding obligation, according to Aarhus, Article 3(7), to introduce Aarhus principles into international negotiations on the environment and have recognised the essential role of national leadership over policy formation in the OECD Paris Declaration on Aid Effectiveness of 2005, they might be expected to create appropriately inclusive policy fora. The fact that they have failed to do so in the case of the new Climate Investment Fund (for example) is an obstacle to the development of sound and legitimate policies to address climate change, and this might be expected eventually to attract the attention of human rights advocates.

Framework created to evaluate funding applications introduced governance criteria requiring indicators that, at the time they were introduced, did not exist for at least twenty-three of the countries most in need of adaptation funding. Mace (2005), 243–5.

[25] McGray *et al.* (2007), 34.

[26] The five largest shareholders to the IBRD each appoint one of twenty-four executive directors on the Bank's Board (the rest are elected). In December 2007, the top five contributors commanded almost 40 per cent of the Bank's voting powers (they are: France (4.3 per cent), Germany (4.49 per cent), Japan (7.86 per cent), the United Kingdom (4.3 per cent) and the United States (16.38 per cent). See the Bank's website at: http://go.worldbank.org/11PWB3RTM0 and http://go.worldbank.org/O9S0U0IOA0.

The Aarhus principles are equally relevant to international mitigation policies, particularly given their cooperative dimension. Because emission caps under the Kyoto Protocol are restricted to Annex I countries, discussions of mitigation strategies have also largely centred on Annex I actors (both public and private). As a result, the most significant long-term mitigation structures – notably the new emissions markets – have gone ahead as though they were of limited interest or relevance for the poorest non-participating states. They are not, of course, given that the markets in question distribute a (henceforth) strictly limited global commodity of which those states may have dire need in future. A broad combination of actors from Annex I countries, both public and private, have been actively constructing this market for over a decade, with the result that the trading regime has become increasingly complex, well resourced and jargon laden. The immense significance that emissions trading will have for the long-term economic prospects of non-Annex I countries has barely been registered or researched in many of these countries, with the exception of the larger 'middle income' countries whose involvement is critical to the market's success. Over time, an initial information gap has led to limited participation by the poorest (or 'least developed') countries, and diminishing appreciation of finer policy detail for those outside the loop, particularly among the general public in those countries. The result is *de facto* exclusion from discussions of carbon trading of very many whose futures will be directly affected.

The CDM has led to the active involvement of some larger developing countries in the trading scheme. Indeed, the CDM is the first and most obvious area where an effectively global regime will impact directly upon development in poorer countries. As outlined above, however, depending on how it is ultimately structured, the trading regime may ultimately price the poorest countries out of the market for access to the carbon dump. It would seem important, then, to ensure that parties with an interest are adequately informed and equipped for full participation, in order to head off such an outcome. Because developing countries, particularly those with the fewest resources, focus on the immediate threats posed by climate change and the urgency of adaptation, mitigation negotiations ought to be transparent and information about policies and programmes should be efficiently communicated. At present, discussions of emissions trading treat their relevance to non-Annex I countries as a matter of CDM prizes, on the one hand, and of the eventual universal adoption of emissions caps, on the other hand. Large host countries, such as the PRC, can enter this trading regime on their own terms, but for smaller countries the stakes of the global market rather concern their long-term development capacity – an issue that has so far suffered relative neglect.

Human rights as thresholds

To look ahead to future eventualities, unprecedented in nature and effect and framed as probabilities rather than known quantities, does not fit easily within the ordinary language of human rights. At first sight, human rights scholars and lawyers appear to be concerned with the past rather than the future (a violation is something that has actually happened or is happening) and with facts rather than probabilities. Nevertheless, forward thinking does, in fact, play an important role in human rights protection. Law enforcement and judicial systems are themselves deterrence mechanisms, warding off future violations via the threat of punishment. The distinction between facts and probabilities is perhaps, then, one of degree: the probability of a given human rights violation taking place in future can – somewhat like a predicted climate change impact – increase or diminish over time according to the relative robustness of the institutions designed to prevent it.

But there is a difference. Whereas human rights prevention mechanisms are tested and familiar, those needed to prevent damage from climate change are still largely innovative or speculative. The means of prevention are almost as hypothetical as the impacts they must prevent and both are subject to the unpredictable feedback effects of the interventions themselves. This has contributed an inherent dynamism in climate change forecasting, reliant as it is on multiple feedback loops and scenarios. Predicted impacts are constantly readjusted to take account of varying or changing assumptions. Innumerable mitigation, adaptation and development paths can be forecast, each with different baseline assumptions and impact ranges. Tweaking any one aspect of a given input – scientific, economic or social – leads to domino alterations elsewhere. Human rights impacts are a relevant aspect of that dynamism, subject to different levels of protection and fulfilment under different scenarios, even if they have not, to date, been factored explicitly into calculations. To mobilise the policy value, and indeed the legal force, of human rights in the construction of a climate change regime, therefore, requires the injection of likely human rights impacts and outcomes into the dynamic forecasting that already characterises climate change scenario construction.

One way to organise data collection and modelling of this kind might be to think in terms of human rights thresholds, as discussed by Simon Caney in his contribution to the present volume (see Chapter 2). These would be levels of protection for individual rights which can be regarded as the minimum acceptable outcome under a given climate change impact or policy scenario. A requirement that basic threshold levels should not be breached – either as an effect of climate change itself or as a by-product of a given mitigation or adaptation policy – ought not to be controversial, especially as such a goal is also a legal requirement for most relevant (public) actors.[27] In addition, as Caney makes clear, such a goal is modest. It does not require large-scale social engineering or assume equal and universal access rights to the atmosphere – as contraction and convergence arguably might. Nor does it involve epic calculations across vast datasets.

Viewing climate change impacts in terms of human rights thresholds will raise a number of questions that have barely been touched upon to date. Elsewhere Caney notes, for example, the level at which global warming becomes 'dangerous', given that impacts are not uniform but touch some persons more severely than others.[28] The common view that an average rise of no more than 2°C from pre-industrial levels is acceptable may appear reasonable from an aggregate perspective, but will appear much less so to those for whom such an increase involves irretrievable losses to livelihood and culture, or those living in places likely to experience warming at higher levels than average.[29] Further questions arise once it is acknowledged that average global warming is, in fact, unlikely to remain below the 'dangerous' 2°C. The pool of individuals certain to be affected grows with each incremental increase in the agreed level of global warming. Should all those caught in this pool be compensated? If so, by whom? Will they have viable claims? Or will it

[27] The General Comments of the UN's Committee on Economic Social and Cultural Rights are relevant here. See, for example, UN Doc. E/C.12/2002/11, General Comment No. 15 (2002), The right to water (Articles 11 and 12 of the International Covenant on Economic, Social and Cultural Rights), para. 8. This highlights the link between 'environmental hygiene', safe drinking water and health, stating among other things that 'States parties should monitor and combat situations where aquatic eco-systems serve as a habitat for vectors of diseases …'.

[28] See UNFCCC, Article 2.

[29] It is a further irony that on many predictions, many effects of a rise between 2°C and 3°C, although devastating in some parts of the world, particularly small island states, may actually be beneficial (on balance) in some OECD countries. Such predictions might presumably delay the urgency to act in countries better equipped to handle the rise.

make more sense, having identified those most at risk, to channel resources in advance toward actions that will ward off their future predicament? In either case, there is a solid argument for identifying as far in advance as possible the likely victims and the mechanisms needed to protect their rights.

Embedding human rights thresholds into policy objectives might first involve reviewing existing climate change scenarios to identify specific human costs across time and in different places, and then asking how countries – provincial and local governments, and eventually communities – are equipped to respond (socially, financially, technologically and institutionally). Down the road, real-time monitoring would need to be supplemented by predictive forecasting of human rights threshold levels under a series of scenarios. Such scenario building would presumably help to guide both adaptation and mitigation policies. Since it is widely recognised that some impacts are now inevitable, adaptation measures are already required for those countries likely to be hardest hit. Yet existing adaptation plans and funding have run into numerous obstacles and continue to advance with difficulty. Identifying likely transgressions of human rights thresholds should refocus attention on the human priorities that ought to drive policy. At the same time, building human rights assessments into long-term mitigation and adaptation scenarios would refine and improve policies, and provide criteria for their adoption or rejection.

Defragmentation of international law?

In a report published in May 2006, the International Law Commission (ILC), a Geneva-based UN body, examined whether international law had become overly 'fragmented'.[30] The question came before the ILC for a number of reasons. On the one hand, international tribunals had not always seemed in agreement on mutually related areas of international case law. On the other hand, certain disputes were being brought simultaneously before different tribunals in an apparent attempt to find the most favourable forum. In addition, with a proliferation of regional legal systems (European, African, Inter-American), and 'special regimes' (trade, environment, human rights), fears were expressed that the architecture of international law might prove to be unable to assimilate or accommodate all the demands placed upon it.[31]

30 For a complete account, see: http://untreaty.un.org/ilc/summaries/1_9.htm.

31 See UN Doc. A/CN.4/L.682, Fragmentation of International Law: Difficulties Arising from the Diversification and Expansion of International Law, Report of the Study

Without taking a strong position on the existence or otherwise of 'fragmentation', the ILC emphasised the inherent tendency toward 'systemic integration' in the evolution and interpretation of international law. Since legal regimes cannot cordon off areas of life into discrete boxes marked 'trade', 'environment', 'human rights', and so on, each legal regime will necessarily constantly run into issues relevant to the others. These collisions are largely managed by judges and lawyers, who will likely turn to 'general principles of international law' at such times (the ILC focused particularly on the 1969 Vienna Convention on the Law of Treaties). Even if it fragments, the ILC implied, the natural tendency of international law is to converge or reconverge.

Although the legal worlds of climate change and international human rights have shown an occasional tendency toward mutual accommodation in this way, they remain very different.[32] Areas of apparent convergence include: reference to a right to the environment in a handful of international documents and over 100 national constitutions;[33] some case law (notably in the European Court of Human Rights), linking environmental pollution with the rights to health and to 'privacy and family life'; recognition that environmental factors matter in protecting the right to water (General Comment 15 of the UN Committee on Economic, Social and Cultural Rights);[34] and recognition of the importance of rights to information, participation and effective remedy to environmental conservation in the Rio Declaration and Aarhus Convention. The UN General Assembly and other UN bodies have repeatedly stressed the links between both.[35]

Yet the climate change and human rights regimes remain fundamentally dissimilar.[36] One is a regime of flexibility, compromise, soft principles and differential treatment; the other of judiciaries, policing, formal equality and universal truths. Faced with injustice, one regime tends to

Group of the International Law Commission Finalized by Martti Koskenniemi, 13 April 2006, 10–16.

32 See Sands (2003), 291–307; Shelton (2001), especially 191–4, 231–6; Shelton (1991).

33 The right appears in two binding regional human rights treaties: the 1981 African Charter on Human and Peoples' Rights (Article 24) and the 1988 San Salvador Protocol to the American Convention on Human Rights (Article 11). Sands (2003), 298.

34 E/C.12/2002/11, 26 November 2002, para. 8. 35 Sands (2003), 295.

36 Despite years of incremental progress on the 'right to a clean environment', Philip Alston could still write in 2001 that 'there is very limited support in international law for the existence of such a right seen as a freestanding human right, attaching either to individuals or peoples': Alston (2001), 281.

negotiation, the other to prosecution. But neither on its own seems quite up to the challenge presented by climate change. The climate regime tends too easily to defer action on the worst injustices (and may even perpetuate them, as we have seen above). Human rights advocates seek hard-and-fast boundaries in a context where they simply cannot be drawn. It may be that the justice claims generated by climate change are simply too large and unsettling to be effectively treated by either regime alone. Or perhaps there is scope for learning to combine the strengths of each – the flexibility and equity of the climate regime, the accountability and rectitude of human rights – with a view to forging an increased capacity for justice in an interdependent world.

After all, as noted earlier, the distinction between facts (in human rights) and probabilities (in climate change) is one of degree: a significant portion of human rights advocacy is concerned with hypotheticals; that is, calling for new laws, reforming judiciaries, training police officers are all means of preventing human rights abuses or at least reducing the *probability* of future occurrence – accountability mechanisms, in short. Looking ahead, if the climate regime is to function effectively, it too cannot continue long without robust accountability mechanisms, both private, as Newell (Chapter 4) indicates in his contribution to this book, and public, as Mutter and Barnard (Chapter 10), as well as Hunt and Khosla (Chapter 8), emphasise in theirs. Key terms, such as equity and vulnerability will need further fleshing out – as Philippe Cullet describes in detail in Chapter 6 – and human rights principles may offer a means of doing so. At the inter-state level, as Dinah Shelton makes clear in Chapter 3, the differing obligations upon sovereign states in these two regimes may dovetail in such as way as to be mutually supportive. The substantive areas of climate change action – adaptation, REDD, biofuels, an emissions market – will gain from incorporating human rights criteria and safeguards, as Seymour (Chapter 7) and Barnett (Chapter 9) among others point out.

None of these pragmatic areas of potential overlap and learning need distract from the wider questions that climate change raises for human rights, questions approached in different ways by Cullet, Adelman (Chapter 5) and Caney in this volume. The phenomenon of climate change, and the way in which it has been tackled to date, highlight the weaknesses of the international system and in particular the gap between the moral statements and aspirations of human rights and the legal norms intended to back them up. In the longer run, climate change will transform much of the world's legal and political, as well as its physical and geographical landscape. Where we are faced, as at present, with a new property rights regime that may, in the longer run, prove detrimental to human rights, it too may be challenged in the language of human rights. And there is also scope to start thinking more broadly about how to rectify, amend and extend rights protections throughout a world transforming under the pressures of a changing climate and a global society and economy adapting in response.

References

Alston, Philip, 'People's Rights: Their Rise and Fall' in Philip Alston (ed.), *Peoples' Rights* (Oxford University Press, 2001).

Baer, Paul, Tom Athanasiou and Sivan Kartha, *The Right to Development in a Climate Constrained World* (Heinrich Böll Foundation, 2007).

Carothers, Thomas, *Promoting the Rule of Law Abroad: In Search of Knowledge* (Carnegie Endowment for International Peace, 2006).

Humphreys, Stephen, 'Theatre of the Rule of Law: An Inquiry into Transnational Rule of Law Promotion' (2008), unpublished doctoral thesis.

IPCC, *Climate Change 2007: Mitigation of Climate Change, Contribution of Working Group III to the Fourth Assessment Report of the IPCC* (Cambridge University Press, 2007).

Lohmann, Larry, *Carbon Trading: A Critical Conversation on Climate Change, Privatization and Power* (Dag Hammarskjold Foundation, 2006).

Mace, M. J., 'Funding for Adaptation to Climate Change: UNFCCC and GEF Developments Since COP-7', *Review of European Community and International Environmental Law* (RECIEL), 14 (2005), 225.

McCarthy, James, J. Osvaldo, F. Canziani, Neil A. Leary, David J. Dokken and Kasey S. White (eds.), *Climate Change 2001. Impacts, Adaptation and Vulnerability, Contribution of Working Group II to the Third Assessment Report of the IPCC* (Cambridge University Press, 2001).

McGray, Heather, Anne Hammill, Rob Bradley with Lisa Schipper and Jo-Ellen Parry, *Weathering the Storm: Options for Framing Adaptation and Development* (World Resources Institute, 2007).

Pogge, Thomas, 'Recognized and Violated by International Law: The Human Rights of the Global Poor' (2006), available at: http://www2.ohchr.org/english/issues/poverty/expert/docs/Thomas_Pogge_new.pdf (revised version of *Leiden Journal of International Law*, 18 (2005), 717).

Pogge, Thomas, *World Poverty and Human Rights* (Polity Press, 2002).

Sands, Philippe, *Principles of International Environmental Law* (Cambridge University Press, 2003).

Shelton, Dinah, 'Environmental Rights' in Philip Alston (ed.), *Peoples' Rights* (Oxford University Press, 2001).

'Human Rights, Environmental Rights and the Right to Environment', *Stanford Journal of International Law*, 28 (1991), 103.

Stern, Nicholas, *Stern Review on the Economics of Climate Change*, HM Treasury, Office of Climate Change (Prepublication) (2006).

Stiglitz, Joseph, 'A New Agenda for Global Warming', *The Economists' Voice*, 3:7, article 3 (2006).

Trubek, David and Alvaro Santos (eds.), *The New Law and Economic Development* (Cambridge University Press, 2006).

APPENDIX: CLIMATE CHANGE IMPACTS ON HUMAN RIGHTS

Human rights impacts of climate change by region

Excerpts from *IPCC, AR4, WG II, Summary for Policymakers*, 13–15 and 59–63; and from *The Stern Review*, 103–5 (references excised).

Africa

IPCC AR4

- By 2020, between 75 million and 250 million people are projected to be exposed to increased water stress due to climate change. If coupled with increased demand, this will adversely affect livelihoods and exacerbate water-related problems.
- Agricultural production, including access to food, in many African countries and regions is projected to be severely compromised by climate variability and change. The area suitable for agriculture, the length of growing seasons and yield potential, particularly along the margins of semi-arid and arid areas, are expected to decrease. This would further adversely affect food security and exacerbate malnutrition in the continent. In some countries, yields from rain-fed agriculture could be reduced by up to 50 per cent by 2020.
- Any changes in the primary production of large lakes are likely to have important impacts on local food supplies. For example, Lake Tanganyika currently provides 25 to 40 per cent of animal protein intake for the population of the surrounding countries, and climate change is likely to reduce primary production and possible fish yields by roughly 30 per cent. The interaction of human management decisions, including over-fishing, is likely to further compound fish off-takes from lakes.
- Local food supplies are projected to be negatively affected by decreasing fisheries resources in large lakes due to rising water temperatures, which may be exacerbated by continued over-fishing.

- Toward the end of the twenty-first century, projected sea level rise will affect low lying coastal areas with large populations. The cost of adaptation could amount to at least 5–10 per cent of GDP. Mangroves and coral reefs are projected to be further degraded, with additional consequences for fisheries and tourism.
- New studies confirm that Africa is one of the most vulnerable continents to climate variability and change because of multiple stresses and low adaptive capacity. Some adaptation to current climate variability is taking place; however, this may be insufficient for future changes in climate.

Stern Review

- Africa will be under severe pressure from climate change. Many vulnerable regions, embracing millions of people, are likely to be adversely affected by climate change, including the mixed arid–semi-arid systems in the Sahel, arid–semi-arid range-land systems in parts of eastern Africa, the systems in the Great Lakes region of eastern Africa, the coastal regions of eastern Africa and many of the drier zones of southern Africa.
- Tens of millions of additional people could be at risk of malaria by the 2080s. Previously unsuitable areas for malaria in Zimbabwe could become suitable for transmission with slight temperature and precipitation variations, whist in South Africa the area suitable for malaria may double with 7.8 million people at risk by 2100.
- Water pressures may be intensified as rainfall becomes more erratic, glaciers retreat and rivers dry up. While there is much uncertainty about flow of the Nile, several models suggest a decrease in river flow, with nine recent climate scenario impacts ranging from no change to more than 75 per cent reduction in flows by 2100. This will have a significant impact on the millions of people who have competing claims on its supplies.
- Many large cities in Africa that lie on or very close to the coast could suffer severe damages from sea level rise. According to national communications to the UNFCCC, a 1-m sea level rise (a possibility by the end of the century) could result in the complete submergence of Banjul, Gambia, and losses of more than US$470 million in Kenya for damage to three crops (mangoes, cashew nuts and coconuts).

Asia

IPCC AR4

- Climate change is projected to impinge on the sustainable development of most developing countries of Asia, as it compounds the pressures on natural resources and the environment associated with rapid urbanisation, industrialisation and economic development.
- Endemic morbidity and mortality due to diarrhoeal disease primarily associated with floods and droughts are expected to rise in east, south and southeast Asia due to projected changes in the hydrological cycle associated with global warming. Increases in coastal water temperature would exacerbate the abundance and/or toxicity of cholera in south Asia.
- A 1-m rise in sea level would lead to a loss of almost half of the mangrove area in the Mekong River delta (2,500 km^2), while approximately 100,000 ha of cultivated land and aquaculture area would become salt marsh. Coastal areas, especially heavily populated megadelta regions in south, east and southeast Asia, will be at greatest risk due to increased flooding from the sea and, in some megadeltas, flooding from the rivers. For a 1-m rise in sea level, 5,000 km^2 of the Red River delta and 15,000 to 20,000 km^2 of the Mekong River delta are projected to be flooded, which could affect 4 million and 3.5 to 5 million people, respectively.
- Glacier melt in the Himalayas is projected to increase flooding, and rock avalanches from destabilised slopes, and to affect water resources within the next two to three decades. This will be followed by decreased river flows as the glaciers recede. Tibetan Plateau glaciers of under 4 km in length are projected to disappear with a temperature increase of 3°C and no change in precipitation. If current warming rates are maintained, Himalayan glaciers could decay at very rapid rates, shrinking from the present 500,000 km^2 to 100,000 km^2 by the 2030s.
- Around 30 per cent of Asian coral reefs are expected to be lost in the next 30 years, compared with 18 per cent globally under the IS92a emissions scenario, but this is due to multiple stresses and not to climate change alone.
- It is estimated that 120 million to 1.2 billion and 185 to 981 million people will experience increased water stress by the 2020s and the 2050s, respectively. The per capita availability of fresh water in India is

expected to drop from around 1,900 m^3 currently to 1,000 m^3 by 2025 in response to the combined effects of population growth and climate change. More intense rain and more frequent flash floods during the monsoon would result in a higher proportion of run-off and a reduction in the proportion reaching the groundwater.

- Freshwater availability in central, south, east and southeast Asia, particularly in large river basins, is projected to decrease due to climate change which, along with population growth and increasing demand arising from higher standards of living, could adversely affect more than a billion people by the 2050s. Agricultural irrigation demand in arid and semi-arid regions of east Asia is expected to increase by 10 per cent for an increase in temperature of 1°C.
- It is projected that crop yields could increase up to 20 per cent in east and southeast Asia, while they could decrease up to 30 per cent in central and south Asia, by the mid-twenty-first century. Taken together and considering the influence of rapid population growth and urbanisation, the risk of hunger is projected to remain very high in several developing countries.
- The frequency and extent of forest fires in northern Asia are expected to increase in the future due to climate change and extreme weather events that would likely limit forest expansion.

Stern Review

Temperatures will increase for all months. Consequently, during the dry pre-monsoon months of April and May, the incidence of extreme heat is likely to increase, leading to greater mortality. Changes in the intensity of rainfall events, and the cycles of the monsoon – combined with an increased risk of critical temperatures being exceeded more frequently – could significantly change crop yields. For example, mean yields for some crops in northern India could be reduced by up to 70 per cent by 2100. This is set against a background of a rapidly rising population that will need an additional 5 million tons of food production per year just to keep pace with the predicted increase in population to about 1.5 billion by 2030.

Latin America

IPCC AR4

- In the future, the frequency and intensity of hurricanes in the Caribbean basin are likely to increase.

- As a result of climate change, rice yields are expected to decline after the year 2020, while increases in temperature and precipitation in south-eastern South America are likely to increase soybean yields if CO_2 effects are considered.
- The number of additional people at risk of hunger under (one) emissions scenario is likely to attain 5, 26 and 85 million in 2020, 2050 and 2080, respectively, assuming little or no CO_2 effects. Cattle productivity is very likely to decline in response to a 4°C increase in temperatures.
- By mid-century, increases in temperature and associated decreases in soil water are projected to lead to gradual replacement of tropical forest by savanna in eastern Amazonia. Semi-arid vegetation will tend to be replaced by arid land vegetation. There is a risk of significant biodiversity loss through species extinction in many areas of tropical Latin America.
- In drier areas, climate change is expected to lead to salinisation and desertification of agricultural land. Productivity of some important crops is projected to decrease and livestock productivity to decline, with adverse consequences for food security. In temperate zones soya bean yields are projected to increase.
- Sea level rise is projected to cause increased risk of flooding in low lying areas. Increases in sea surface temperature due to climate change are projected to have adverse effects on Mesoamerican coral reefs, and cause shifts in the location of south-east Pacific fish stocks.
- Changes in precipitation patterns and the disappearance of glaciers are projected to significantly affect water availability for human consumption, agriculture and energy generation. By the 2020s between 7 million and 77 million people are likely to suffer from a lack of adequate water supplies, while for the second half of the century the potential water availability reduction and the increasing demand, from an increasing regional population, would increase these figures to between 60 and 150 million.
- Some countries have made efforts to adapt, particularly through conservation of key ecosystems, early warning systems, risk management in agriculture, strategies for flood, drought and coastal management and disease surveillance systems. However, the effectiveness of these efforts is outweighed by lack of basic information, observation and monitoring systems, lack of capacity building and appropriate political, institutional and technological frameworks, low income and settlements in vulnerable areas, among others.

Stern Review

- Countries in Latin American and the Caribbean are significantly affected by climate variability and extremes. The region's economy is strongly dependent on natural resources linked to climate, and patterns of income distribution and poverty exacerbate the impacts of climate change for specific subregions, countries and populations. Living conditions and livelihood opportunities for millions of people may be affected. By 2055 subsistence farmers' maize production (the main source of food security) in the Andean countries and Central America could fall by around 15 per cent on average. The potential die-back, or even collapse, of the Amazon rainforest presents a great threat to the region. The Amazonian forests are home to around 1 million people of 400 different indigenous groups, and provide a source of income and medical and pharmaceutical supplies to millions more.

Middle East and North Africa

Stern Review

- The region is already very short of fresh water and faces difficulty in meeting the needs of fast-growing populations. Most, if not all, of the region may be adversely affected by changing rainfall patterns as a result of climate change. An additional 155 to 600 million people may suffer an increase in water stress in North Africa with a 3°C rise in temperature according to one study. Yemen is particularly at risk given its low income levels, rapidly growing population and acute water shortages today. Competition for water within the region and across its borders may grow, carrying the risk of conflict.
- Reduced water availability combined with even modestly higher temperatures will reduce agricultural productivity and in some areas may make crops unsustainable. Maize yields in North Africa, for example, could fall by between 15 and 25 per cent with a 3°C rise in temperature according to one recent report.
- Some parts of the region – notably the Nile Delta and the Gulf coast of the Arabian peninsula – are in addition vulnerable to flooding from rising sea levels which could lead to loss of agricultural land and/or threats to coastal cities. Others are vulnerable to increased desertification.

Small islands

IPCC AR4

- Small islands, whether located in the tropics or higher latitudes, have characteristics which make them especially vulnerable to the effects of climate change, sea level rise and extreme events.
- Deterioration in coastal conditions, for example, through erosion of beaches and coral bleaching, is expected to affect local resources, for example, fisheries, and reduce the value of these destinations for tourism.
- Sea level rise is expected to exacerbate inundation, storm surge, erosion and other coastal hazards, thus threatening vital infrastructure, settlements and facilities that support the livelihood of island communities.
- Climate change is projected by mid-century to reduce water resources in many small islands, for example, in the Caribbean and Pacific, to the point where they become insufficient to meet demand during low rainfall periods.
- International airports on small islands are mostly sited on or within a few kilometres of the coast, and the main (and often only) road network runs along the coast. Under sea level rise scenarios, many of them are likely to be at serious risk from inundation, flooding and physical damage associated with coastal inundation and erosion.
- Without adaptation, agricultural economic costs from climate change are likely to reach between 2 and 3 per cent and 17 and 18 per cent of 2002 GDP by 2050, on high terrain (for example, Fiji) and low terrain (for example, Kiribati) islands.
- Outbreaks of climate-sensitive diseases such as malaria, dengue, filariasis and schistosomiasis can be costly in lives and economic impacts. Increasing temperatures and decreasing water availability due to climate change is likely to increase burdens of diarrhoeal and other infectious diseases in some small island states.
- Studies so far conducted on adaptation on islands suggest that adaptation options are likely to be limited and the costs high relative to GDP. Recent work has shown that, in the case of Singapore, coastal protection would be the least-cost strategy to combat sea level rise under three scenarios, with the cost ranging from US$0.3–5.7 million by 2050 to US$0.9–16.8 million by 2100.

Human rights impacts of climate change by affected right

Excerpts from *IPCC, AR4, WGII, Technical Summary*, 44–7; and the *Stern Review*, 62–76 (references excised).

Right to water

IPCC AR4

- Water volumes stored in glaciers and snow cover are very likely to decline, reducing summer and autumn flows in regions where more than one-sixth of the world's population currently live.
- Drought-affected areas will probably increase, and extreme precipitation events, which are likely to increase in frequency and intensity, will augment flood risk. Increased frequency and severity of floods and droughts will have implications for sustainable development.
- Up to 20 per cent of the world's population live in river basins that are likely to be affected by increased flood hazard by the 2080s in the course of global warming.
- Many semi-arid areas (for example, Mediterranean basin, western United States, southern Africa and north-eastern Brazil) will suffer a decrease in water resources due to climate change.
- The number of people living in severely stressed river basins is projected to increase from 1.4–1.6 billion in 1995 to 4.3–6.9 billion in 2050 (according to one scenario).
- Sea level rise will extend areas of salinisation of groundwater and estuaries, resulting in a decrease in fresh water availability for humans and ecosystems in coastal areas.
- Groundwater recharge will decrease considerably in some already water-stressed regions, where vulnerability is often exacerbated by the rapid increase in population and water demand.
- Higher water temperatures, increased precipitation intensity and longer periods of low flows exacerbate many forms of water pollution, with impacts on ecosystems, human health and water system reliability and operating costs.
- Areas in which run-off is projected to decline will face a reduction in the value of services provided by water resources. The beneficial impacts of increased annual run-off in other areas will be tempered by the negative effects of increased precipitation variability and seasonal run-off shifts on water supply, water quality and flood risks.

Stern Review

- People will feel the impact of climate change most strongly through changes in the distribution of water around the world and its seasonal and annual variability. Areas that are already relatively dry, such as the Mediterranean basin and parts of southern Africa and South America, are likely to experience further decreases in water availability, for example, several (but not all) climate models predict up to 30 per cent decrease in annual run-off in these regions for a 2°C global temperature rise and 40–50 per cent for 4°C rise.
- The effects of rising temperatures against a background of a growing population are likely to cause changes in the water status of billions of people. Considerably more effort and expense will be required on top of existing practices to meet people's demand for water.
- Climate change will have serious consequences for people who depend heavily on glacier meltwater to maintain supplies during the dry season, including large parts of the Indian sub-continent, over quarter of a billion people in the PRC and tens of millions in the Andes. In the Himalaya–Hindu Kush region, meltwater from glaciers feeds seven of Asia's largest rivers, including 70 per cent of the summer flow in the Ganges, which provides water to around 500 million people. In the PRC, 23 per cent of the population (250 million people) lives in the western region that depends principally on glacier meltwater. Virtually all glaciers are showing substantial melting in the PRC, where spring stream flows have advanced by nearly one month since records began. In the tropical Andes in South America, the area covered by glaciers has been reduced by nearly one-quarter in the past 30 years. Some small glaciers are likely to disappear completely in the next decade given current trends. Many large cities such as La Paz, Lima and Quito and up to 40 per cent of agriculture in Andean valleys rely on glacier meltwater supplies. Up to 50 million people in this region will be affected by loss of dry season water.

Right to food

IPCC AR4

- Future climate change is expected to put close to 50 million extra people at risk of hunger by 2020 rising to an additional 132 million and 266 million by 2050 and 2080, respectively.

- Projected changes in the frequency and severity of extreme climate events, together with increases in risks of fire, pests and disease outbreak, will have significant consequences on food and forestry production, and food insecurity, in addition to impacts of projected mean climate.
- Smallholder and subsistence farmers, pastoralists and artisanal fisherfolk will suffer complex, localised impacts of climate change.
- Global food production potential is likely to increase with increases in global average temperature up to about 3°C, but above this it is very likely to decrease.
- Local extinctions of particular fish species are expected at edges of ranges. (*Stern Review*: 'About one billion people worldwide [one-sixth of the world's population] rely on fish as their primary source of animal protein.')
- Food and forestry trade is projected to increase in response to climate change, with increased food import dependence of most developing countries.
- In mid- to high-latitude regions, moderate warming will benefit cereal crops and pasture yields, but even slight warming decreases yields in seasonally dry and tropical regions. Further warming has increasingly negative impacts in all regions.

Stern Review

- Around 800 million people are currently at risk of hunger (approximately 12 per cent of world's population), and malnutrition causes around 4 million deaths annually, almost half in Africa. Once temperatures increase by 3°C, 250–550 million additional people may be at risk – over half in Africa and western Asia, where: (1) the declines in yield are greatest; (2) dependence on agriculture is highest; and (3) purchasing power is the most limited.
- In tropical regions, even small amounts of warming will lead to declines in yield. In higher latitudes, crop yields may increase initially for moderate increases in temperature but then fall. Higher temperatures will lead to substantial declines in cereal production around the world.
- With a 4°C increase, entire regions may be too hot and dry to grow crops, including parts of Australia. Agricultural collapse across large areas of the world is possible at even higher temperatures (plus 5 or 6°C), but clear empirical evidence is still limited. The impacts will be

strongest across Africa and western Asia (including the Middle East), where yields of the predominant regional crops may fall by 25–35 per cent, or 15–20 per cent once temperatures rise by 3 or 4°C. Maize-based agriculture in tropical regions, such as parts of Africa and Central America, is likely to suffer substantial declines.

Right to health

IPCC AR4

- The projected relative risks attributable to climate change in 2030 show an increase in malnutrition in some Asian countries.
- Later in the century, expected trends in warming are projected to decrease the availability of crop yields in seasonally dry and tropical regions. This will increase hunger, malnutrition and consequent disorders, including child growth and development, in particular in those regions that are already most vulnerable to food insecurity, notably Africa.
- By 2030, coastal flooding is projected to result in a large proportional mortality increase; however, this is applied to a low burden of disease so the aggregate impact is small. Overall, a two- to three-fold increase in population at risk of flooding is expected by 2080.
- Estimates of increases of people at risk of death from heat differ between countries, depending on the place, ageing population and adaptation measures in place. Overall, significant increases are estimated over this century.
- Mixed projections for malaria are foreseen: globally an estimated additional population at risk between 220 million and 400 million has been estimated. In Africa, estimates differ from a reduction in transmission in south-east Africa in 2020 and decreases around the Sahel and south-central Africa in 2080, with localised increases in the highlands, to a 16–28 per cent increase in person-months of exposure in 2100 across all scenarios. For the United Kingdom, Australia, India and Portugal, some increased risk has been estimated.
- By 2030 an increase in the burden of diarrhoeal diseases in low income regions by approximately 2–5 per cent is estimated. An annual increase of 5–18 per cent by 2050 was estimated for Aboriginal communities in Australia.
- In eastern North America under (one) climate scenario, a 4.5 per cent increase in ozone-related deaths is estimated. A 68 per cent increase in average number of days per summer exceeding the 8-hour regulatory

standard is projected to result in a 0.1–0.3 per cent increase in non-accidental mortality and an average 0.3 per cent increase in cardio-vascular disease mortality.
- By 2085 it is estimated that the risk of dengue from climate change increases to include 3.5 billion people.

Stern Review

- Climate change will amplify health disparities between rich and poor parts of the world. WHO estimates that climate change since the 1970s is already responsible for over 150,000 deaths each year through increasing incidence of diarrhoea, malaria and malnutrition predominantly in Africa and other developing regions. Just a 1°C increase in global temperature above pre-industrial levels could double annual deaths from climate change to at least 300,000 according to the WHO. At higher temperatures, death rates will increase sharply, for example, millions more people dying from malnutrition each year.

INDEX